Marcel Van de Voorde, Gunjan Jeswani (Eds.)
Handbook of Nanoethics

Also of interest

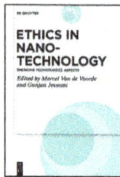

Ethics in Nanotechnology.
Emerging Technologies Aspects
Marcel Van de Voorde, Gunjan Jeswani (Eds.), 2021
ISBN 978-3-11-070181-4, e-ISBN 978-3-11-070188-3

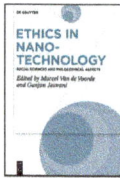

Ethics in Nanotechnology.
Social Sciences and Philosophical Aspects
Marcel Van de Voorde, Gunjan Jeswani (Eds.), 2021
ISBN 978-3-11-071984-0, e-ISBN 978-3-11-071993-2

Nanoscience and Nanotechnology.
Advances and Developments in Nano-sized Materials
Marcel Van de Voorde (Ed.), 2018
ISBN 978-3-11-054720-7, e-ISBN 978-3-11-054722-1

Environmental Functional Nanomaterials
Qiang Wang, Ziyi Zhong (Eds.), 2020
ISBN 978-3-11-054405-3, e-ISBN 978-3-11-054418-3

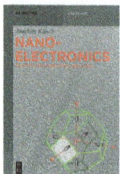

Nanoelectronics.
Device Physics, Fabrication, Simulation
Joachim Knoch, 2021
ISBN 978-3-11-057421-0, e-ISBN 978-3-11-057550-7

Handbook of
Nanoethics

Edited by
Marcel Van de Voorde and Gunjan Jeswani

DE GRUYTER

Editors
Prof. Dr. Marcel Van de Voorde
University of Technology Delft (NL)
Bristol – A, Apt. 31
Rue du Rhodania 5
3963 Crans-Montana
Switzerland

Assoc. Prof. Gunjan Jeswani
Shri Shankaracharya Group of Institution
Department of Pharmaceutics
Bhilai 490020, Chhattisgarh
India

ISBN 978-3-11-066923-7
e-ISBN (PDF) 978-3-11-066928-2
e-ISBN (EPUB) 978-3-11-066947-3

Library of Congress Control Number: 2021933271

Bibliographic information published by the Deutsche Nationalbibliothek
The Deutsche Nationalbibliothek lists this publication in the Deutsche Nationalbibliografie;
detailed bibliographic data are available on the Internet at http://dnb.dnb.de.

© 2021 Walter de Gruyter GmbH, Berlin/Boston
Cover image: piranka/iStock/Getty Images Plus
Typesetting: Integra Software Services Pvt. Ltd.
Printing and binding: CPI books GmbH, Leck

www.degruyter.com

Nano-Covid-19 Vaccine is a great innovation but also a nano-ethical problem. Vaccine is a matter of privilege to the rich countries, who take the right to serve themselves first and regrettably treat the rest of the world as second-class people. Vaccine nationalism takes precedence over world health, human rights, and solidarity. The rich will regret it once!

Prof. Marcel Van de Voorde, doctor honoris causa

Preface: Converging technologies from the nanoscale require enhanced ethics

Reaching the atomic and molecular levels with investigative and transformative tools has been a historic turning point not only for human knowledge but also for philosophy, manufacturing, and human well-being. The rapid progress of nanotechnology has been mirrored by increased concerns about unknown material behavior, powerful transformative tools, human-technology coevolution, and other societal implications.

Developments at the nanoscale have awakened other foundational science and technology fields such as modern biology, information technology, wireless technologies, artificial intelligence (AI) systems, and cognitive technologies. Together, these are transforming ways of life and societal trends. A new,converging science and technology global platform– nano-bio-info-cogno-AI –has been created worldwide. This platform built from the nanoscale is accelerating economic and human development. Risks of secondary effectson "environmental, health and safety (nano-EHS)" and "ethical, legal and societal implications (nano-ELSI)" must be identified and mitigated from the beginning of any large project, in concurrence with the science and technologyprogress of the respective projects.

Significant nanoscale science and technology research programs continue around the world;once President Bill Clinton announced the U.S. National Nanotechnology Initiative(NNI)in January 2000, national programs sprang up in 80 other economies within just a few years. These programs are pushed by successive nanoscale breakthroughs and convergence with other emerging science and technology fields. Programs also are pulled by their extensive and multifaced societal benefits, which were recognized early; nanotechnology stakeholdersconvened on the topic within the first month of NNI,leading tothe report, "Societal Implications of Nanoscience and Nanotechnology" (Kluwer, now Springer 2001). The NNI'scumulative research and development funding has reachedabout $28 billion in 2019. Between 2000 and 2019, the average annual growth rate of nanotechnology publications has been about 14% (faster in the first decade, decelerating slightly after 2010) and of nanotechnology applications of about 25% (accelerating after 2010). Societal implications grew with several kinds of applications, such as those related to human and environmental safety and ethical and legal aspects. Global revenue of products having nanotechnology as the key competitive factor is estimated to reach about $3 trillion in 2020, of which about one quarteris from the U.S. New areas of research and engineering such as metamaterials and plasmonics have emerged, and new uses appear in emerging technologies such as molecular manufacturing, regenerative medicine from the nanoscale, and smart phone platforms. Recent R&D programs are directed to future nanosystem architectures and converging technology platforms for production, health, infrastructure, and services.Today, responsible governance focuses attention on the confluence of nanotechnology with other domains,

https://doi.org/10.1515/9783110669282-202

its role on social sustainability, and synergistic effect from integration with other emerging technologies ("Converging Knowledge, Technology and Society: Beyond Nano-Bio-Info-Cognitive Technologies", Springer 2013, available on www.wtec.org/NBIC2/).

As part of this worldwide effort, an international community has been established for nano-EHS and nano-ELSI. This has beenfacilitated by the creation of large networks in the U.S., Europe, and Asia, beginning with the networks on "Nanotechnology in Society" (centered at the UC-Santa Barbara and Arizona State University) and "Centers for Environmental Implications of Nanotechnology" (at UCLA and Duke University). After an initial focus on the toxicity of nanomaterials themselves, scientific interest in nanodevices and exposure of industrial and incidental nanoproducts to consumer has risen and led to essential improvements in evaluating and governing the risk. Significant research and regulatory activities have addressed the environmental, health and safety implications of nanotechnology applications, especially for passive as well as chemically and biologically active nanoparticles. Responsible development of nanotechnology increasingly addresses ethical, legal, and other societal issues, including ecological and human development aspects. Ethical questions have been raised about who will receive the benefits of nanotechnology and who will be affected by its secondary effects. Ethics need to be included in the vision of an R&D program in an emerging area. Based on my experience in planning research initiatives, on average about a third of the overall program's success depends on the initial topic selection, research directions, and consideration of societal implications, including ethical aspects.

The next decade will bring several main advancements to nanotechnology-inspired science and technology fields. Larger and multifunctional nanostructures and devices already are being built with more atoms, complex networking, and information content, increasing the EHS and ELSI risks. They are increasingly used in advanced manufacturing, quantum information systems, AI systems, the bioeconomy, neurotechnology, sustainable society, and advanced wireless S&T. Such technologies will converge to better serve people and contribute to human development, but new safety concerns may arise. Nanoscale science and technology evolve closer together, with more partnerships between academia and the private sector. Translational nanotechnology takes a center stage. The necessary infrastructure becomes more complex and integrated with the needs of the global emerging technology system. A flexible nanotechnology workforce education and training are essential requirements for safe and sustainable progress and along-term challenge strongly connected to changes in the human-technology ecosystem and other societal trends.

With the prospect of these and other emerging, convergent areas of nanotechnology before us, the need to incorporate ethics – for both program success and positive societal impacts – is clear and crucial

The "Handbook on Nanoethics" assembled by Marcel Van de Voorde and Gunjan Jeswani bring together a rich range of topics and various perspectives from five continents. The book presents highlights of ethics in nanotechnology (Chapter 1) and ethical reflections of nanotechnology (Chapters 2–6). Essential aspects of Health, environmental and industrial nanoethics are systematically presented (Chapters 7–11). Nanotechnology governance including societal and legal aspects are discussed in the last part of the book (Chapters 12–17). The co-editors provide guiding remarks and recommendations to nanotechnology providers and users. The collection of opinions presented in this volume is a remarkable selection of research contributions concerning the ethics of nanotechnology research.

This book brings together some of the most vigorous minds active in nanotechnology ethics in the world. Converging and emerging technologies from the nanoscale bring new health, environmental, ethical, and legal concerns that are highlighted in these volumes. We encourage all research, education, and industry experts to closely read these state-of-the-art contributions on ethics in nanoscale science and technology fields.

Mihail C. Roco
National Science Foundation and National Nanotechnology Initiative
Alexandria, Virginia, May 3, 2021

Forewords

I

When I was invited to write the foreword of this book, I thought it would be important to start by giving justice to the visionary work of Nobel laureate R. Feyman[1] on the groundbreaking capabilities of nanotechnology.

In 1959, Prof. Feyman first mentioned the broad possibilities afforded by substances of an extremely small size. Addressing an audience at the annual American Physical Society, he stated: "[. . .] when we have some control of the arrangement of things on a small scale we will get an enormously greater range of possible properties that substances can have, and different things that we can do [. . .]."[2]

Since this speech, the miniaturization described by Prof. R. Feyman has led to the development of, what is called today, nanotechnology. Nanotechnology enables the manufacturing of materials with size ranging from 1 to 100 nm. Due to the extremely small size of the particles, these materials exhibit novel physical and chemical properties. Considered an area of incredible promise, this disruptive technology has delivered over the years many innovative applications in very different domains.

Several everyday products (e.g., kitchenware, batteries, coatings for solar cells, cosmetics, and sunscreens) show improved features, thanks to the nanomaterials they contain. For example, silver nanoparticles, with their antimicrobial attributes, reduce odor-producing bacteria in sport textiles.[3]

Also, in medical devices, coatings in the nanoscale increase biocompatibility and thus improve integration of a variety of medical implants (e.g., stent coating in cardiology)[4] with the surrounding tissues. The application of nanotechnology to medicine includes the use of precisely engineered materials to develop novel therapies that may reduce toxicity as well as enhance the efficacy of treatments. The first nanotechnology-based cancer drugs are already on the market.[5]

However, as is the case for many other technological breakthroughs, the development of nanotechnology is also hindered by several ethical controversies. Today there is a clear split between those who see the unmet potential of the innovation and those who fear a negative impact on the environment and on health, and question how to ensure a responsible use of nanotechnology.

The different views between scientists and the general public on how to address the balance between the risks and the benefits of nanotechnology create mistrust in

---

1 https://www.nobelprize.org/prizes/physics/1965/feynman/biographical/
2 Feyman RP. There's plenty of room at the bottom. *Caltech Engineering and Science* 1959; **23**(5): 149–160.
3 https://ec.europa.eu/health/scientific_committees/emerging/docs/scenihr_o_039.pdf
4 https://www.rivm.nl/publicaties/nanotechnologies-in-medical-devices
5 https://www.cancer.gov/nano/cancer-nanotechnology/current-treatments

https://doi.org/10.1515/9783110669282-203

the deployment of this innovation. This leads to significant barriers to its broad diffusion.

One area in which nanotechnology is facing controversy is the packaged food and beverage sector, a field in which, as chairman emeritus of Nestlé S.A., I am particularly interested. Nanotechnology can have implications, for example, in the reduction of food waste: by using it in packaging we can prevent the food from gases and light, which cause degradation of food.

From a consumer's perspective, product safety is what drives the acceptance of any technology, including nanotechnology, in food. This is where and when regulators step in. Indeed, food applications of nanotechnology fall under stringent general food regulations. They are authorized only when safety requirements, based on detailed risk analysis, are fully met. However, despite this high level of institutional reassurance, concerns on the novel technology in food may persist. Consequently, manufacturers may decide not to make any use of it even if allowed by regulators.

Science and novel technologies are drivers of societal change. They can accelerate progress, leading to greater well-being for all. The ethical questions that are raised on the safety and on the responsible use of nanotechnology are legitimate, and the answers available at this stage may not be reassuring enough for some of us. However, the lack of adequate responses should not limit our endeavor to explore new scientific avenues. We are still at the early stages of the nanotechnology journey: research should be encouraged and should continue.

Scientists, policymakers, and other stakeholders can join forces to help understand the major advantages of this new technology to the general public. With this, and as we progress and learn, we can overcome the existing distrust and uncertainty, and bring more reassurance on the unmet benefits that nanotechnology can provide to humanity.

Peter Brabeck-Letmathe
Chairman Emeritus, Nestlé S.A.
Vevey, 29 April 2020

II

It is a pleasure to write the foreword of this book, because we are witnessing the emergence of the next version of the industrial revolution; hence, there is a need for understanding and practice of ethical standards.

In the era of Industry 4.0, dynamic technology is the main growth platform and is playing a crucial role in raising the standards of the society. Nanotechnology has brought several benefits to humanity. Nevertheless, as technological advancement is profound importance, society is facing newer challenges as well. There is still a significant difference for those who perceive the unfulfilled promise of science and those who fear adverse effects on the climate, safety, and the safe usage of the technology. Consequently, ethical discussions among the thought leaders and technical experts are required in order to address the gray areas.

The nuclear power, weapons, biotechnology, and artificial intelligence are the main areas driven by nanotechnology, which raises several legal and moral concerns. The ethical dilemmas, environmental problems, and work displacement are among the key ethical or moral concerns in nanotechnology and are discussed in different chapters of this book.

The fields like nanomedicines, nano-cosmetics, and nanodevices are discussed at length which are interesting and demonstrates significant enhancement in pharmacokinetics, increased tissue selectivity, and effectiveness due to their size, structure, composition, and design. The use of nanomedicines besides benefits raises ethical challenges regarding probable side effects of nanoparticles on humans and disposal in the environment. The overall risk analysis, research involving human embryonic stem cells, germline gene therapy, and gene editing are of great concern. Currently, in more than 40 countries the use of germline editing for reproduction is prohibited. Nanoparticles of zinc oxide and titanium dioxide used in sunscreen products also run the risk of dangerous entry into the systemic circulation, genotoxicity, and cytotoxicity.

This handbook on nanotechnology ethics would foster discussion and probable solutions for complexities. The chapters are written by experts representing prestigious organizations around the globe.

I genuinely believe thisbook will help intellectuals across the various profession and consumers to develop an understanding of ethical challenges in these areas.

Shailendra Saraf
Vice President, Pharmacy Council of India, New Delhi, India

III

Unprecedented situations need unprecedented solutions.
Amin Maalouf

"There is no any challenge beyond the human distinctive capacity to create," said President J. F. Kennedy in June 1963. This is a very important statement: the impossible today can be possible tomorrow and there is no place in the scientific mind for fanaticism, dogmatism, and fatalism. The past cannot be rewritten. The future is to be invented.

Taking into account the potentially irreversible processes, it is our common responsibility to bear in mind "the ethics of time," acting in a way that will prevent us from taking a path that allows no return. We have to specifically take into account next generations: our legacy cannot condemn them to have a worst standard of living and prevents them from fully exercising the distinctive capacities of human beings.

In this context, nanotechnology such as the study, design, creation, synthesis, manipulation, and application of materials, devices, and functional systems through the control of matter at nanoscale, and the exploitation of phenomena and properties of matter at nanoscale offer a number of possibilities in different fields to help alleviate situations of great need.

Linked to the scientific research developed by the main public institutions of higher education, nanotechnology fosters a model of interdisciplinary collaboration in fields such as the so-called nanomedicine – application of techniques that allow the design of drugs at the molecular level – nanobiology and the development of microconductors.

When matter is manipulated on such a tiny scale of atoms and molecules, it demonstrates completely new phenomena and properties. Therefore, scientists use nanotechnology to create new and inexpensive materials, devices, and systems with unique properties. Nanotechnology must be used as a tool to help alleviate the great challenges we face.

The past has already been written. It must now be accurately described. We must learn from the lessons of the past to be able to invent the future, to be able to provide all required conditions to each and every human being, so they can all have a dignified life.

The time has come to raise our voices with both serenity and resolution. The time has come for the emancipation of citizenship, for "We, the peoples . . ." as stated in the first phrase of the UN Charter. We need peace within each human being, at home, in the villages, in the cities, and across the world.

Our conscience allows us to progressively discover what we are like, what we are composed of and how that is reflected in "health" and in mental instability (physiopathology). We research incessantly to uncover the realities of all living beings, of the planet we inhabit, and of the whole universe. We contemplate and reflect on the cosmos, on the galaxies, and on their colossal dimensions; we do so as well with the smallest particles, the elements that comprise the minimum parts of matter. From the most distant stars to quarks, our curiosity and desire for greater knowledge knows no bounds. And all of this enables us to discover *what we are like* and *where we are*, but it adds little to *who we are*.

I firmly insist that sustainable development and avoid the further deterioration of the environment cannot be postponed, because no return points can be reached. Therefore, the time has come for a great mobilization at the global scale so that people at last take control of their own common destiny. Words are our only "mass construction weapons."

Enough! The time has now come to "rescue" the citizens and, to that end, we must make a quick and courageous shift from an economy of war to an economy of global and sustainable development.

Dare to know! (*sapere aude*) is written in the Oxford County Wall Crest. Yes, but afterward, "no how to dare," how to share, and how to care. This is the *crucial role of the scientific community* today: to be closed – never more dependent – to the governments, parliaments, councils, and civil society associations . . . in order to provide them with science and technology assessment, to advise and anticipate, to be watchtowers, to foresee, and to prevent.

Now, more than in the past, is urgent interweaving the real world with the affordable solutions. On this way, particularly in the moments of deep crisis that we are living – financial, environmental, democratic, ethical – that are opportunities as well, the scientists must become actively involved in a new farsighted vision of otherness and brotherhood taking into account the earth as a whole. There are no frontiers anymore, and should not be privileged citizens (a minority of 20%) when most of them live in very difficult, even inhuman, conditions.

Time has come for action!

Federico Mayor
Former President, UNESCO, Paris, France
March 4, 2020

Acknowledgments

The editors would like to extend special thanks of gratitude to Dr. Ilise Feitshans Director, ESI Safernano, for her valuable contribution to this book by writing some chapters and by reviewing and editing various chapters to ensure high standards. Her deep thoughts and passionate experience along with an intense sense of understanding of vision and mission of nanoethics have greatly contributed to the success of this book.

We are also thankful to Dr. Evrard, University of Manitoba for his valuable contribution by the book by writing a chapter and for his critical comments and suggestions which helped a lot in adding value to the book.

The editors are deeply indebted to Professor Michael FITZPATRICK, Executive Dean, Faculty of Engineering, Environment and Computing and Lloyd's Register Foundation Chair in Structural Integrity and Systems Performances, Coventry University (UK), for his work throughout the book in technical and academic editing to ensure a consistent style and accessibility of the book for experts and non-experts readers, as well as for numerous technical contributions.

https://doi.org/10.1515/9783110669282-204

Contents

Part I: Highlights of ethics in nanotechnology

Part II: Nanotechnology and ethical reflections

Introduction: Overall vision of ethics in nanotechnology developments

Book's objective

The new and emerging nanotechnology is remaking the world at an alarmingly rapid pace with many applications in a wide variety of fields ranging from health care over industrial products to crime prevention and defense. Despite the many impactful benefits of nanotechnology, such as the cultivation of new organs for patients, there are potential hazards and risks involved, such as the toxicity of certain nanoparticles that could cross the blood–brain barrier. Nanotechnology, it should be clear, has the potential to profoundly change society for better or for worse.

The societal implications of nanotechnology encompass many ethical issues such as the violation of privacy, the violation of a person's autonomy, environmental pollution, economic abuse, security and global justice (between developed and developing countries).

The biggest question we are facing is how we should deal with uncertainty and risk in this emerging technology? Uncertainty is one of the major obstacles to the commercialization of nanotechnology – uncertainty about the risks to health and environment and uncertainty about how the governments might regulate nanotechnology in the future. There is an urgent need for answers both in the realm of science and in the realm of law and regulation. In the absence of scientific clarity about the potential health and environment effects of exposure to nanoparticles, the policymakers must be provided with guidance on how to deal with hazards, risks, and controls.

Nanotechnology will affect everyone; thus, all members of society should have a voice in its development and commercialization. The book focuses on the main societal and ethical issues that arise from the development of nanotechnology. The result is a clear set of questions and solutions proposed by leading researchers working in a variety of fields such as applied ethics, bioethics, ethics of science, ethics of technology, and business ethics.

There is an urgent need for constructive interaction between the natural sciences (physics, chemistry, mathematics, biology, engineering) and the human sciences (philosophy, law, sociology and ethics), with the aim of formulating ethical standards for the development of nanotechnology. This book hopes to be an important step in that direction.

This book presents an overview of new and emerging nanotechnologies and their societal and ethical implications. To support sustainable, ethical, and economic nanotechnological development, it is imperative to educate all nanotechnology scientists, engineers and stakeholders about the long term benefits, limitations, and risks of nanotechnology. In the introduction, we provide the reader with an insight into the multiple societal and ethical questions raised by nanotechnology.

https://doi.org/10.1515/9783110669282-206

Convergence of natural and human sciences in nanotechnology

Nanotechnology is a popular term these days, associated with multiple uses and potential applications in the future. It is a dynamic multidisciplinary field. The applications are constantly extending to various areas, and their implementations are impressive. In fact, the main engineer/alchemist offering nanoscale materials appears to be life's unstoppable force, our mother nature. In ancient times, certain practices created nanoparticles through the customary procedures. One of the most acclaimed ancient applications is the Lycurgus cup, which was created from nanoparticles of gold and silver that were inserted in the glass. However, these were not recognized as nanosystems or nanoparticles until the twentieth century.

One nanometer is basically one billionth of a meter, or 10^{-9} of a meter. A newspaper sheet is around 100,000 nm wide, in contrast. Nanotechnology – as we know it – was introduced in 1959 in a talk delivered by Richard Feynman, "There's plenty of room at the bottom." Nevertheless, until 1974, when a Japanese physicist Norio Taniguchi invented and described the word "nanotechnology," the real phrase "nanotechnology" was not coined. Thereafter, the development of scientific instruments such as the scanning tunneling microscope (1981) and the atomic force microscope (1986) generated the impetus for the further development of contemporary nanotechnology. Scientists discovered that atoms and molecules behave differently at the nanoscale. As a result, experts believe that nanotechnology has potential applications in a broad variety of fields with significant consequences for human well-being, the atmosphere, biodiversity, and national security. Although plenty of us do not realize the incredible effect it might have and – to a certain extent – is already having on our everyday lives. A promising indication is the growing amount of capital that policymakers are investing into improving such innovations worldwide.

With the advent of the fourth industrial revolution encompassing robotics, artificial intelligence, machine learning, Internet of things, 3D printing, and so on, the world of nanotechnology will change our lives and our society dramatically. There will be a great number of applications that at present, we cannot even conceive.

Some of the important examples of applications include:

In medicine:
- Nanoparticle-based vaccines may one day provide permanent immunity to the common cold and influenza.
- Artificial replacements for body tissue such as the skin, muscle, tendon, and even organs can be produced by nanoscale fabrication, mimicking natural processes.
- Nanoparticles in pharmaceutical products will smartly deliver chemotherapy drugs to specific cells, such as cancer cells.

In consumer product:
- Filters for producing clean drinking water will remove all viruses and bacteria.
- Nanoparticles used in food packaging will reduce UV exposure to prolong shelf life.
- Faster, smaller, and more powerful computers will consume far less power, with biodegradable longer-lasting batteries.

In industrial applications:
- Nano-strengthened materials will produce lightweight alloys for cars, reducing fuels
- Tires will be fabricated with a better grip in wet conditions.
- Electrode materials for rechargeable batteries will be greatly enhanced through nanotechnology, reducing weight and improving performance in the next generation of hybrid and electric vehicles.

History shows that new science and technology is often met by society with unrest. Prominent examples include the development of synthetic chemicals and nuclear power in the mid-twentieth century, and biotech and genomics in the twentieth century. The same goes for nanotechnology and the potential threats it raises. Therefore, there is an urgent need for a better understanding of the safety aspects of nanotechnology applications. For the nanomaterials, numerous toxicological tests have shown that nanoparticles can damage both humans and the ecosystem. Many argue that we should apply to the precautionary principle to engineered nano products and particles. An expert study published by the Federal Ethics Committee on Non-Human Biotechnology (ECNH) concluded that the presumption of proof of the harmlessness of synthetic nanoparticles will be the duty of the producers because there are good reasons to believe that such particles may seriously endanger humans and the atmosphere under some circumstances. Moreover, given the unpredictability of the future developments in nanotechnology, governmental authorities and consumers are faced with unknown risks.

Nanotechnology, it should be clear, has the power to revolutionize human lives and transform human societies. We must ask ourselves, what kind of world are we creating? The development of nanoscience and nanotechnology should not be left to engineers and nanoscientists alone. It is of utmost importance that experts in the social sciences and humanities and the public at large are involved. Developments in nanotechnology should be transparently communicated and closely monitored. For this purpose, scientists, engineers and industrial players should work closely together with ethicists and social scientists and the public should be informed and consulted.

At present, such a multidisciplinary and ongoing dialogue on the ethical concerns surrounding nanotechnology is, for the most part, lacking. This book hopes to contribute to this urgent task. It addresses the ethical concerns raised by nanotechnology from various perspectives: from a scientific, social, and philosophical point of view. Chapters have been written by experts from various countries across the world, including the United States, Japan, China, Israel, India, Australia, Canada, and European countries, proposing different strategies but reaching the same conclusion: that there is an urgent need for ethical guidelines as well as regulations to ensure safe and healthy production and use of nanoproducts. It is also clear that these regulations and guidelines should have a worldwide dimension.

Our sincere hope that this multiauthored book covering the various aspects of nanotechnology and ethics will be welcomed by the scientific community and philosophers alike. Besides, we sincerely hope that the book will assist and enrich readers to understand the ethical challenges of nanomaterials and the discussed solutions to the safe use of nanomaterials.

The book also highlights perception of nanotechnology and ethical concerns from different angles and proposes a holistic approach towards the perceived dangers of nanotechnology.

We believe this book will be equally relevant to scientists and engineers employed in the area of nanotechnology and philosophers involved in nano or any technology-related ethics. The text is structured in such a manner that each chapter stands on its own and can be read independently.

We express our sincere thanks to Ms Ute Skambraks for her kind help in publication process. Sincere thanks to Dr Karin Sora for continually supporting us through the publication house. We express our sincere gratitude to all the writers who made diligent efforts to write chapters during the time of the COVID-19 pandemic. We are grateful to all the eminent persons: *Mihail C. Roco* – National Science Foundation and National Nanotechnology Initiative; *Peter Brabeck-Lemathe* – Former President NESTLE; *Shailendra Saraf* – Vice President, Pharmacy Council of India and *Federico Mayor Zaragosa*, Former President of UNESCO from respectable authorities for the excellent foreword to this book. We appreciate our respective families, for without their continuous support this work could not have been completed.

Marcel Van de Voorde and Gunjan Jeswani, book editors

Embedding ethics in nanomedicine: Europe acted promptly

Preamble

The *Oxford Dictionary* defines nanotechnology as "the branch of technology that deals with dimensions and tolerances of less than 100 nanometres, especially the manipulation of individual atoms and molecules".[1] Nanotechnology has application in multiple scientific fields, such as surface science, organic chemistry, molecular biology, semiconductor physics, energy storage, microfabrication, and molecular engineering. It is a technology capable of creating new materials and devices with a potential vast range of applications, such as in medicine (hence, the term "nanomedicine"), electronics (hence, "nanoelectronics"), biomaterials energy production and consumer products. As innovative technologies often do, however, nanotechnology applications have raised an intense debate in the circles of scientists, lawyers, and policymakers, including on ethical issues.

In the early 2000s, controversies emerged regarding the definitions and potential implications of nanotechnology, including the safety (for health and/or the environment) of the very first products originating from nanotechnology such as silver nanoparticles, carbon fiber strengthening using silica nanoparticles and carbon nanotubes. In particular, nanomedicine has acquired a progressive importance on the societal debate on the ethical dimension of this technology and the expectations and concerns on its use.

Nanomedicine: What is possible to do?

A European technology platform on nanomedicine has been created and in 2005 issued its first vision paper.[2] Its activities included diagnosis, drug delivery, and triggering of biomaterials. This activity in nanomedicne is continuing.[3] With no

---

1 https://www.lexico.com/definition/nanotechnology
2 http://ec.europa.eu/research/industrial_technologies/pdf/nanomedicine-visionpaper_en.pdf
3 https://etp-nanomedicine.eu/

Disclaimer: The information and views set out in this chapter are those of the authors and do not necessarily reflect the official opinion of the European Commission.

Renzo Tomellini, Head of Unit "Science Policy, Advice and Ethics", European Commission, Brussels, Belgium
Maurizio Salvi, Unit "Science Policy, Advice and Ethics", European Commission, Brussels, Belgium

https://doi.org/10.1515/9783110669282-207

Application area	Expected benefits
Drug delivery	Nanotechnology provided the possibility of delivering drugs to specific cells using nanoparticles.
Cancer	Nanoparticles can be used against tumor cells and accumulate at tumor sites.
Imaging	Nanoparticles may be used for *in vivo* imaging (cardiovascular imaging or oncology imaging)
Sensing	Research on nanoelectronics could lead to tests for the diagnosis, and treatment of cancer.
Blood purification	Magnetic micro particles are research instruments for the treatment of systemic infections and they may also provide alternatives to traditional dialysis methods.
Tissue engineering	Nanotechnology may be used to reproduce, repair, or reshape damaged tissue using suitable nanomaterial-based scaffolds.[5]
Medical devices	Nanoscale enzymatic biofuel cells for nanodevices have been developed that uses glucose from biofluids including human blood and watermelons.

pretention of completeness, some applications of nanotechnology in the medical fields include:[4]

In addition to this set of current applications of nanomedicine, the community of scientists has also advocated the possibility of engineering molecular assemblers (hence, the term "nano-robots") which could reorder matter at a molecular or atomic scale and then be implanted into the systems or organisms, including the human body, to repair or detect damages and infections.

The safety issues

Because nanomedicine may affect citizens, both directly (trials) and indirectly (possible exposure to free nanoparticles into the environment), it is important to underline the safety of this technology in order to protect their rights and their aspirations. A central consideration in assessing the legitimacy of medical technologies therefore refers to their safety.[6]

---

4 See also https://www.britannica.com/science/nanomedicine

5 Nanoparticles such as graphene, carbon nanotubes, molybdenum disulfide and tungsten disulfide are being used as reinforcing agents to fabricate mechanically strong biodegradable polymeric nanocomposites for bone tissue engineering applications.

6 However, even if it is hard to define a precise borderline between the two dimensions, a distinction needs to be done between risks for the patients undergoing an application of nanomedicine and risks associated with the toxicological and eco-toxicological effects of nano-components.

Safety issues of nanotechnology and nanomedicine have been addressed in several reports across the world. The Scientific Committee on Emerging and Newly Identified Health Risks[7] (SCENIHR) report and the White Paper *Nanotechnology Risk Governance*[8] published in June 2006 by the International Risk Governance Council are two examples of reports on risk governance issues of nanotechnology. While using different approaches and methods, the above reports stress the lack of data (in particular, long-term data) on possible risks associated with nanotechnology with regard to both the human health and the ecological consequences of nanoparticles accumulating in the environment. In addition (to quote the UNESCO report on the ethics and politics of nanotechnology[9]), the issue of the safe and responsible use of nanomedicine and nanotechnology raises "two concerns: the hazardousness of nanoparticles and the exposure risk. The first concerns the biological and chemical effects of nanoparticles on human bodies or natural ecosystems; the second concerns the issue of leakage, spillage, circulation, and concentration of nanoparticles that would cause a hazard to bodies or ecosystems." Concerns are also raised by the difficulties of identifying, estimating, and managing risks in an area where there are considerable uncertainties and knowledge gaps, and when the short-term and long-term risks may be different.

The existing framework on ethics and human rights

While the above considerations show the difficulty to assess the safety of biomedical products, it is important to refer to the existing framework on ethics and human rights that constitutes a reference point for the analysis of the ethical dimension of nanotechnology. Let us then consider some international references having a legal or moral authority in this analysis (soft and hard law).

(a) The *Council of Europe* has issued the Oviedo Convention – Convention on Human Rights and Biomedicine.[10] Its main purpose is to "preserve human dignity, rights and freedoms, through a series of principles and prohibitions against the misuse of biological and medical advances." The Convention also concerns equitable access to health care, professional standards, protection of genetic heritage, and scientific research. It contains several detailed provisions on informed consent. A number of additional protocols supplement the Convention.[11]

---

7 Scientific Committee on Emerging and Newly Identified Health Risks (SCENIHR).
8 See the White Paper *Nanotechnology Risk Governance* published in June 2006 by the International Risk Governance Council and the references in that report.
9 http://unesdoc.unesco.org/images/0014/001459/145951e.pdf.
10 https://www.coe.int/en/web/bioethics/oviedo-convention
11 http://www.coe.int/t/e/legal_affairs/legal_cooperation/bioethics/texts_and_documents/1Treaties_COE.asp#TopOfPage.

(b) The *Universal Declaration on the Human Genome and Human Rights*,[12] adopted by UNESCO's General Conference in 1997 and subsequently endorsed by the United Nations General Assembly in 1998, deals with the human genome and human rights. Since the Declaration was drafted in 1997, it does not refer *explicitly* to nanomedicine, but modifications that are targeted to DNA may fall within its scope.[13] The Declaration also contains provisions on the informed consent principle. The *Universal Declaration on Bioethics and Human Rights* (adopted by UNESCO on 19 October 2005) also contains specific provisions on ethical issues related to medicine, life sciences and associated technologies, and advocates several ethical principles, including human dignity, consent, autonomy and responsibility, privacy, equity and justice, solidarity, and benefit sharing.[14]

(c) The *European Charter of Fundamental Rights*[15] emphasizes that the Union is founded on the indivisible and universal values of human dignity, freedom, equality, and solidarity and on the principles of democracy and the rule of law. It contributes to the preservation of these common values while respecting the diversity of the cultures and traditions of the peoples of Europe, as well as the national identities of the Member States and the organization of their public authorities. The Charter formulates a common set of basic shared values at EU level.[16] Respect for human dignity, a ban on human reproductive cloning, respect for people's autonomy, non-commercialization of biological components derived from the human body, prohibition of eugenic practices, protection of people's privacy, freedom of science are examples of values enshrined in the Charter, which was adopted at the Summit of Nice in 2001.

(d) *The precautionary principle*, according to the Commission Communication of February 2000, is the basic constituent of the precautionary principle and the prerequisites for its application, are the existence of a risk, the possibility of harm, and scientific uncertainty concerning the actual occurrence of this harm.[17] Although

---

12 https://en.unesco.org/themes/ethics-science-and-technology/human-genome-and-human-rights

13 The Declaration asserts, "Dignity makes it imperative not to reduce individuals to their genetic characteristics and to respect their uniqueness and diversity".

14 http://portal.unesco.org/shs/en/file_download.php/46133e1f4691e4c6e57566763d474a4d BioethicsDeclaration_EN.pdf

15 Approved on 28 September 2000 and proclaimed by the European Parliament, the Council and the Commission on 7 December 2000.

16 For example Article 1 (respect for human dignity), Article 3 (ban on human reproductive cloning, respect for people's autonomy, non-commercialisation of biological components derived from the human body, prohibition of eugenic practices), Article 8 (data protection issues), Article 13 (freedom of science).

17 The precautionary principle does not necessitate impassable boundaries or downright bans. It is a general risk management tool, which was originally restricted to environmental matters. In the Commission's Communication of February 2000, it is stated, "The precautionary principle is not

this principle is not explicitly mentioned in the treaty except in the environmental field, its scope is far wider and covers those specific circumstances where scientific evidence is insufficient, inconclusive, or uncertain and there are indications through preliminary objective scientific evaluation that there are reasonable grounds for concern that the potentially dangerous effects on the environment, human, animal, or plant health may be inconsistent with the chosen level of protection.[18]

(e) *The innovation principle* as understood and applied by the Commission[19] promotes smart and future-oriented regulation, which is able to encourage new discoveries and solutions to address the most pressing social and environmental issues.

Bioethical questions

In its opinion on ethics of nanomedicine, the European Group of Ethics[20] has indicated a list of concerns that need to be faced to assess the ethics of this technological sector, such as: "How is it possible to give information about future research possibilities in a rapidly developing research area and to make a realistic risk assessment in view of the many unknowns and the complexities? What are the implications of nanomedicine for problems raised in cases where the information obtained by refined nanomedical diagnostic methods is used by third parties, in particular insurance companies and employers? How can the development of nanomedicine and nanotechnology be tailored to the benefit of the public? How can societies remain at least partly autonomous in their decisions, when the development of nanomedicine is closely connected to the economic prosperity of a given society and plays a part in international competition on the global market?"

---

defined in the Treaty, which prescribes it only once – to protect the environment. But in practice, its scope is much wider, and specifically where preliminary objective scientific evaluation indicates that there are reasonable grounds for concern that the potentially dangerous effects on the environment, human, animal or plant health may be inconsistent with the high level of protection chosen for the Community" (Communication Summary paragraph 3). Accordingly, the Commission believes that "the precautionary principle is a general one" (i.e., a general principle) (Section 3 of the Communication), whose scope goes beyond the EU – as shown by several international instruments starting with the Declaration on Environment and Development adopted in Rio de Janeiro in 1992.

18 https://eur-lex.europa.eu/legal-content/EN/TXT/PDF/?uri=CELEX:52000DC0001&from=EN

19 https://ec.europa.eu/info/news/innovation-principle-makes-eu-laws-smarter-and-future-oriented-experts-say-2019-nov-25_en

20 Group of Ethics examined the ethical aspects of medical applications related to nanotechnology. Some of the basic ethical values include the principle of respect for dignity; the principle of individual autonomy; the principle of justice and of beneficence; the principle of freedom of research; and the principle of proportionality. https://op.europa.eu/en/publication-detail/-/publication/4d7d9c99-2129-42e1-993e-c815b91f256b/language-en/format-PDF/source-77404425, pp. 39–41

The above questions relate to the ethical dimension of nanomedicine. As other technological mediums they do not induce univocal response, but they refer to a number of values/principles that need to be assessed with regard to each different use of nanomedicine is concerned. Case-by-case analyses are therefore inherent to the ethics of nanomedicine, also reflecting the ethical pluralism of modern society. In whatever views we therefore want to focus the ethical analysis of nanomedicine, whether following utilitarianism, Kantianism, virtue theory, the principles of autonomy, beneficence, non-maleficence, and justice, it is important that specific elements that are embodied in the international ethics frame described earlier are taken into account.

The protection offered by international declarations and guidelines applies to both health care and medical research; it includes the obligation to obtain free and informed consent from patients and participants in research and specifies the measures to be taken when patients and participants in research are for various reasons (minors, mentally incapacitated, etc.) unable to give consent. The principles stated in the above declarations and guidelines specify the obligations to protect individuals and societies against unpredictable risks based on the precautionary principle and a risk-benefit analysis, which includes also long-term risks and benefits. The principles mentioned above are also applied to health-related risks of nanotechnology, not only in the medical contexts, which are in focus here, but also in other contexts where nanotechnology is used. To summarize, the bioethics issue of nanomedicine that deserve specific attention include:

- **_Informed consent_**
  Informed consent requires the information to be understood. In view of the knowledge gaps, and the complexity of the matter, concerning the long-term effects of nanomedical diagnostic and therapeutic tools, it may be difficult to provide adequate information concerning a proposed diagnosis, prevention, and therapy needed for informed consent.
- **_Diagnostic complexity and increased personal responsibility_**
  Nanomedicine offers new diagnostic possibilities, where the results will be available with high speed, magnitude, and precision at the molecular level. The results may be complex and difficult to interpret. The increased complexity of diagnostic potentials, affects the level of responsibility by the medical community to properly interpret diagnostic results and propose therapeutic actions based on the above provisions.
- **_Medical and non-medical uses_**
  The fine line between medical and non-medical uses of nanomedical methods for diagnostic, therapeutic and preventive purposes is often hard to define, but it is possible to give examples of both. Non-medical applications include _intentional_ changes in, or to, the body due to what a person wants, when these wants are not related to medical needs, even if such medical needs are difficult to define clearly.

– *Access from an individual perspective*

Access to health-care and new medical technologies is often seen as a challenge for health-care systems and then opening issues related to fairness. Individuals may struggle to gain access to nanomedical innovations, even taking on considerable financial costs. If they cannot afford new diagnostics, drugs, or therapies offered to them, they might feel left behind or even as second-class citizen.

As stated before, what is described here is a non-exhaustive frame of references points that need to be confronted with the specific applications of nanomedicine. This method applies to case-by-case analyses and aims to reflect the ethical pluralism of modern society.

In the next part of this chapter, we describe how the approach chosen in the EU, intrinsically linking ethical considerations (both individual and social ethics) into the policy frame of nanomedicine and nanotechnology.

Embedding ethics in nanomedicine: The policy frame adopted in the EU

When this technology was developed, the European Commission had to face its controversy, not only in terms of public acceptance but also in terms of safety and in terms of adequacy of regulatory frame for nanotechnology-based products. Differently than other technology, items the Commission decided to *anticipate* the debate on both the safety and social acceptability of this medium and the risk management approach were built in terms of an unprecedented transparency, inclusiveness, and anticipatory nature.[21] In its strategic document on nanosciences and nanotechnology, the Commission indicated that nanotechnology products must comply with the high levels of consumer, worker, and environmental protection set in Community regulations.[22] In June 2008, the Commission adopted the Communication "Regulatory Aspects of Nanomaterials,"[23] fulfilling a commitment made in the Action Plan.[24] The Communication was accompanied by a Staff Working Document providing a summary of legislation in relation to health, safety, and environmental aspects of nanomaterials, and outlining regulatory research needs and related measures.[25] This regulatory review concluded that existing Community regulatory frameworks cover *in principle* the potential health, safety and

---

21 https://ec.europa.eu/research/industrial_technologies/pdf/policy/action_plan_brochure_en.pdf

22 https://www.europarl.europa.eu/registre/docs_autres_institutions/commission_europeenne/com/2009/0607/COM_COM(2009)0607_EN.pdf.

23 *Regulatory Aspects of Nanomaterials*, COM(2008)366.

24 https://ec.europa.eu/research/industrial_technologies/pdf/policy/action_plan_brochure_en.pdf.

25 SEC(2008)2036.

environmental risks related to nanomaterials. Without excluding regulatory change in the light of new information, the Commission stressed that the protection of health, safety, and the environment needed to be enhanced mainly by improving the implementation of existing legislation. In addition to supporting research on risk assessment, the Commission has worked in several regulatory areas to improve implementation, assess the adequacy of existing legislation, and consider whether regulatory changes on specific aspects were necessary.[26]

The above Communication was examined by both the European Parliament[27] and the European Economic and Social Committee.[28] The European Parliament in particular questioned whether, in the absence of explicit provisions for nanotechnology in Community law, legislation can be deemed adequate to cover the risks related to nanomaterials. Given the lack of appropriate data and assessment methods, the Parliament asked that existing regulations be carefully reviewed (not only nanomedicine but also the use of nanotechnology in cosmetics, novel food and food additives). As planned, the Commission presented an updated regulatory review in 2011, paying particular attention to the points raised by the European Parliament and the European Economic and Social Committee.

The Commission has therefore supported innovation in nanotechnology through different policies and actions. The main initiatives related to nanotechnology included increased emphasis on applications in the research funded under the research and innovation framework programs; a continued commitment to regulatory and standardization activities; and the creation of a nanotechnology observatory, ObservatoryNANO,[29] to study opportunities and risks in various technology sectors. In this context, special attention was also paid to SMEs and start-ups.

Several actions were equally undertaken in pursuit of the general objective of taking people's expectations and concerns into account. In February 2008, the Commission adopted the recommendation for a "Code of Conduct for responsible nanosciences and nanotechnologies research,"[30] which provides guidelines favoring a responsible and open approach. As called for by the Council in September 2008,[31] the Commission regularly monitor the Code, and revise it every two years in order to take into account developments in nanotechnology and their integration in European society.

All research project proposals that were considered for funding (by the European Commission) and raise ethical issues underwent a thorough ethical review. This

---

26 For example, the working group for nanomaterials under REACH has made progress and published initial results: http://ec.europa.eu/environment/chemicals/reach/pdf/nanomaterials.pdf.
27 Resolution of 24 April 2009 on regulatory aspects of nanomaterials (2008/2208(INI))
28 Opinion of 25 February 2009 on the Communication on Regulatory Aspects of Nanomaterials, INT/456; http://eesc.europa.eu/documents/opinions/avis_en.asp?type=en
29 www.observatorynano.eu
30 *Code of Conduct for responsible nanosciences and nanotechnologies research*, C(2008)424
31 12959/1/08 REV 1 (2891st Council Meeting Competitiveness)

included many nanotechnology proposals. Such proposals had discussed the ethical dimension of the research to be undertaken and meet community ethical requirements, such as the EU Charter of Fundamental Rights, in addition to national requirements.

Several outreach projects have been funded under the European Commission research programs. These suggest that there is a need for a more permanent public deliberation on nanotechnology in its broad societal context. The Commission has pursued an active policy of engagement and consultation with stakeholders, in particular through their continuous involvement in Commission working groups in charge of coordinating the implementation of regulation; and in the annual nanotechnology "Safety for Success Dialogue" workshops. The call for dialogue and engagement in the action plan has also been reflected in various other initiatives organized by industry, in European Technology Platforms, and in special interest forums such as consumers' groups. The existence of diverse fora indicates a need to monitor the debates at national, European, and international levels, for instance, with support from future European Commission activities, in order to consistently convey messages from public debates to policymakers. The Commission has also published a wide range of informative materials in many languages and for various age groups. In addition, a specific entry for nanotechnology on the Commission's Europa website has helped the public to follow all its nanotechnology activities.

As far as the science for policy aspects of this example are concerned, what it was relevant from the Commission was not only to address the scientific assessment of the nanotechnology products in specific application domains (e.g. medicine and food – SCENIHR), but to ask the European Group on Ethics to assess the governance aspects of this technology. This request was done at European Commission's top level (President of the Commission) and aimed to analyze this technological sector under the remit of this independent advisory body: issuing a set of policy recommendations based on the scientific, legal, social, and ethical implications that may rise from the use of nanotechnology. This work was also done in consultation of key experts in the field, decision-makers, and relevant stakeholders.[32] In addition, a coordination group (ISG/ Inter Services Group) clustering 20 commission services was established to check the consistency of actions taken by different Commission Directorates in this specific field. The strategic choice by the Commission was to have scientific advice not only on technical aspects of nanotechnology but also on a strategy to endorse for a transparent, inclusive, responsible, and socially acceptable use of this technology and the products based on its use. In fact, this strategy has showed being successful and nanotech-based products are part of many innovative industrial productions across the EU.

---

32 https://ec.europa.eu/archives/bepa/european-group-ethics/archive-activities/activities-2005–2010/index_en.htm and https://ec.europa.eu/archives/european_group_ethics/archive/2005_2010/activities/docs/roundt_nano_21march2006_final_en.pdf

Conclusions

Pluralism is a characteristic of the EU,[33] mirroring the richness of its traditions and adding the need for mutual respect and tolerance. Respect for different philosophical, ethical/moral, or legal approaches and for diverse cultures is implicit in the ethical dimension of building a democratic Europe. Social and ethical pluralism requires that a culture of debate and communication needs to be established wherever and whenever wide-ranging changes to the lives of individuals, or in social practices, take place or are liable to take place in the future.

The above is relevant also for the controversies prompted by nanomedicine. Such issues have been addressed upfront in Europe, both in the European Commission Communication "Towards a European Strategy for Nanotechnology"[34] and stimulating international cooperation such as with the workshop "Nanotechnology: Revolutionary Opportunities & Societal Implications" co-organized with the US National Science Foundation[35] and launching an open "International Dialogue on Responsible Research and Development of Nanotechnology."[36]

The mentioned European strategy spelled clearly out how ethical principles must be respected and, where appropriate, enforced through regulation. These principles are embodied in the European Charter of Fundamental Rights[37]and other European and international documents[38] The EU has furthermore taken several steps in the development of policy design of nanomedicine *within* the constraints of the principle of respect for the rights of individuals, respect for multiculturalism, dialogue, and tolerance. The proposed approach did not try to fix ethics rules for nanomedicine, but rather stimulate a flexible approach to the governance of nanomedicine where ethical evaluation, safety, legal clarification (including data protections and patenting), and social debate were all equally requested for the design and use of nanomedical application and research in nanomedicine.

Renzo Tomellini and Maurizio Salvi

---

33 Respect for pluralism is in line with Article 22 of the European Charter of Fundamental Rights, on "Cultural, religious and linguistic diversity" and with Article 6 of the Amsterdam Treaty, which ensures the protection of fundamental rights at EU level, based in particular on international instruments as well as common constitutional traditions, while also stressing respect for the national identity of all Member States.
34 https://ec.europa.eu/research/industrial_technologies/pdf/policy/nano_com_en_new.pdf
35 https://www.nsf.gov/mps/dmr/lecce_workshop.pdf
36 https://ec.europa.eu/research/industrial_technologies/pdf/policy/report-third-international-dialogue-2008_en.pdf
37 See http://www.europarl.eu.int/charter/default_en.htm
38 See http://europa.eu.int/comm/research/science-society/ethics/legislation_en.html

List of contributing authors

Preface, Foreword and Conclusion
Mihail C. Roco
National Science Foundation Engineering
2415 Eisenhower Avenue
22314 Alexandria
Virginia
USA
mroco@nsf.gov

Peter Brabeck-Letmathe
Nestlé SA
Chairman Emeritus
Av. Nestlé 55
1800 Vevey
Switzerland
peter.brabeck-letmathe@nestle.com

Shailendra Saraf
University Institute of Pharmacy
Pt. Ravishankar Shukla University
Raipur
Chhattisgarh
India

Federico Mayor Zaragosa
Former President, UNESCO, Paris, France
Ciudad Universitaria de Cantoblanco,
Pabellón C - C/Einstein 13,
28049 Madrid
Spain
info@fund-culturadepaz.org

Renzo Tomellini
Head of Unit "Science Policy,
Advice and Ethics"
European Commission
DG Research & Innovation
ORBN 3/124
1049 Brussels
Crans – Montana
Belgium
renzo.tomellini@ec.europa.eu

Maurizio Salvi
Unit "Science Policy,
Advice and Ethics"
European Commission
DG Research & Innovation
ORBN 3/126
1049 Brussels
Belgium
maurizio.salvi@ec.europa.eu

Marcel Van de Voorde
Bristol – A, Apt. 31
Rue du Rhodania 5
3963 Crans-Montana
Switzerland

Gunjan Jeswani
Faculty of Pharmaceutical Sciences, Shri
Shankaracharya
Group of Institutions, SSTC
Department of Pharmaceutics
Junwani
490020 Bhilai, Chhattisgarh
India
gunjanjeswani@gmail.com

Chapter 1
Frans W.A. Brom
Netherlands Scientific Council for
Government Policy (WRR)
P.O. Box 20004
NL-2500 EA
The Hague
The Netherlands
brom@wrr.nl

Rinie van Est
Rathenau Instituut, the Dutch parliamentary
TA organization
P.O. Box 95366
NL-2509 CJ
The Hague
The Netherlands
q.vanest@rathenau.nl

https://doi.org/10.1515/9783110669282-208

Bart Walhout
National Institute for Public Health and the
Environment (RIVM)
P.O. Box 1
NL-3720 BA
Bilthoven
The Netherlands
bartwalhout@gmail.com

Chapter 2
Bengt Fadeel
Karolinska Institutet
Institute of Environmental Medicine
13 Nobels väg
171 77 Stockholm
Sweden
bengt.fadeel@ki.se

Phil Sayre
NanoRisk Analytics, LLC
5 Terrace Court
95603 Aubur, CA
United States
phil.sayre@verizon.net

Chapter 3
Marcel Van de Voorde
Bristol – A, Apt. 31
Rue du Rhodania 5
3963 Crans-Montana
Switzerland

Michael Vlerick
1. Tilburg University
2. University of Johannesburg
Philosophy
44 Broekkantstraat
9051 Afsnee
Belgium
michaelvlerick@gmail.com

Chapter 4
Dave H. A. Blank
University of Twente
MESA+ institute
P.O.Box 217
7500AE Enschede
The Netherlands
d.h.a.blank@utwente.nl

Chapter 5
Karolina Jurczyk
Department of Reconstructive Dentistry
and Gerodontology
University of Bern
School of Dental Medicine
Bern
Switzerland
karolajur@gmail.com

Urs Braegger
Department of Reconstructive Dentistry
and Gerodontology
University of Bern
School of Dental Medicine
Bern
Switzerland

Mieczyslaw Jurczyk
Poznan University of Technology,
Institute of Materials Science and
Engineering,
M. Sklodowska-Curie 5 Sq.,
60–965 Poznan
Poland

Chapter 6
Daisuke Fujita
National Institute for Materials Science
Research Center for Advanced
Characterization and Measurement
1 chome 2-1 Sengen
305-0047 Tsukuba, Ibaraki
Japan
fujita.daisuke@nims.go.jp

Chapter 7
Thomas H. Brock
German Social Accident Insurance Institution
for the raw materials and chemical industry
Department of Hazardous Substances
and Biological Agents
62 Kurfürsten-Anlage
69115 Heidelberg
Germany
thomas.brock@bgrci.de

Chapter 8
Ajey Lele
Manohar Parrikar Institute For Defence
Studies and Analyses
Strategic Technologies
1, Development Enclave
Rao Tula Ram Marg
110010 New Delhi
India
ajey.lele@gmail.com

Kritika Roy
Hertie School Centre for International
Security
180 Friedrichstraße
10117 Berlin
Germany
kritikaroy.26@gmail.com

Chapter 9
Thomas Reuter
University of Melbourne
Asia Institute
24 Faulbacherstrasse
56206 Kammerforst
Germany
thor2525@gmail.com

Chapter 10
Jean Pierre Massué
—Deceased – November 2020
– Member of the Senate of the European
Academy of Sciences and Arts,
– Member of the Senate of the European
Materials Research Society: EMRS
– Former Executive Secretary of the EUR-OPA
Major Hazard Agreement of the Council of
Europe.

Chapter 11
Oluwatosin Ademola Ijabadeniyi, Ph.D
Durban of University of Technology
Department of Biotechnology and Food
19 Steve Biko
4001 Durban, KZN
South Africa
oluwatosini@dut.ac.za

Chapter 12
Ilise L. Feitshans
Director, ESI Safernano,
Fellow in international law of
Nanotechnology,
European Scientific Institute Archamps
France (ESI) and Guest Researcher
Center for Biomedical Law,
Faculty of Law University of Coimbra
Portugal and Invited Professor
ISTerre University of Grenoble
France
ilise.feitshans@univ-grenoble-alpes.fr

Chapter 13
Armin Grunwald
Karlsruher Institut für Philosophie
Institute for Technology Assessment and
Systems Analysis (ITAS)
11 Karlstr.
76133 Karlsruhe
Germany
armin.grunwald@kit.edu

Chapter 14
Albert Ed. Evrard
University of Manitoba
School of Law (visiting)
224 Dysart Road
R3T2N2 Winnipeg, Manitoba
Canada
albert.evrard@unamur.be

Praveen Martis
Associate Professor
Postgraduate department of Chemistry
Principal, St Aloysius College
(Autonomous)
Mangalore, India
praveenmartis@gmail.com

Chapter 15
Michael Vlerick
1. Tilburg University
2. University of Johannesburg
Philosophy
44 Broekkantstraat
9051 Afsnee
Belgium
michaelvlerick@gmail.com

Chapter 16
Nobuyuki Haga
Ishinomaki Senshu University
Biological Science
1 Minamizakai
986-8580 Ishinomaki, Miyagi
Japan
haga@isenshu-u.ac.jp

Chapter 17

Rune Nydal
Norwegian University of Science
and Technology
Programme for Applied ethics, Department of
Philosophy and Religious studies
40 Dravoll alle
7491 Trondheim
Norway
rune.nydal@ntnu.no

About the editors

Prof. Dr. Ing. ir. Dr. h. c. Marcel Van de Voorde has many years' experience in European Research Organizations, including CERN- Geneva and the European Commission research. He was involved in research, research strategies and management. He is emer. professor at the University of Technology, Delft (NL), holds multiple visiting professorships and is doctor honoris causa. He has been a member of numerous Research Councils and Governing Boards: e.a. CSIC (F), CNR (I), CSIC (E), NIMS (JP), of science and art academies, of the Science Council of the French Senate and the National Assembly, in Paris, and Fellow of multiple scientific societies. He has been honored by the Belgian King and received an award for European merits in Luxemburg by the former President of the European Commission. He is author of multiple scientific and technical publications and books.

Gunjan Jeswani is associate professor and head of the department of Pharmaceutics, Faculty of Pharmaceutical Sciences at Shri Shankaracharya Technical Campus, Shri Shankaracharya Group of Institution, Bhilai, Chhattisgarh, India. She is an innovative researcher in the field of pharmaceutical nanotechnology. Her research interest involves advanced polymeric dosage forms for cosmeceutical and parenteral drug delivery. She has developed novel long-acting dosage forms, including transdermal creams and particulate-based systems for the delivery of synthetic and natural drug molecules. She has published review and research articles in high impact national and international journals, co-authored book chapters in international books published by reputed publishers, and also authored a few textbooks of pharmaceutical sciences. Along with ten years of teaching experience, she is also the recipient of the "Young Scientist Award-2018" by Chhattisgarh Council of Science and Technology (CGCOST), India. She has additionally received various scholarships and grants from the Indian government for qualifying national exams and organizing national conferences.

https://doi.org/10.1515/9783110669282-209

Part I: **Highlights of ethics in nanotechnology**

Frans W.A. Brom, Rinie van Est and Bart Walhout

1 Nanoethics: Giving orientation to societal reflection

Abstract: We are now looking back at over 20 years of social reflection on nano-technology. Gaining legitimate trust is vital for the nanotechnology community. That is why nanotechnologists engage in public discussions, social engagement, and interdisciplinary nanoethical research. Ensuring safety is an important ele-ment in ensuring trust. The transformative potential of nanotechnology, however, becomes apparent in its convergence with other emerging technologies.

The notion of converging technologies helps to explore this potential. It opens a new perspective for nanoethics and helps to identify societal challenges that need to be on the agenda of societies. The ethical issues raised by nanobiology, nanomedicine, and nanoelectronics illustrate these challenges: from nanobiology to the discussion on synthetic biology, from nanomedicine to the discussion on human enhancement, and from nanoelectronics to the discussion on artificial in-telligence (AI). In all three fields, we will show a similar structure: (1) development in these fields is impossible without nanotechnology, (2) these fields raise funda-mental and relevant normative issues, and (3) a nanoethics cannot confine itself to the direct and strict impact of nanotechnology itself, but needs to open up to the broader questions raised by the interaction between nanotechnology and other emerging technological fields. This shows us the challenge for nanoethics: it needs to orient societal reflection on the impact of nanotechnology, and in doing so it cannot confine itself to the direct consequences of specific nanotechnologies. Nanoethics needs to orient the societal reflections and discussions on where to go (direction), what to protect, and whom to empower in order to protect themselves (protection), and on practical ways to govern these developments (organization).

Key Words: Public deliberation, Technology Assessment, NBIC-converging, Human Enhancement, Synthetic Biology, Artificial Intelligence, Trust, Intimate technology

Frans W.A. Brom, Netherlands Scientific Council for Government Policy (WRR), NL-2500 EA, The Hague, The Netherlands; Ethics Institute, Faculty of Humanities, Utrecht University, Utrecht, The Netherlands
Rinie van Est, Rathenau Instituut, The Dutch Parliamentary TA Organization, The Hague, The Netherlands; Technology, Innovation and Society, Industrial Engineering and Innovation Sciences, Eindhoven University of Technology, Eindhoven, The Netherlands
Bart Walhout, National Institute for Public Health and the Environment (RIVM), Ministry of Health, Welfare and Sport, (RIVM), Bilthoven, The Netherlands

https://doi.org/10.1515/9783110669282-001

1.1 Introduction

Nanoethics emerged in the debate on nanotechnology. Nanotechnology is new, and nanoethics therefore has two goals: to give orientation to the development of nanotechnology and to structure and orient the social reflection on the future of nanotechnology. Nanoethics can thus be seen as a form of "reflexivity": it is part of social reflection and it looks at ways to stimulate and structure this social reflection. It is important to take this reflexivity into account, because it emphasizes the double task of nanoethics: to support the ethical development of nanotechnology and to support societal reflection on the ethical development of nanotechnology [1]. In this brief introduction, we will combine both tasks. We will use the content of the first task – giving orientation to the development of nanotechnology – as an agenda for the second task – fostering societal reflection. In this way, we develop three fields where nanoethics orients societal reflection: (1) direction, (2) protection, and (3) organization. These three fields revolve around three normative questions: (1) where to go and thus what to stimulate; (2) what to protect and whom to empower to actively protect; and (3) how to organize stimulation and protection in practical terms. These three normative questions require normative answers. Our main message is that, in order to fulfill its goals and to orient societal reflection on the impact of nanotechnology, nanoethics cannot confine itself to the direct impact of specific nanotechnologies. Nanoethics – just like nanotechnology – needs a broad perspective.

In this introductory chapter, we will neither try to give an overview of the ethical issues that nanotechnology raises, nor present all the various methods and approaches that are being developed in the field of nanotechnology. Instead, an in-depth analysis of specific issues in nanoethics will be given in the following chapters. We will give a thematic introduction to the field of nanoethics, showing the challenges for nanoethics in orientating societal reflection in the future of nanotechnology. We start by looking backward. The social reflection on nanotechnology has a history of over 20 years. A core issue in the engagement of scientists and engineers working in nanotechnology in these debates was (and is) the need to gain justified societal trust in the development of nanotechnology (Section 1.2). Ensuring safety is an important element in ensuring trust. That is why we continue this thematic introduction with an inquiry into an issue that is specifically raised by nanotechnology: the safety of nanomaterials (Section 1.3). Next, we will look at the societal and ethical issues that are raised when we consider nanotechnology as an element in the broader development of technology. For this purpose, we explore the notion of converging technologies. The convergence of nanotechnology, biotechnology, information technology, and cognitive science (NBIC convergence) involves increasing interaction between the life sciences and the natural or engineering sciences. This convergence opens a new perspective for nanoethics (Section 1.4). We illustrate this development with an inventory of ethical discussions in nanobiology, nanomedicine, and nanoelectronics. These sections illustrate how for society the transformative potential of nanotechnology becomes apparent in its convergence with other emerging technologies. Developing a nanoethics from this "converging technologies" perspective helps to identify societal challenges that need to be on the agenda of societies. In each section, we use an actual societal field to illustrate this claim: from nanobiology to the discussion on synthetic biology (Section 1.5), from nanomedicine to the discussion on human enhancement (Section 1.6), and from nanoelectronics to the discussion on AI (Section 1.7). In all three fields, we show a similar structure: (1) development in these fields is impossible without nanotechnology, (2) these fields raise fundamental and relevant normative issues, and (3) a nanoethics cannot confine itself to the direct and strict impact of nanotechnology itself, but needs to open up to the broader questions raised by the interaction between nanotechnology and other emerging technological fields. This brings us to the conclusion of this introduction and the challenge for this book: nanoethics needs to orient societal reflection on the impact of nanotechnology, and in doing so it cannot confine itself to the direct consequences of specific nanotechnologies. Nanoethics needs to orient the societal reflections and discussions on where to go, what to protect, and whom to empower in order to protect themselves, and on practical ways to govern these developments.

1.2 From gaining societal trust to engaged interaction

The social reflection on nanotechnology has a history going back over 20 years. The potential societal impact of nanotechnology was already being discussed in the comprehensive foresight study coordinated between 1996 and 1998 by the Netherlands Study Centre for Technology Trends [2]. And from the start of the twenty-first century, the discussion on the social and ethical implications of nanotechnology [3] got underway, not only in academia but also in public discourse. In 2003, the report "The Big Down" by the ETC group (a Canadian NGO) played an important role in the public debate, because it posed the legitimate question as to what the benefits and risks of this new technology would be for society, and pointed to the many uncertainties with regard to the health impact of nanoparticles [4].

For many scientists, the ETC group's criticism had a familiar flavor, reminding them of the debate on genetically modified organisms (GMOs). They saw GMO technology as one of many promising and useful technologies that had been unable to connect successfully to the broader society, due to negative campaigning by environmental groups such as Greenpeace. As a commentator wrote in *Nature*, "Nanotechnology is set to be the next campaign focus of environmental groups. Will scientists avoid the mistakes made over genetically modified food, and secure trust for their research?" [5]. One could say that the debate on nanotechnology followed a recurring type of argumentation in which the various societal actors play a pre-scripted role in the debate [6]. As Swierstra and Rip indicate, a debate of this kind takes on a predictable pattern in which different arguments "hang together" in the sense that they provoke each other into existence. The tropes and the "storylines" in the argumentative patterns have become a repertoire that is available in late-modern societies, both as a framing of how actors view issues and expect others to view them, and as a kind of toolkit that can be drawn upon in concrete debates" [7]. Ethical reflection aims to open up these ossified patterns.

For the good development of nanotechnology, a more responsive relationship with society was seen as necessary. This relationship is important because those working in the field of nanotechnology mostly have the honest belief that it can be an important aid to meeting societal challenges. Nanotechnological developments bring hope with regard to fighting famine, helping the transition to non-fossil energy, or developing new treatments for diseases such as cancer. That is why securing adequate societal trust in nanotechnology through a more responsive relationship with society is important. At least two elements are important in this responsive relationship: systematic consideration of the societal and ethical implications of nanotechnology, and more and better communication between the nanotech community and society.

The wish to develop a better understanding of the implications of nanotechnology and better processes of social reflection led to various research programs and organized societal deliberations on nanotechnology in the years 2003–2010. In the Netherlands, for instance, the national research consortium NanoNed decided in 2002 to include research into the societal and ethical implications in their research program and to link these implications with nanoresearch and innovation [8]. In the UK, several "upstream" activities [9] were organized in 2005–2006 to include citizens in the discourse on nanotechnology and its development [10, 11]. And in Germany, the Office of Technology Assessment at the German Bundestag "was already commissioned by the research committee of the German Bundestag to carry out a TA study on nanotechnology as early as 2000. The results were presented to the committee and published in 2003" [12].

As the German example makes clear, a third element is needed to secure societal trust: adequate and democratically legitimized oversight and governance. From our perspective, the search for a more responsive relationship between nanoscience and society can be seen as a search for technology assessment. Technology assessment is the institutionalized practice of studying and publicly deliberating on the broad societal impact of new and emerging technologies [13]. Technology assessment is not just academic thinking; it includes organized societal discussion and support for democratic political decision-making on new and emerging technologies [14]. For us, technology assessment "combines an awareness about potential negative and positive effects of technological change with the belief or hope that one can anticipate these effects" [15].

Nanoethics has developed against the background and as a part of the critical reflection on nanotechnology. Like other fields in applied ethics (e.g., bioethics, agricultural ethics, or public health ethics), nanoethics is not pure philosophical ethics but an interdisciplinaryendeavor "to advance the examination of ethical and social issues surrounding nanotechnologies in a philosophically rigorous and scientifically informed manner" (in the words of the first editorial of the journal *Nanoethics* in 2007) [16]. And in doing so, nanoethics does not need to be a separate field of ethics, nor is it necessary to identify specific "unique" nanotechnology questions. In nanoethics, general ethical issues are linked to specific nanotechnological developments.

Various interdisciplinary approaches have been developed in nanoethics: methods regarding the engagement of citizens in the discussions on the impacts of nanotechnologies under the heading of upstream engagement [17], methods for bringing normative considerations into the development of the technology itself under the heading of value-sensitive design [18], methods for analyzing and weighing the societal aspects of risk [15], and finally comprehensive approaches to implementing responsibility in research and innovation under the heading of responsible research and innovation [19]. All these approaches have one thing in common: they seek for ways that help engineers, citizens, scientists, members of parliament, and people in governments, funding agencies, and NGOs to start a sensible conversation on nanotechnology and its impacts.

A peculiar aspect of ethics is that it is not neutral and descriptive when it comes to fostering a conversation of this kind. Ethics, and hence nanoethics, is a normative and prescriptive discipline. It is not primarily interested in describing the way things are empirically, but in prescribing how things ought to be [20]. The core of doing ethics is the belief that it is relevant to reflect on and argue about the way things ought to be. As stated in the introduction, the three functions of nanoethics (direction, protection, and organization) revolve around three normative questions: where to go and what to stimulate, what and whom to protect and empower, and how to organize. The normative answers to these questions are not "just" given by the way the world happens to be. When discussing possible answers, people put forward arguments to support certain choices. If we are to have a sensible conversation about these choices, facts are important but not enough: exchanging and reflecting on the different arguments is at least as important. And we might differ, of course, as regards the answers to these questions. Doing nanoethics, however, only makes sense if – without being blind to the reality of power structures, lobbying, and other influences on the way citizens and society make their choices – we strive for choices that are supported by the best possible arguments. Doing ethics thus means looking for the best possible arguments, including a critical analysis of those arguments, their structure, and their persuasiveness.

1.3 From nanosafety to innovation

Ensuring safety is an important element in gaining trust. From the moment nano-technology development started to be organized in large-scale publicly funded re-search programs, one of the most pressing issues has been the question of whether the "nano" in nanotechnology (i.e., engineering material properties at the nano-scale) would itself be safe for human health and the environment. Nanomaterials exhibit unique properties compared with "normal" materials made of the same chemical substances. For example, whereas normal carbon does not allow for electrical conductivity, very small carbon nanotubes can be used as a new genera-tion of electronic switches in ever smaller and faster computer chips. In longer lengths, these tubes can be used as lightweight but superstrong materials. How-ever, in such lengths, carbon nanotubes might have the same properties as rigid fibers like asbestos. Bearing in mind that asbestos was once also touted as a mira-cle material and its large-scale use caused many early deaths, it is no wonder that the introduction of nanomaterials sparked a global debate among experts and NGOs on nanosafety and about what a precautionary approach to the safety of nanomaterials would look like.

Addressing the potential safety issues of nanomaterials is far from straightfor-ward, however. First of all, carbon nanotubes comprise only a very small fraction of the many nanomaterials being developed, and only the long versions of these carbon nanotubes may pose the risk that small fibers exhibit in general when being processed. Other nanomaterials also have totally different uses, going as far as the protection of human health (e.g., nanoparticles for new cancer therapies). Both the novelty of nanomaterial characteristics and the many shapes and sizes in which nanomaterials can be used make the development of adequate safety test-ing a daunting task, bringing other discussions in its wake – for example, the large number of animal tests that would be needed for regulatory approval. Uncer-tainties about safety thus bring questions of direction, protection, and organiza-tion directly to the fore.

One such area is the organization of protection. There is a clear need to *consoli-date* relevant knowledge from the growing number of scientific papers on nanosafety into regulatory requirements. At the same time, *innovation* in risk assessment method-ologies is needed, to enable us to cope with the rapidly increasing range of materials to be assessed and to respond to the calls to reduce animal testing. However, the same innovations unsettle the process of standardization and consolidation. Furthermore, instead of "end-of-pipe" testing just before market introduction, both industry and the authorities would prefer early screening and the application of safe-by-design princi-ples in R&D processes as far as possible [21]. The development of such approaches here requires close collaboration between the "innovators" and the "regulators." Such a *transformation* of public–private relationships brings with it new challenges in terms of building trust, sharing costs, checks and balances, and intellectual property protection. Given that resources are limited, ethics with regard to risk is thus a question not only of safety levels for individual nanomaterials in specific contexts but also of the practical choices involved in trying to regulate current generations of nanomaterials, while anticipating future ones. The boundaries of existing institutional structures therefore need to be renegotiated and transformed.

Another area is that of choices related to balancing the various values that drive innovation. For example, society needs to transition to non-fossil energy production. That is why substantial budgets are being spent on nanomaterial research to enable the transition. Core challenges include the safety of these nanomaterials. Increasing the energy efficiency of solar panels or improving battery lifetime, for example, often involves the use of rather toxic materials, which present pressing safety problems when these products reach the end of life. Here, new value articulations (e.g., the transition to a circular economy, which demands that waste be turned into new feedstock) open new perspectives on the trade-offs between safety and sustainability [22]. There is consequently an increased need for holistic approaches, broadening the scope from Safe-by-Design (SbD) to Safe-and-Sustainable-by-Design (SSbD) principles and linking protection and direction in approaches for responsible research and innovation.

As this example makes clear, practical choices in innovation show the importance of opening up risk assessment and risk management practices. Proper governance of innovative processes and the balancing of safety, sustainability, and practical use need to incorporate expert-driven risk assessment in broader deliberative processes. Special consideration, therefore, needs to be given to issues of risk appraisal and decision-making under conditions of complexity, uncertainty, and ambiguity [15]. This sketch of an ethics of nanorisks shows that ethical reflection and societal discussion on the risk governance of nanomaterials needs to broaden out beyond sheer "protection" to include the dimensions of direction and organization. An ethics of nanosafety cannot confine itself to the responsibilities of experts to protect human and environmental safety; it needs to include the empowerment of citizens with regard to safety issues, as well as the concepts that guide innovation (e.g., the "cradle-to-cradle" approach) and practical issues regarding the organization of safety governance, citizen empowerment, and innovation for societal challenges.

1.4 From nanotechnology to NBIC convergence

Combining technologies in novel ways often drives innovation. Over the past few decades, the technological convergence between four technological revolutions has been creating a new technological wave. Known as "NBIC convergence" nanotechnology, biotechnology, information technology, and cognitive technology are propelling each other.

NBIC convergence signifies an increasing interaction between the life sciences, which have traditionally studied living organisms, and the natural or engineering sciences, which have traditionally studied and built non-living systems [23]. This merger is reflected in two bioengineering megatrends: "biology is becoming technology" and "technology is becoming biology" [24].

Inherent in the "biology is becoming technology" trend is the promise that living systems (e.g., genes, cells, organs, and brains) might be engineered in much the same way as non-living systems (e.g., bridges and electronic devices). This expectation is driven by the fact that the nanosciences provide more and more ways to measure, analyze, and intervene in living organisms. "Biology becoming technology" promises a major increase in new types of interventions in living organisms, including the human body and brain, as illustrated by cultured heart valves, bacteria with complete synthetic genomes, human germline editing, and deep brain stimulation.

The "technology becoming biology" trend entails the ambition of engineering properties we associate with living organisms (e.g., self-assembly, self-healing, reproduction, and intelligence) into technology. The "technology becoming biology" trend embodies a future increase in new types of bio-, cogno-, and socio-inspired artifacts, which will be used in our bodies and brains and/or intimately integrated into our social lives. Examples of bio-inspired artifacts are stem cells, hybrid artificial organs, 3D-printed artificial blood vessels, and synthetic cells. Humanoid or android robots, avatars, emotion detection, and AI are examples of cogno-inspired artifacts.

This new wave of technologies enabled by NBIC convergence radically broadens the bio-debate [25]. Technologies such as human germline editing and deep brain stimulation fuel the debate on human enhancement. Those types of interventions in the human body and brain force policymakers to anticipate new issues in the field of safety, privacy, bodily and mental integrity, and informed consent. Secondly, intelligent machines – from smartphones to automated cars, service robots, and augmented reality glasses – are able to detect and portray social and emotional behavior and shape our behavior. These intimate technologies, which are increasingly penetrating our privacy and social life, lead to new safety, privacy, data ownership and liability issues, and questions regarding the limits to the simulation of, for example, friendship and violent behavior [26]. Besides raising all kinds of societal and ethical issues, NBIC convergence challenges some of the basic concepts we use to understand ourselves and the world we live in [27]. The "biology is becoming technology" and "technology is becoming biology" trends are slowly but surely blurring the boundaries between living and non-living; sickness and health; technology and nature; human and machine intelligence; human and machine agency; and science, technology, and society. These types of conceptual uncertainties are putting the regulation of numerous existing social practices under pressure. How to regulate AI, the development of killer robots, human germline editing, or virtual reality in a humane way?

If nanotechnology is indeed an enabler for technological integration, then it is clear that nanoethics needs to take this enabling role of nanotechnology in the convergence of technologies as its point of departure. We will illustrate the implications for nanoethics with the aid of three short examples.

1.5 From nanobiology to synthetic biology

One of the ways converging gets real is by combining nanotechnology with genomics, which leads to nanobiology. This is not the place to discuss the technical issues involved in this combination, but it seems clear that, combined with an engineering perspective on biology, it leads to the idea of synthetic biology. The essential feature of synthetic biology is the notion that the biological world is no longer a "given" but becomes an object of design, engineering, and human creativity [28].

This change of perspective opens a Pandora's box of ethical issues. It raises questions regarding the direction of synthetic biology: should we confine synthetic biology to smart biological systems for the specific production of important materials, or should we consider our ecosystem as an object for technical improvement [29]? Systematic agriculture and traditional breeding already give us enormous influence over the ecosystem; hence, it might be worthwhile to look at it more systematically and with more and better tools from a sustainability perspective.

Questions regarding protection follow directly from this perspective. It is not only important to protect humans against biological risks created by synthetic biology; the protectability of nature, ecosystems, endangered species, or biodiversity also becomes an object of deliberation and reflection. Fundamental philosophical questions regarding the moral status of non-human entities go hand in hand with practical ethical questions regarding the management of biological dangers and the role of a precautionary principle.

Synthetic biology intensifies the need for a discussion on the ethics of organization. Changing the biological outlook of the planet is by definition not a private issue. It is clear that at least some decisions regarding the development of synthetic biology go beyond the territory of individual freedom. How to reconcile this with the idea of academic freedom, or with a free market? How should we organize collective decision-making? What institutions can be helpful in setting collective goals and protecting individual freedoms and the environment?

If we take the role of nanotechnology in the development of synthetic biology seriously, it is clear that there are very few boundaries to the field of nanoethics.

1.6 From nanomedicine to human enhancement

A combination of nanotechnology with medical technologies is being exploited in nanomedicine. Issues in nanomedicine range from delivering precise quantities of medicine to precise sites in the body to replacing diseased organs with artificial (improved) artifacts. These developments can raise the same ethical questions as high-tech medical technologies in general, such as fairness and equity with respect to high therapy costs, or the treatment of so-called welfare diseases. Here, however, we focus on more radical questions involved in the step from nanomedicine to human enhancement, which follow on from the idea that there is no medical necessity to limit the goals of nanomedicine to restoring normal functioning. We focus on human enhancement because it is an extreme perspective that helps to clarify the issues at stake in nanomedicine. The essential feature of human enhancement is the notion that, with the aid of growing biomedical understanding, the human mind and body can be transformed into objects of design, engineering, and human creativity [30].

Human enhancement can be seen as the use of biomedical technology to achieve goals other than the treatment or prevention of disease. There are many different definitions of "human enhancement," however [31]. The discussion on these definitions is not "just" a game of words; it is a discussion on the proper framing of the debate, and the various definitions pre-structure different perspectives on the social and political appraisal of human enhancement [32]. The question of direction would seem to be hidden in these definitional discussions. The direction of human enhancement – improving human performance – is not new. "Existing enhancement technologies, like dietary supplements and hearing aids, are relatively uncontroversial. Other examples prompt more public discussion: cosmetic surgery, the use of drugs beyond their original medical settings, narcotics and doping in sports. These technologies in combination with prospected future enhancement technologies spur public debate on human enhancement" [30]. The discussion on the direction of human enhancement is also a discussion on what it means to be human. Is what it means to be human fixed, or is it open to our collective creativity?

But if "being human" is at stake in the discussion on human enhancement, then this raises the question of how to protect human vulnerability. There are two sides to protect human vulnerability in this context. The first is protecting individuals against harms and risks caused by the negative consequences of developing technologies. Being in control of the human body might be the dream of human enhancement, but for the foreseeable future, control is out of reach. Unintended, unforeseen collateral damage due to enhancement techniques that are still under development might pose risks for their users and can cause real harm. The medical profession and the health system generally have a strict professional ethic and governance in which the "primum non nocere" principle (above all, do no harm) is pivotal. A vital consideration when looking for and developing new medical techniques is the idea of proportionality: positive and negative chances need to be in balance. Balancing the risks and benefits for people who are seriously ill and need treatment seems to be of a different order from balancing the positive chances of enhancement against the risks to someone's currently uncompromised health. But besides the questions of "Is human enhancement safe?" and "How safe should it be in order to be allowed?" there are also societal values at stake that need to be protected. A significant discussion that human enhancement provokes is that of justice [33]. How can human equality be protected if some have access to technologies that help them to enhance their intelligence, strength, or vitality? Improving individual performance is valued in our society, but not at all costs and not by any means. Doping is banned from competitive sports because it is considered unfair. But what about the use of drugs that are intended to improve focus and attention (e.g., Ritalin) by students during their exams? And do these practices create peer pressure: will improved results achieved by users pressure other students or co-workers into using similar drugs? A critical discussion on human enhancement relates to the idea that unnecessary changes to the body could be considered as a violation of human dignity or as a lack of respect for human life. A more positive perspective emphasizes the continuity between self-development and the quest for self-improvement through education, sport, or a healthy lifestyle and the use of human enhancement techniques. The goal of self-improvement is deeply embedded in human development, and the use of new means – in itself – changes little.

Finally, there is the question of organization. How should we organize the possible implementation of human enhancement? Is applying human enhancement technologies an individual choice by individual consumers, or should the choice be made more collectively? Collective decision-making is difficult because – as former colleagues of the Rathenau Instituut have written – "State-promulgated policies and religious movements that have been claiming a monopoly in defining 'humanity' have a rather dubious track-record. Still, new technological developments can have collective effects. To discuss and to re-think the consequences of enhancement in terms of their meaning for our self-perception is a necessary condition for a well-functioning democracy. But what space is left to start such a collective conversation?" [34]. They claim that, in a liberal society, the basic setup for decision-making on such a dazzling topic as human enhancement therefore seems to consist of two components: state-secured safety and individual choice. And although they realize that "this picture is a bit of a caricature. Nevertheless, it serves to make our main point, namely that notwithstanding the great benefits of having a liberal society, in its basic structure it might have insufficient space for public discourse to address questions on collective issues that cannot be dealt with on a satisfactory level by either state regulation or individual choice" [34]. And as we have argued elsewhere, it is doubtful that "the fora for ethical reflection and debate that are institutionally linked to the practices of professional health research and health care will be able to take up the challenge to explore the ethical issues raised by the developments outside the medical domain" [35]. Thus, nanoethics cannot evade the fundamental questions of political philosophy: democracy, individual freedom, public goods, and collective decision-making.

The role of nanotechnology in nanomedicine in the development of human enhancement technologies confirms the conclusion of the section on synthetic biology (Section 1.5). If we take the development of nanotechnology as an enabling technology in the context of emerging technologies seriously, then there are very few boundaries to the field of nanoethics.

1.7 From nanoelectronics to artificial intelligence

Artificial intelligence (AI) already dates back to the 1950s, but it has gained momentum over the past two decades, especially in the form of machine learning and deep learning. This is due mainly to the increased collection and storage of "big data" and the growing processing power of computers. Both developments are driven by great advances in information and communication technologies, and in particular micro- and nano-electronics. AI is not just a single technology: it can better be understood as a "cybernetic system" that observes, analyzes (thinks), and acts, and can learn from doing so. It cannot therefore be separated from other digital technologies such as robotics, the Internet of things, digital platforms, biometrics, virtual and augmented reality, persuasive technology, and big data. AI can be seen as the "brain" behind various "smart" applications such as self-driving cars, military drones, and cognitive assistants. It therefore plays a role in many intelligent machines and is strongly related to the broad trend of digitization as cybernetization [36].

The development of digitization involves a wide range of social and ethical issues. Besides privacy and security, issues such as the control of technology, justice, human dignity, and unequal power relationships play a role. Human rights such as non-discrimination and the right to a fair trial can come under pressure as a result of the digitization (among other things using AI) of judicial decisions [36, 37]. The above-mentioned issues are all being identified and discussed in the worldwide public debate that has arisen about AI in recent years. In the wake of and through the AI discussion, the public debate about digitization – and therefore about nanoelectronics as the driving force behind it – has broadened and reached the political agenda. Thus, nanoethics – as an ethics of a crucial enabling technology – cannot evade questions of direction and how research funding can encourage responsible innovation.

One way to discuss responsible innovation in AI is by formulating orienting principles. Interestingly, the principles that have been formulated for AI in recent years are similar to important biomedical principles, such as beneficence, non-maleficence, autonomy, and justice [38]. In biomedical practice, these principles aim to guide medical interventions in the human body and brain. It seems that awareness has grown that digital technology, in particular AI, intervenes in our lives in a fundamental way. As a result of the technological convergence we discussed above, digital technology is gaining fundamental impact on human life. And parallel to the need for bioethical reflection in the case of medical technology, AI needs digi-ethical reflection. Principles, however, are not sufficient to encourage responsible innovation in AI, because the practices in which AI is gaining importance differ fundamentally from physician–patient relationships in biomedical practice. An example is the discussion on the concept of meaningful human control. Is meaningful human control of autonomous AI systems possible, when these systems potentially make critical decisions (as in the case of killer drones or artificial judicial decisions) [39]? Another issue specific to AI is the discussion on the "right to meaningful human contact." In practices where human contact and interaction are crucial – for example, raising and educating children, caring for elderly people, and the interaction between governments and citizens – "meaningful human contact" is a key value that needs protection. Here, we see the importance of the second question on our nanoethics agenda: what to protect and who should be empowered to enable them to be protected or to protect themselves? Which features of human–human interaction constitute meaningful human contact need protection? How can we empower vulnerable groups such as children and elderly citizens against digital deception [37]?

As said above, principles alone are not enough. And there is no shortage of guiding principles. In recent years, we have seen a flood of ethics codes in the field of AI. The strength of these codes is that they are a result of collective awareness processes in various parts of society. Their weakness, however, is that the individual codes do not constitute enforceable rules. That is why the ethical question of organization is also important for these developments. In a democracy based on the rule of law, legislation ought to play a crucial role in the implementation of AI. The use of AI must comply with fundamental rights, including constitutional rights, and specific legislation such as the EU's General Data Protection Regulation or sector-specific legislation in fields such as health care or the transport market. The connection between innovation and public values must therefore also be pursued far more emphatically [40]. Innovation means not only technological innovation but also social, economic, and legal innovation.

1.8 Conclusion: The need for a broad nanoethics agenda

We are now looking back at over 20 years of social reflection on nanotechnology. Gaining legitimate trust is vital for the nanotechnology community. That is why nanotechnologists engage in public discussions, social engagement, and interdisciplinary nanoethical research. Justified trust needs to be built upon information and an open exchange of arguments. Looking at these interactions from an nanoethical perspective means looking for the best possible arguments, including a critical analysis of those arguments, their structure, and their persuasiveness. This sets the context for the challenge to nanoethics in orientating societal reflection on the future of nanotechnology.

Ensuring safety is an important element for ensuring trust. That is why we looked more closely at the agenda-setting questions raised by the safety issues that nanomaterials evoke. But nanotechnology is much more than the creation of nanomaterials. Nanotechnology is a crucial enabling technology in many technological and scientific developments. It is, for instance, crucial in what has been called "NBIC convergence." NBIC convergence signifies increasing interaction between the life sciences and the natural or engineering sciences, between the perspective on studying and cultivating living organisms and the perspective on studying and building non-living systems. Looking at nanoethics from the NBIC point of view broadens our perspective. The transformative power of nanotechnology is such that there are very few boundaries to the field of nanoethics.

In this introductory chapter, we have argued that nanoethics involves reflection in three areas: direction (including what to encourage to achieve that direction), protection (including whom to empower in order to protect themselves), and organization. The first area, "direction," relates to our hopes and fears, which structure our reflections on the future. This is especially important for the ethics of research agendas: in what direction is research going? What problems need to be solved? This involves such things as the importance of new technologies in fighting famine, supporting health, or helping to achieve more sustainable ways of living.

As we have seen, new technologies emphasize the need to protect what is vulnerable (rights, identities, autonomy). Societal reflection on this topic is urgently needed when innovations and inventions enter the market: for example, the way new devices or structures influence human autonomy and developmental freedom, and how solidarity and human welfare can come under pressure from large-scale surveillance. Protection includes the empowerment of those that need to be protected. The examples of the importance of human contact and interaction in fundamental human practices such as raising and educating children, caring for elderly people, and interaction between citizens and government show that "meaningful human contact" is a key value that needs protection. But this value also shows the importance of the follow-up question: who needs to be empowered to actively protect this value? We need to reflect on ways in which humans in vulnerable situations can be empowered against digital deception.

The third issue that nanoethics needs to take up is that of organization. This focuses on the way we organize innovation and the way we institutionalize the ethical debates about innovation (meta-governance). Nanoethics reflects on the institutions and organizations that are responsible for setting the direction (funding agencies) and are designed to govern responsibilities (legal and voluntary regulation) and the way we legally structure the distribution of costs and benefits (liability, intellectual and other property rights, and the role of companies). As we saw in the example of nanosafety, ethical reflection cannot confine itself to the responsibilities of experts to protect human and environmental safety; it needs to include the concepts that guide innovation, for example, the "cradle-to-cradle" approach. Safety issues are not confined to the actual risks of certain nanomaterials; they include safety problems when these products reach the end of life. Governance structures aim to transform the "end-of-pipe" safety question into an innovation challenge.

Nanotechnology has enormous transformative power. That is why doing nanoethics is not just making a calculation of the risks and benefits, and the distribution of those risks and benefits, somewhere in the development of a new product, device, or process. Doing nanoethics is developing orientation for continuous societal reflection and structures for societal dialogue about the direction, protection, and organization of nanotechnology. This is what has been done for more than 20 years, and during those two decades nanoethics has created and encouraged new perspectives on the relationship between science, society, technology, and governance.

References

[1] Brom FWA, Chaturvedi S, Ladikas M, Zhang W. Institutionalizing Ethical Debates in Science, Technology and Innovation Policy: A Comparison of Europe, India and China. In: Ladikas M, Chaturvedi S, Zhao Y, Stemerding D, eds. Science and Technology Governance and Ethics. A Global Perspective from Europe, India and China. Cham etc, Springer, 2015, 9–23.

[2] Ten Wolde A, ed. Nanotechnology. Towards a Molecular Construction Kit. Den Haag, Netherlands Study Centre for Technology Trends (STT), 1998.

[3] Roco MC, Sims Bainbridge W, eds. Societal Implications of Nanoscience and Nanotechnology. Dordrecht, Kluwer, 2001.

[4] Van Est R, Van Keulen I. 'Small technology – big consequences: Building up the Dutch debate on' nanotechnology from the bottom. Technikfolgenabschätzung: Theorie und Praxis 2004, 13, 3, 72–79.

[5] Brumfield G, A little knowledge . . . Nature, 2003, 424 (17 July), 246–48.

[6] Swierstra T, Rip A. Nanoethics as NEST-ethics: Patterns of moral argumentation about new and emerging science and technology. Nanoethics 2007, 1, 3–20.

[7] Rip A, Van Lente H. Bridging the gap between innovation and ELSA: The TA program in the dutch NanoR&D program nanoned. Nanoethics 2013, 7, 7–16.

[8] Wilsdon J, Willis R. See-through Science: Why Public Engagement needs to Move Upstream. London, Demos, 2004.

[9] Mampuys R, Brom FWA. Ethics of dissent: A plea for argumentative restraint in the scientific debate about the safety of GM crops. J Agric Environ Ethics 2015, 28, 903–24.

[10] Rogers-Hayden T, Pidgeon N. Moving engagement "upstream"? Nanotechnologies and the royal society and royal academy of engineering's inquiry. Public Underst Sci 2007, 16, 345–64.

[11] Rogers-Hayden T, Pidgeon N. Developments in nanotechnology public engagement in the UK: 'Upstream' towards sustainability? J Clean Prod 2008, 16, 1010–13.

[12] Grunwald A. Ten Years of Research on Nanotechnology and Society – Outcomes and Achievements. In: Zülsdorf TB, et al, eds. Quantum Engagements. Heidelberg, AKA Verlag, 2011, 41–58.

[13] Grunwald A. Technology Assessment in Practice and Theory. London, New York, Routledge, 2019.

[14] Joss S, Bellucci S, eds. Participatory Technology Assessment: European Perspectives. London, Centre for the Study of Democracy, 2002.

[15] Van Est R, Brom FWA. Risk and Technology Assessment. In: Roeser S, Hillerbrand R, Sandin P, Peterson M, eds. Handbook of Risk Theory – Epistemology, Decision Theory, Ethics, and Social Implications of Risk. Dordrecht, Springer, 2012, 1067–109.

[16] Weckert J. Editorial. Nanoethics. 2007, Vol. 1, 1–2.

[17] Rogers-Hayden T, Mohr A, Pidgeon N. Introduction: Engaging with nanotechnologies – engaging differently? Nanoethics 2007, 1, 123–30.

[18] Van de Poel I. Translating Values into Design Requirements. In: Michelfelder D, McCarthy N, Goldberg D, eds. Philosophy and Engineering: Reflections on Practice, Principles and Process. Philosophy of Engineering and Technology, Vol. 15. Dordrecht, Springer, 2013, 253–66.

[19] Ruggiu D. Anchoring European governance: Two versions of responsible research and innovation and EU fundamental rights as 'normative anchor points'. Nanoethics 2015, 9, 217–35.

[20] Beekman V, Brom FWA. Ethical tools to support systematic public deliberations about the ethical aspects of agricultural biotechnologies. J Agric Environ Ethics 2007, 20, 3–12.

[21] Soeteman-Hernandez L, et. al. Safe innovation approach: Towards an agile system for dealing with innovations. Mater Today Commun 2019, 20, 100548.

[22] Beekman M, et al. Coping with Substances of High Concern in a Circular Economy. Bilthoven, RijksinstituutvoorVolksgezondheiden Milieu (RIVM), 2020.

[23] Arthur WB. The Nature of Technology: What it is and How it Evolves. London, Allen Lane, 2009.

[24] Van Est R, Stemerding D. eds. European governance challenges in bio-engineering – Making perfect life: Bio-engineering (in) the twenty-first century. Final report. Brussels: European Parliament/STOA, 2012.

[25] Van Est R, Stemerding D, Van Keulen I, Geesink I, Schuijff I, eds. Making Perfect Life: Bio-engineering (in) the Twenty-first Century. Monitoring report. Brussels, European Parliament/STOA, 2010.

[26] Van Est R. Intimate Technology: The Battle for our Body and Behaviour. The Hague, Rathenau Instituut, 2014.

[27] Swierstra T, Boenink M, Van Est R. Taking care of the symbolic order. How converging technologies challenge our concepts. Nanoethics 2014, 3, 269–80.

[28] De Vriend H. Constructing Life. Early Social Reflections on the Emerging Field of Synthetic Biology. The Hague, Rathenau Instituut, 2006.

[29] Mampuys R, Brom FWA. The quiet before the storm: Anticipating developments in synthetic biology. Poiesis & Praxis Int J Ethics Sci Technol Assess 2010, 7, 3, 151–68.

[30] Van Est R, Klaassen P, Schuijff M, Smits M. Future Man – No Future Man. Connecting the Technological, Cultural and Political Dots of Human Enhancement. The Hague, Rathenau Instituut, 2008.

[31] Savulescu J, Sandberg A, Kahane G. Well-Being and Enhancement. In: Savulescu J, Ter Meulen R, Kahane G, eds. Enhancing Human Capacities. Chichester, Wiley-Blackwell, 2011, 3–18.

[32] Schuijff M, Brom FWA. The Dynamics of Citizen Deliberation Regarding Human Enhancement. In the Netherlands. In: Vedder A, Lucivero F, eds. Beyond Therapy v. Enhancement? Multidisciplinaryanalyses of a Heated Debate. Robolaw Series. Pisa, Pisa University Press, 2013, 143–61.

[33] Luby Š, Lubyová M. Nanoethics – A Way of Humanization of Technology for the Common Benefit. In: Kelso J, ed. Learning To Live Together: Promoting Social Harmony. Cham, Springer, 2019, 189–203.

[34] Staman J, Dijstelbloem H, Smits M. Afterword. In: Zonneveld L, Dijstelbloem H, Ringoir D, eds. Reshaping the Human Condition: Exploring Human Enhancement The Hague: (in Collaboration with the British Embassy, Science and Innovation Network and the Parliamentary Office of Science & Technology). Rathenau Instituut 2008, 149–57.

[35] Van Est R, Stemerding D, Rerimassie V, Schuijff M, Timmer J, Brom FWA From Bio to NBIC convergence – From Medical Practice to Daily Life. Report written for the Council of Europe, Committee on Bioethics, The Hague, Rathenau Instituut 2014.

[36] Royakkers L, Timmer J, Kool L, Van Est R. Societal and ethical implications of digitization. J Ethics Inf Technol 2018, 20, 127–42.

[37] Van Est R, Gerritsen J, Kool L. Human Rights in the Robot Age: Challenges due to the use of Robotics, Artificial Intelligence, Virtual and Augmented Reality. Expert Report Written for the Council of Europe, Parliamentary Assembly, Committee on Culture, Science, Education and Media. The Hague, Rathenau Instituut, 2017.

[38] Floridi L, et. al. AI4People: An ethical framework for a good AI society: Opportunities, risks, principles, and recommendations. Minds Mach 2018, 28, 689–707.

[39] AI Now Report. The social and economic implications of Artificial Intelligence technologies in the near-term: A summary of the AI Now public symposium, hosted by the White House and New York University's Information Law Institute, 7 July 2016. (Accessed June 26, 2020 at https://ainowinstitute.org/AI_Now_2016_Report.pdf)

[40] Kool L, Dujso E, Van Est R. Directed Digitalisation: Working towards a Digital Transition Focused on People and Values – The Dutch Approach. Den Haag, Rathenau Instituut, 2018.

Part II: **Nanotechnology and ethical reflections**

Bengt Fadeel and Phil Sayre

2 Toward a revitalized vision of ethics and safety for the revolutionary nanotechnologies

Abstract: The nanotechnologies offer the potential for breakthrough applications in numerous sectors of society. However, there is a potential for both good and bad, and there is a need to consider the cost–benefit of new and sophisticated materials and technologies. There is also the all-important question of safety. Indeed, if nanotechnology is an enabling technology in terms of boosting innovation and industrial competitiveness and growth, then safety assessment of nanomaterials and nano-enabled products with respect to human health and the environment is the path to successful and sustainable implementation of this technology. Here the authors take a pragmatic view of ethics and safety of nanomaterials and nano-enabled products and discuss the reliability of safety assessment methods and results, the importance of a risk governance framework for nanotechnology, and the key role of promoting a dialogue between all the relevant societal stakeholders. Nanosafety research has reached adolescence, and as we look back at the first 15 years, there are important lessons to be learned with respect to nanomaterials as well as other advanced materials.

Keywords: ethics, nanotechnology, risk governance, safety assessment, sustainability

2.1 New wine into old wineskins: Nano and ethics

The biblical parable of new wine and old wine bottles (Luke 5:36–39) is well known. The question is: does nanotechnology pose a new problem, or would it be safe to assume that the existing risk management framework is sufficient? On the other hand, perhaps nanotechnology is an entirely new entity and the existing guidelines and regulations cannot contain it, leading the new wine to burst the old wineskins?

Acknowledgments: Bengt Fadeel (BF) wishes to acknowledge the generous support of the Swedish Foundation for Strategic Environmental Research through the MISTRA Environmental Nanosafety program, and the European Commission through the Horizon2020 consortium, BIORIMA (biomaterial risk management) (grant agreement no. 760928). BF is chair of the expert panel of the national nanosafety platform, SweNanoSafe, commissioned by the Swedish Ministry of the Environment. Phil Sayre (PS) serves on the advisory board of the BIORIMA project and works for the Gov4Nano consortium. PS also works for Wiley Law, LLP (Washington, DC).

Bengt Fadeel, Institute of Environmental Medicine, Karolinska Institutet, Stockholm, Sweden
Phil Sayre, NanoRisk Analytics, LLC, Auburn, CA, USA

https://doi.org/10.1515/9783110669282-002

It has been argued that science has entered a new era in which the quest for truth ("pure" science) has been gradually replaced by science as a source of economic power and, by extension, political power, leading some authors to the conclusion that contemporary science can be viewed as being "post academic" [1]. Furthermore, nanotechnology was suggested as a prime example of this transition from basic science to a more applied or "post academic" science. The question, then, is whether this requires a reframing of ethical or other societal concerns? Indeed, does nanotechnology pose new or unique ethical problems or are the concerns familiar and shared with other emerging technologies? [2]. This topic has been extensively covered in other chapters in the present volume. Here we discuss safety assessment of engineered nanomaterials along with the reliability of the methods used. We posit that a clear understanding of *safety* is a prerequisite not only for adequate risk assessment of nanomaterials – it is also required to realize the full potential of the revolutionary nanotechnologies.

It is worth noting that while we are still debating whether nanomaterials raise new safety (or ethical) concerns, several mRNA vaccine candidates delivered via lipid nanoparticles have recently been developed to manage the current pandemic [3]. More complex nanomaterial-enabled products will likely continue to be developed, as science races ahead of the development of methods to assess safety.

Safety to humans and to the environment is a tangible part of commercializing nanomaterials [4]; other tangible factors to consider are the potential benefits of a material in the sense of whether or not it cures or prevents a disease. However, beyond such tangible and easily measured aspects of judging the safety and ethics of the nanotechnologies, factors such as the ability of a material to provide high-quality jobs, the public's acceptance of a new material (cf. genetically modified organisms), and the ability of the material to contribute to clean energy or environmental remediation should also be considered. Such factors, in total, contribute to the overall ability of society to achieve a better and more sustainable future for all, while also protecting our planet, in line with the United Nations (UN) sustainable development goals (SDGs), "a blueprint for a better future for all." [5].

The intent of this chapter is thus to focus on how to address the environmental health and safety assessment of nanomaterials both from a fundamental science and a regulatory perspective: this is still a key priority that will ultimately contribute to the sustainable deployment of nanotechnologies. We discuss risk governance and elaborate a set of suggestions for the implementation of safety assessment in a harmonized manner in order to enable developed and less developed countries to assess nanomaterials and other new materials using common standards.

2.2 Advanced tools for nanosafety assessment

How safe are nanomaterials? The answer to this simple question is not simple [6]. Even the question of how one should define a "nanomaterial" is surprisingly difficult, not least from a regulatory perspective [7]. We know that both the chemical and physical properties of a material change as we approach the nanoscale. Indeed, it has been suggested that nanomaterials exist in a realm where the properties are governed by "a complex combination of classical physics and quantum mechanics" [8]. Therefore, it is reasonable to assume that the biological effects may also change [9].

Auffan et al. [10] argued in an essay published more than a decade ago that evidence for novel size-dependent properties rather than particle size *per se* should be the main criterion in any definition of nanoparticles (from a safety perspective). This makes a great deal of sense; after all, materials are produced at the nanoscale because we are interested in novel (size-dependent) properties. The question is whether we are in a position to draw conclusions regarding size-dependent toxicities? The answer is: yes and no. But the truth is that there is little evidence of any general structure–activity relationship (SAR) with respect to "nanotoxicity" even though progress is being made with the use of high-throughput screening approaches [11]. The problem may be due, in part, to the fact that nanomaterials are often bundled into one single material category when this is clearly not the case; for some materials, it is the tendency to undergo dissolution with the release of toxic metal ions that drives the toxicity, while for other materials it is the fiber-like dimensions and the high degree of biopersistence that drives the biological responses, and so on. The situation is compounded by the fact that nanomaterials (indeed, all biomaterials) rapidly adsorb proteins and other biomolecules in a living system, and this means that the biological responses are dependent both on the synthetic "identity" defined by the material intrinsic properties and by the context-dependent biological "identity" of the material [12]. This biological "identity" (also referred to as the bio-corona) may differ between *in vitro* and *in vivo* conditions, but it is not something that can be ignored.

Appropriate and scientifically rigorous risk assessments of nanomaterials are key to the sustainable use of nanotechnology products. However, nanomaterials are challenging due to the many different variations not only in terms of the core composition of the materials but also in terms of size, shape, surface properties, degree of dissolution, and so on. At the same time, reducing animal tests by introducing alternative and/or predictive *in vitro* and *in silico* methods has become a priority [13]. Considerable progress has been made in recent years with respect to mechanism-based hazard assessment, including the use of high-throughput screening platforms to speed up safety assessment, and the application of omics-based systems biology or systems toxicology approaches to probe the underlying mechanisms [14]. The question is not whether scientists are willing to embrace these new technologies, but if regulatory bodies consider these alternative methods (i.e., alternatives to animal testing) to be reliable and predictive of what happens in a living organism including in humans. As pointed out previously [15], even though great strides have been made, nanotoxicology still faces a number of challenges not associated with traditional toxicology (of chemicals). One of the more pressing needs has to do with improving *in vitro* to *in vivo* predictivity [15]. Ten years ago, it was argued that accurate evaluation of the pulmonary hazard of nanomaterials using *in vitro* or *in silico* (modeling) approaches is a "myth" [9]. However, provided that *in vitro* assays are validated (admittedly, a long and arduous process) they may serve as valuable predictive screening tools to assess hazard potency of nanomaterials, resulting in simpler, faster, and less expensive assessments than the corresponding animal tests [16]. It is instructive to note that the so-called lung-on-a-chip assay for nanomaterial testing were also reported 10 years ago [17]; the authors found that mechanical strain accentuated the cytotoxic and inflammatory responses to silica nanoparticles. Such human "breathing" microfluidic devices hold great promise for hazard screening and perhaps, pending validation, in predictive safety assessment of nanomaterials as well.

2.3 Reliability of safety data and assays/methods

As pointed out previously, "one would not expect good experimental data from badly designed experiments" [18]. Furthermore, even if an experiment is carefully designed, good (i.e., reliable and reproducible) data cannot be expected if the test material is not carefully characterized. Indeed, it is important to know not only the chemical composition of the nanomaterial but also how the material interacts with its biological surroundings (the "nano-bio" interface) [12]. The intrinsic physicochemical properties may be viewed as the "synthetic identity" of the nanomaterial. However, nanomaterials may also acquire a "biological identity" upon introduction into a living system, defined by the adsorption of biomolecules including proteins or lipids, or both, onto the nanomaterial surface [19]. This also means that in order to fully appreciate and decipher the impact of nanomaterials on a living system, multiple scientific disciplines need to collaborate: the nano-bio field and by extension the field of nanotoxicology is a playground for biologists, chemists, and physicists – an interdisciplinary adventure.

The number of publications in the field of nanotoxicology has increased exponentially during the last 10 years (Figure 2.1). However, some experts have cautioned that many (early) nanotoxicology studies are flawed in terms of experimental design and that they suffer from, among other things, a lack of rigorous characterization of the test materials, and the use of excessively high doses of the materials in question [20]. In an effort to remedy this, the publicly funded communication project called DaNa ("data and knowledge on nanomaterials"; in German: Daten und Wissen zu Nanomaterialien) provides curated information on nanomaterials and their safety assessment (www.nanoobjects.info). Importantly, this project has defined a checklist

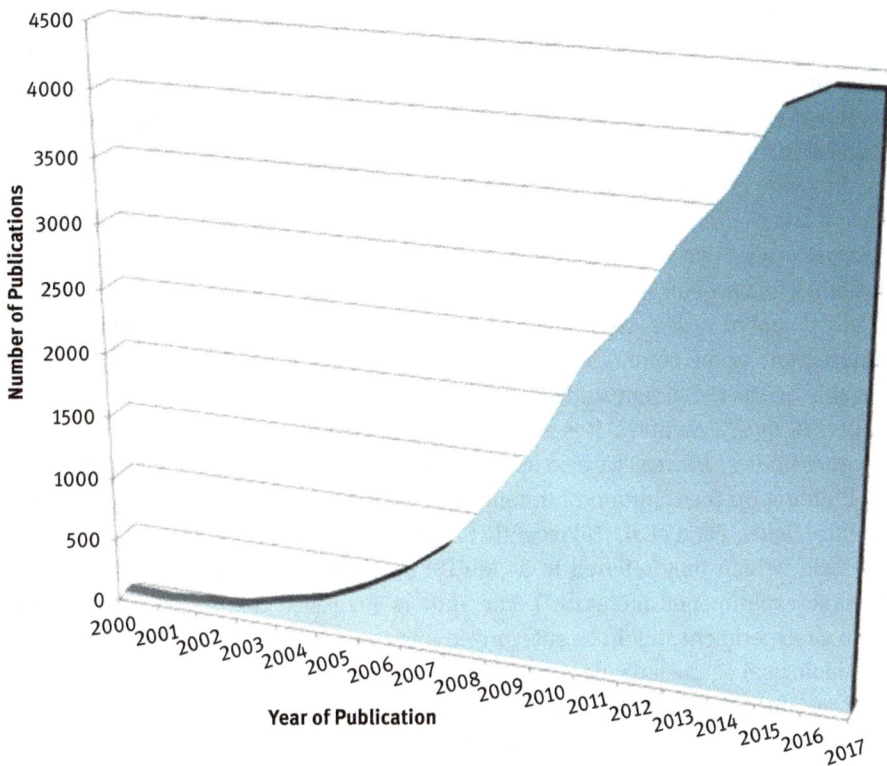

Figure 2.1: The number of published studies in the field of nanotoxicology during the period 2000–2017. Reproduced from Krug [20], with permission from Elsevier.

for publications that need to be fulfilled in order to judge a published study as "acceptable" for the database or "knowledge base" developed by the project. The DaNa knowledge base currently provides information on 26 types of nanomaterials used in various products already on the market (as of February 2018) [20]. Even though not strictly an "ethical" issue, we believe that it is vitally important to ensure that published studies on nanomaterials are of high quality, and we concur that it is important that no adverse effect studies are also published [21], in order not to skew the perception of nanomaterials as inherently "good" or "bad."

Interference of nanomaterials with conventional toxicity assays can occur via several mechanisms [22–24] and yet despite this well-known fact, many researchers persist in using the same assays. This is problematic as it may skew the results (and conclusions) of hazard assessments of nanomaterials. A few years ago, Ong et al. [25] conducted a literature survey and found that 95% of papers from 2010 using biochemical techniques did not account for potential nanomaterial interferences. Clearly, more awareness of this problem is needed, and new developments including label-free methodologies for toxicity testing are also needed [26]. Other, nontrivial issues include the stability of the dispersions, and the dose metric used (i.e., dose based on mass, particle number, or surface area; refer to the recent WHO-IPCS report on immunotoxicity testing of nanomaterials) [27].

However, not all sources of variation in nanomaterial safety assessment are due to technical inconsistencies; some variations are simply explained by differences in biology. In a very instructive example, Stoudmann et al. [28] systematically analyzed published studies on nanocelluloses. The authors identified 16 cell lines (mostly of human or mouse origin) that had been tested against crystalline nanocelluloses (CNC), and 13 for fibrillar nanocelluloses (CNF) in altogether 22 studies. The results showed that the toxicity potential varied greatly depending on the cell line (Figure 2.2). Moreover, and perhaps more to the point, for some commonly used cell lines, the results varied considerably between different studies. This could perhaps be explained by the use of different CNC or CNF samples. It is also pertinent to note that all the studies were acute exposure studies, whereas we also need to consider chronic, low-dose exposure [28].

Building on the tradition of minimum reporting standards in the biological and chemical fields, Faria et al. [29] recently proposed a reporting standard for the "nano-bio" field, which they referred to as MIRIBEL ("minimum information reporting in bio-nano experimental literature"). The authors suggested that information pertaining to an experiment should be categorized into three sections: material characterization, biological characterization, and details of experimental protocols; or, to put it in more succinct terms: the standard requires that information on the materials, models, and methods is provided in every publication, using a checklist approach. The aim of the MIRIBEL standard is to provide a guide, rather than a barrier to research [30]. This is an important step toward improving communication across various disciplines in "nano-bio" research, if such standards are endorsed by the scientific community and by journals [31].

In our view, safety assessment of nanomaterials should fulfill the following three criteria, which we refer to as the R.I.P. guidelines for nanotoxicity testing: (1) R = relevant (meaning realistic and relevant *in vitro* or *in vivo* models that generate results relevant for risk assessment); (2) I = integrated (or intelligent, as in integrated or tiered approaches to safety assessment, starting with acellular or *in vitro* tests); (3) P = predictive (as in *in vitro* results that are predictive of *in vivo* outcomes, but also results that allow for grouping of nanomaterials on the basis of predictive screening).

2.4 Nanomaterial regulations and risk governance

There has been much debate as to whether we need to reinvent regulations for nanomaterials [8]. However, while engineered nanomaterials may exhibit hazard or exposure properties that differ from those of traditional chemicals, it is generally thought

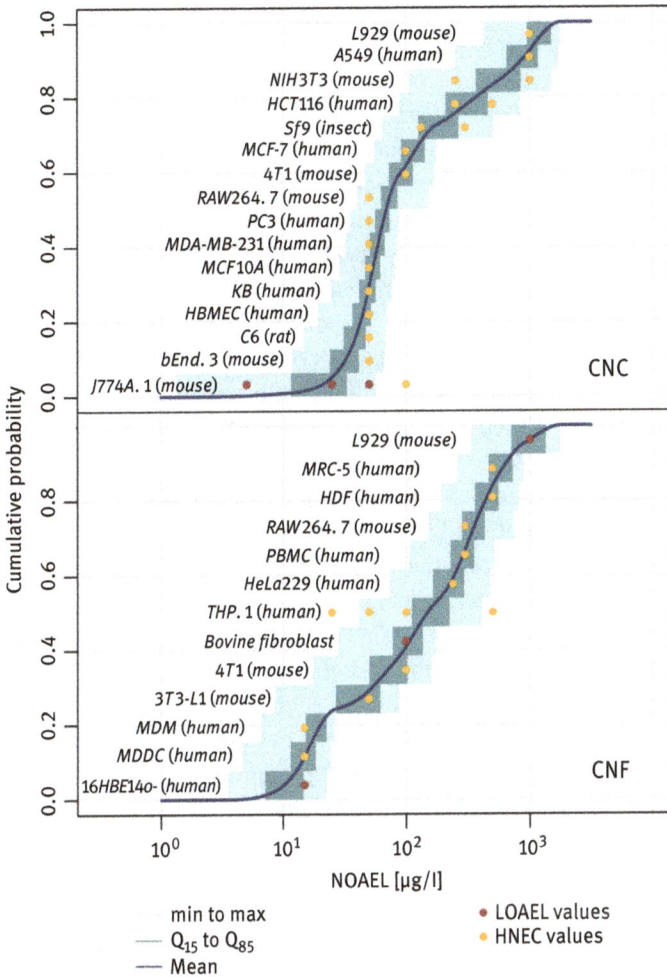

Figure 2.2: Different strokes for different folks: the sensitivity of various cell lines to cellulose nanocrystals (CNC) (top) and cellulose nanofibers (CNF) (bottom). HNEC, highest no-effect concentration (values); NOAEL, no observed adverse effect level. From Stoudmann et al. [28], with permission from Taylor & Francis.

that the basic risk assessment paradigm for chemicals also applies to nanomaterials [32]. The EU-funded project, NanoREG, and its sister project, ProSafe, have published a series of publications describing the reliability of methods and data for regulatory assessment of nanomaterial risks [32, 33], with specific attention to the standardization and validation of methods used to characterize nanomaterials [34] and the *in vitro* and *in vivo* assays used for hazard assessment of nanomaterials [16, 35]. In their provocative analysis, Miller and Wickson [36] identified several major barriers to reliable risk assessment of nanomaterials, including a lack of nanospecific regulatory requirements and validated methods for safety testing. The authors argued that "the [nanotechnology] Emperor is naked" in the sense that reliable risk analysis of nanotechnology is unattainable. Other experts have sounded a less pessimistic note [37], and in a recent paper, the theoretical basis for the development and implementation of an "effective, trustworthy, and transparent" risk governance framework for nanomaterials is laid out [38], though the authors concede that operationalization of such a risk governance framework remains a formidable challenge. Notwithstanding, it is clear that the nanosafety community has come a long way in the past 10–15 years [39, 40].

The term "governance" has a broader connotation than government and refers to the interaction between the formal institutions of a country and those in civil society. The following definition of "governance" is provided by the World Bank [41]: "Good governance is epitomized by predictable, open and enlightened policymaking, a bureaucracy imbued with a professional ethos acting in furtherance of the public good, the rule of law, transparent processes, and a strong civil society participating in public affairs." Three risk governance projects funded under the Horizon 2020 program of the European Union (Gov4Nano, NanoRIGO, and RiskGONE) are working closely together to ensure a "future-proof" risk governance model for nanotechnology in Europe. One of the overarching objectives is to establish a Risk Governance Council. The governance council mission is still being developed, yet in addition to monitoring progress and identifying key needs, it also aims to bring together stakeholders such as industry and regulators to work toward setting priority goals (such as the development of test guidelines, fostering cooperation through dialogue, and promoting funding/research to address key regulatory needs for nanomaterials). Inspired by the International Risk Governance Council's Risk Governance Framework, [42] the projects also seek to develop an ethical impact assessment module for nanomaterials and nano-enabled products [43].

Both regulation and governance of nanomaterials begin with the *definition* of a nanomaterial. This definition has varied, but the most common definition of a nanomaterial refers to materials that have at least one dimension in the size range of approximately 1–100 nm. For example, the US National Nanotechnology Initiative (NNI) Program (www.nano.gov) defines nanotechnology as "the understanding and control of matter at dimensions between approximately 1 and 100 nm, where unique phenomena enable novel applications." The definition was originally intended to spur research and innovation to develop commercial materials in this size range. However, as regulatory authorities cast about for a definition of a nanomaterial that would serve to identify nanomaterials with a higher potential risk, this same NNI definition based on size was adopted by some regulatory agencies such as the US EPA (under the Toxic Substances Control Act (TSCA)) to subject these materials to closer scrutiny than traditional chemicals: the premise was that certain physicochemical properties in this size range could be associated with higher, or unforeseen, risks. A similar, but more refined, size-driven definition has emerged within the EU that is based on this size range [44]. This has been further refined to include other physicochemical descriptors to more narrowly define a nanomaterial under the registration, evaluation, authorization, and restriction of chemicals (REACH) regulation. Hence, the size distribution for materials between 1 and 100 nm (and smaller) must also be described in terms of their surface functionalization or treatment, morphological characterizations (including crystallinity, shape, and aspect ratio), and surface area. These additional parameters segregate nanomaterials into narrower "nanoforms," thus driving the hazard/fate/risk assessments of very specific nanomaterials. Another distinct set of legislations that does not align with the "traditional" 1–100 nm size cutoff relates to nanomedicines: both the U.S. Food and Drug Administration (FDA) and the European Medicines Agency (EMA) view nanomedicines as materials up to, but also larger than, 100 nm (up to possibly 1 µm in size); this is driven by the fact that unique properties of nanomaterials can arise at sizes greater than 100 nm (U.S. FDA) [45]. Therefore, with this in mind, and in light of what is surmised to be the well-defined human exposures for nanomedicines [46], larger size materials are scrutinized more closely than under the broader statutes for industrial chemicals (such as TSCA and REACH). A third factor to consider in these nanomedicine regulations is that some statutes, such as those under which FDA and EMA operate, include the fact that the material must be intentionally engineered to exhibit properties (including physical or biological properties, or biological effects) that are attributable to dimensions up to 1,000 nm. Hence, from these somewhat divergent approaches, different types of products employing nanomaterials are captured for nanomaterial review under different statutes. Such differing approaches may be justified by the nature of the different statutes under which nanomaterial safety reviews are conducted. However, a single definition for a nanomaterial may be preferable across statutes and nations to encourage uniformity of assessments. A proposal for such a single definition could be built on components of these existing divergent definitions to result in a *single regulatory nanomaterial definition*. The original NNI definition (see above) is not supported by increased concern for unexpected properties and/or increased hazard/fate/biodistribution attributes. Furthermore, FDA has shown that there may be increased concerns for nanoparticles in size ranges of several hundred nanometers (for instance, with regard to *in vivo* biodistribution). The EPA recently attempted to identify unique properties of nanomaterials that would trigger reporting requirements under TSCA [47], but this effort proved problematic. However, certain physicochemical properties (such as shape and crystallinity) clearly give rise to increased safety concerns. Therefore, a new proposal for a nanomaterial that would subject the material to a safety review with nano-specific concerns in mind could be comprised of those particles that are up to 1,000 nm in size (not 100 nm). Such nanomaterials would need to be defined in terms of their particle size distribution, surface functionalization, or treatment, morphological characterizations (including crystallinity, shape, aspect ratio), and surface area (as has been done for REACH nanomaterial reporting, and adopted also by other

EU agencies such as the European Food Safety Authority (EFSA)). To avoid determining whether the particles have unique properties due to their size, this criterion would be dropped from the unified definition. The advantages of such an
approach include application of the same definition across all legislations, thus
simplifying reporting and reducing confusion, good scientific support for the size
range selected, little ambiguity regarding whether nanoscale properties are exhibited, and the inclusion of additional physicochemical parameters to correlate
with assessments of safety. The disadvantage, or challenge, is that more materials would be subject to regulatory review, increasing reporting, and assessment
burdens.

2.5 Toward a sustainable (nano)future: a proposal

To meet the very pressing needs for harmonized and regulatory-accepted test guidelines for nanomaterials, regulatory authorities rely on the work of organizations
such as the Organisation for Economic Co-operation and Development (OECD), especially the OECD Working Party on Manufactured Nanomaterials. However, due to
the complexity of nanomaterials, the consensus-building needed, and the heavy
cost and time burdens of validating a new test guideline, not many new OECD test
guidelines and guidances are available at this time to address the specific needs of
nanomaterials. The currently listed efforts now underway in the OECD Working
Group of the National Coordinators for the Test Guidelines Programme to address further nanomaterial needs can be found in the OECD 2020 Workplan (www.oecd.org).
However, there is potential to spur the needed development of nano-specific or nano-
adapted regulations and guidances worldwide in order to provide a uniform set of
standards to which all countries could adhere. We propose that such an effort could
be spearheaded through the UN, at least on a broader policy basis (including mechanisms for setting standards worldwide), while specific test methods and tools could
continue to be developed by the OECD. The 17 SDGs were adopted by all Member
States in 2015, as part of the 2030 Agenda for Sustainable Development which set out
a 15-year plan to achieve the goals (www.un.org). As noted in the beginning of this
chapter, these SDGs may provide a good overall structure for the ethical deployment
of nanomaterials. This meshes well with, for instance, the EU's Green Deal initiative
(where "no one will be left behind") [48].

The European Commission has recently (October 2020) set out its own goals of
how it sees implementation of a "chemicals strategy for sustainability toward a
toxic-free environment [49]." The strategy already contains specific goals for the
commercialization of nanomaterials. Many of these goals could be added to provide more nano-specific goals to a similar, yet global UN goal-setting initiative for
the ethical and sustainable use of nanomaterials and nanotechnologies. It is also

pertinent to mention training: a set of education goals to inform academic scientists, industry, and regulators could be helpful in making scientists across different disciplines (not only in (nano)toxicology) aware of risk issues so that they could be more easily mitigated, ideally at an early stage in the development process. The governance council concept could also be employed at the UN level also to promote dialogue between scientists worldwide, thus enabling more discussions on not just educational needs but on other topics as well (governance needs, resource needs, etc.). In particular, it would be helpful to have such a governance council focusing on those pressing structural needs that we identified here.

2.6 Mind the gap: The importance of dialogue

Significant global investments have been made in nanosafety research over the past 10–15 years, not least in the frame of European Commission-funded collaborative research projects in FP7 and H2020 (refer to the EU Nanosafety Cluster) (www.nano safetycluster.eu) as well as National Institute of Environmental Health Sciences-funded consortia in the United States (refer to the NNI) (www.nano.gov). In addition to multiple academic research programs, other coordinated international efforts including the decade-long program on the testing and assessment of manufactured nanomaterials under the auspices of the OECD have provided insights regarding the applicability of existing test guidelines [50]. However, risk assessment of nanomaterials and nano-enabled products remains a challenge. Teunenbroek et al. [51] argued that this is due mainly to the fact that research in the field has been (basic) science oriented and not "regulatory oriented," and the authors pointed out that standardized test methods are a prerequisite for risk assessment in a regulatory context. We believe that the answer to this problem will also require a *dialogue* on benefits and safety concerns of the nanotechnologies between different stakeholders. Indeed, it is time for researchers to descend from their ivory towers and engage in a dialogue with society including regulatory and other government bodies, industry representatives, and the public. We need to communicate what we know and do not know about the risks of engineered nanomaterials. For example, in Sweden, a publicly funded nanosafety platform called SweNanoSafe was launched to address this need. The Swedish nanosafety "story" started back in 2006 with the organization of the first national meeting on nanotoxicology, soon after the first international nanotoxicology conference [52]; 10 years later, a national platform, supported by the Ministry of the Environment, was inaugurated. The main goal of the platform is to serve as a forum for cooperation and communication concerning health and environmental safety of nanomaterials for scientists and other societal stakeholders (www.swenano safe.se). The platform has launched a national network of nanosafety researchers and organizes annual workshops in the frame of this network. The platform also comprises a scientific expert panel whose members have complementary expertise in the field of nanosafety, including material characterization, exposure assessment, (eco)toxicology/hazard assessment, and risk assessment. In addition, the platform is currently mapping both research and educational activities relevant for nanosafety (at the national level), and in September 2020 hosted a virtual meeting of other similar initiatives across Europe.

The US–EU Nanotechnology Communities of Research provide a platform for scientists to collaboratively identify and address key research needs (www.us-eu.org). This bilateral platform has organized workshops since 2011. Additionally, international nanotoxicology conferences have been held biannually during the past 15 years, alternating between different geographic regions; the 2014 meeting in Antalya, Turkey, gathered close to 600 participants from 40 countries. The meetings in Beijing (2012) and Boston (2016) were also a great success.

There is a tremendous interest in nanotechnologies for various applications, from clean energy to precision medicine. Especially nanomedicine is a rapidly expanding field [53], as evidenced by the increasing numbers of papers and patents (Figure 2.3). However, it is important not to lose sight of the fact that nanotechnologies can also be harnessed to address some of the world's most critical development problems [54]. Indeed, it has been argued that nanotechnology can be deployed to address SDGs of the UN [55]. UNESCO, an intergovernmental organization with a global vantage point, may play a key role in promoting the dialogue between all relevant stakeholders. Importantly, building "nano-literacy" starts in the classroom.

What does the public know about nanomaterials, and what do experts say? The European Chemicals Agency (ECHA) has launched a website called European Union Observatory for Nanomaterials that provides information about existing nanomaterials on the EU market (www.euon.echa.europa.eu). In November 2020, ECHA published the results of a survey that was carried out in five selected EU countries on their website. The results showed that EU citizens demand better labeling of products containing nanomaterials [56]. The authors of the report concluded that it is necessary to raise awareness of nanomaterials/nanotechnologies and their benefits and risks to ensure that the public can make informed choices. In another recent study, Larsson et al. [57] conducted a survey among expert stakeholders; the study sample included 237 individuals working in the fields of nanotechnology innovation and regulation in Sweden. The study showed that experts in government organizations favored stronger regulation and perceived higher risks and they were also more concerned with the ethical implications of nanotechnologies than their counterparts from industry. The authors concluded that the attitudes toward regulation are influenced by perceived risk, a result that accords well with previous research [58].

Foss Hansen et al. [59] argued that "a new technology will only be successful if those promoting it can show that it is safe." We believe that those using a

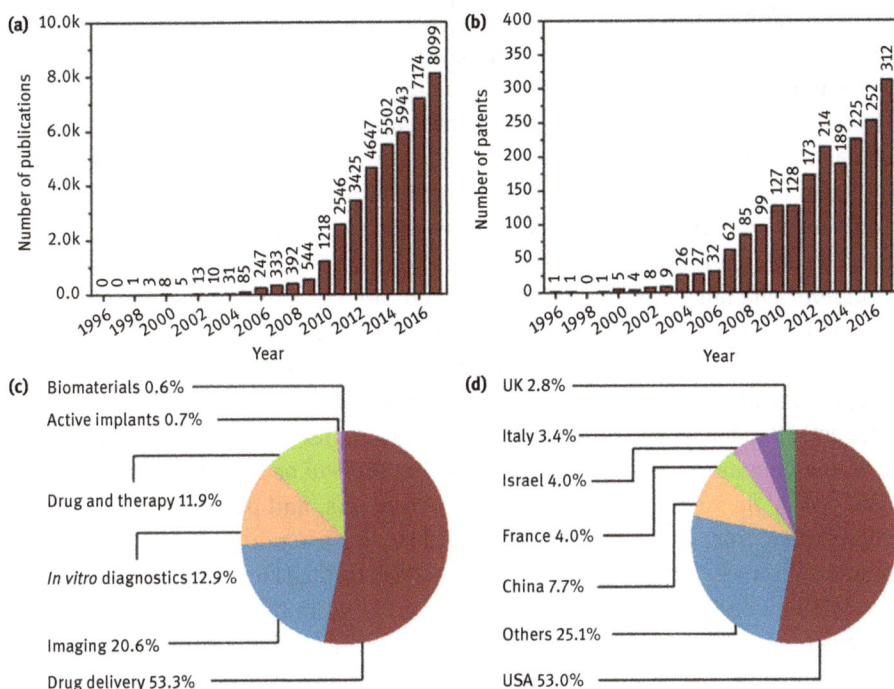

Figure 2.3: Global activities in nanomedicine: (a) publications, (b) patents, (c) sectorial breakdown of nanomedical applications, and (d) breakdown of clinical trials per country. From Yan et al. [53], with permission from Wiley-VCH Verlag GmbH & Co. KGaA.

technology or in some way benefiting from a technology also must be included in the discussion. Nanosafety research should not be exercised in isolation from society; a dialogue between scientists/experts and other societal stakeholders is critically important. Scientists are good at findings answers, but it is equally important to understand what the questions are from a regulatory and/or innovation perspective. Indeed, what we need is an international ecosystem of nanosafety covering all the processes along the value chain and all relevant stakeholders.

2.7 Concluding remarks and a personal perspective

Taking a broad view of the field of nanosafety research, it appears that we managed to spend the better part of the first decade arguing over the definition of "nano" along with a trial-and-error approach to nanomaterial testing, until the community finally figured out that nanotoxicology begins and ends with careful characterization of the physicochemical properties of the test material, and realized that the transformation of nanomaterials in the environment and in the organism also plays a key role [60]. This was followed by a new phase during the past 5 years or so, in which more and more comprehensive studies using advanced methods emerged [14, 61]. Ten years ago, we asked whether REACH was out of reach with respect to regulation of nanomaterials [62], but with the new revisions (January 1, 2020) of the European Union's chemical legislation, more clarity and coherence has been provided in this regard [63]. So, where do we stand, and where do we go from here? Does nanotechnology pose an ethical problem, in addition to admittedly being a regulatory challenge? No, in our view, there are no new issues with regard to nano-ethics when compared with other new and emerging technologies, but there is an obvious need to ensure that safe nano-enabled products or nanotechnologies are brought to market, and this hinges on governance of nanomaterials and nanotechnologies and attendant risk(s) both for human health and the environment. Hence, in our view, the impetus to deliver new products that provide benefits to society must be balanced by the need to protect workers, consumers, and patients, as well as the environment from harm. To be perfectly clear, we are not arguing in favor of a moratorium until "nano" has been proven safe; instead, we believe that product development and safety assessment should go hand in hand. This is, in essence, the meaning of safe by design. We also believe that in order to ensure safe and sustainable development of the nanotechnologies and of nano-enabled products of ever-increasing sophistication, there is a need to develop the science of safety assessment (i.e., hazard, exposure, and risk assessment) of nanomaterials and other advanced materials. Indeed, as famously pointed out more than a decade ago, "there is almost no other scientific field in which the core experimental protocols have remained nearly unchanged for more than 40 years" [64]. In other words, an overhaul of (regulatory) toxicology is needed, and new methods including the so-called alternative (animal-free) testing strategies should be adopted to promote the assessment of hazard and risk of new substances [13]. Risk assessment of new materials is not about putting out fires, it is a scientifically informed approach whereby potentially harmful substances are weeded out. Notably, risk assessment and regulation of new materials reside at the intersection of law and science, and it is at this intersection that progress is needed.

C. P. Snow argued that science and humanities had become divided into "two cultures" and that this division was a major handicap in solving the world's problems. We caution that the division of toxicology (especially, nanotoxicology) into basic, mechanism-based science on the one hand, and regulatory toxicology on the other, is detrimental to both. Indeed, screening ought to be mechanism based, and new materials may necessitate the development of new assays. Furthermore, basic toxicology and regulatory toxicology face similar challenges including, for instance, how to handle the tsunami of "big data" and how to properly exploit such data by using *in silico* (modeling) approaches; in addition, how to best model the real-life (human) situation: surely, we are now approaching an era when biomedical scientists as well as toxicologists and regulators are getting ready to leave behind their unholy obsession with animal models [65].

As this essay draws to an end, a few additional reflections: first, it appears that the term "nanotechnology" is frequently used to describe purposely engineered nanoscale objects or particles (in fact, any material in the range of 1–100 nm) when, in reality, nanotechnology is a term that describes the manipulation of matter at the atomic scale, and the exploitation of distinct properties and phenomena at that scale [66]. However, a large number, if not a majority of papers in nanotoxicology or, for that matter, in the field of nanomedicine, are not concerned with the manipulation of matter, but merely with the study of matter that happens to exist on a nanoscale; this distinction is important if we are ever going to realize the full potential of the revolutionary nanotechnologies. Second, and not withstanding the fact that nanotoxicology does not always address nanoscale phenomena, nanotoxicology as a scientific discipline may be viewed as a lesson in interdisciplinarity [67], and may serve as a template for future studies of other emerging materials at the interface between the material sciences and biology. It has been argued that the public discourse around nanotechnology may provide useful lessons for other fields such as synthetic biology [68].

References

[1]   Jotterand F. The politicization of science and technology: Its implications for nanotechnology. J Law Med Ethics 2006, 34, 4, 658–66.
[2]   Kuiken T. Nanomedicine and ethics: Is there anything new or unique? Wiley Interdiscip Rev Nanomed Nanobiotechnol 2011, 3, 2, 111–18.
[3]   Shin MD, Shukla S, Chung YH, Beiss V, Chan SK, Ortega-Rivera OA, Wirth DM, Chen A, Sack M, Pokorski JK, Steinmetz NF. COVID-19 vaccine development and a potential nanomaterial path forward. Nat Nanotechnol 2020, 15, 8, 646–55.
[4]   Hjorth R, van Hove L, Wickson F. What can nanosafety learn from drug development? The feasibility of "safety by design". Nanotoxicology 2017, 11, 305–12.
[5]   United Nations: Sustainable Development Goals (www.un.org/sustainabledevelopment). Last accessed May 3, 2021.
[6]   Valsami-Jones E, Lynch I. Nanosafety. How safe are nanomaterials? Science 2015, 350, 6259, 388–89.
[7]   Considerations on a Definition of Nanomaterial for Regulatory Purposes. JRC. (2010). See: https://ec.europa.eu/jrc/en/publication/reference-reports/considerations-definition-nano material-regulatory-purposes. Last accessed May 3, 2021.
[8]   Renn O, Roco MC. Nanotechnology and the need for risk governance. J Nanopart Res 2006, 8, 2, 153–91.
[9]   Warheit DB. Debunking some misconceptions about nanotoxicology. Nano Lett 2010, 10, 12, 4777–82.
[10]   Auffan M, Rose J, Bottero JY, Lowry GV, Jolivet JP, Wiesner MR. Towards a definition of inorganic nanoparticles from an environmental, health and safety perspective. Nat Nanotechnol 2009, 4, 10, 634–41.

[11] Collins AR, Annangi B, Rubio L, Marcos R, Dorn M, Merker C, Estrela-Lopis I, Cimpan MR, Ibrahim M, Cimpan E, Ostermann M, Sauter A, Yamani NE, Shaposhnikov S, Chevillard S, Paget V, Grall R, Delic J, de-Cerio FG, Suarez-Merino B, Fessard V, Hogeveen KN, Fjellsbø LM, Pran ER, Brzicova T, Topinka J, Silva MJ, Leite PE, Ribeiro AR, Granjeiro JM, Grafström R, Prina-Mello A, Dusinska M. High throughput toxicity screening and intracellular detection of nanomaterials. Wiley Interdiscip Rev Nanomed Nanobiotechnol 2017, 9, 1, e1413.

[12] Fadeel B, Fornara A, Toprak MS, Bhattacharya K. Keeping it real: The importance of material characterization in nanotoxicology. Biochem Biophys Res Commun 2015, 468, 3, 498–503.

[13] Nel AE, Malloy TF. Policy reforms to update chemical safety testing. Science 2017, 355, 6329, 1016–18.

[14] Fadeel B, Farcal L, Hardy B, Vázquez-Campos S, Hristozov D, Marcomini A, Lynch I, Valsami-Jones E, Alenius H, Savolainen K. Advanced tools for the safety assessment of nanomaterials. Nat Nanotechnol 2018, 13, 7, 537–43.

[15] Hussain SM, Warheit DB, Ng SP, Comfort KK, Grabinski CM, Braydich-Stolle LK. At the crossroads of nanotoxicology *in vitro*: Past achievements and current challenges. Toxicol Sci 2015, 147, 1, 5–16.

[16] Drasler B, Sayre P, Steinhäuser KG, Petri-Fink A, Rothen-Rutishauser B. *In vitro* approaches to assess the hazard of nanomaterials. NanoImpact 2017, 8, 99–116.

[17] Huh D, Matthews BD, Mammoto A, Montoya-Zavala M, Hsin HY, Ingber DE. Reconstituting organ-level lung functions on a chip. Science 2010, 328, 5986, 1662–68.

[18] Parr D. Will nanotechnology make the world a better place? Trends Biotechnol 2005, 23, 8, 395–98.

[19] Lynch I, Ahluwalia A, Boraschi D, Byrne HJ, Fadeel B, Gehr P, Gutleb AC, Kendall M, Papadopoulos MG. The bio-nano-interface in predicting nanoparticle fate and behavior in living organisms: Towards grouping and categorizing nanomaterials and ensuring nanosafety by design. BioNanoMaterials 2013, 14, 3–4, 195–216.

[20] Krug HF. The uncertainty with nanosafety: Validity and reliability of published data. Colloids Surf B Biointerfaces 2018, 172, 113–17.

[21] Krug HF, Wick P. Nanotoxicology: An interdisciplinary challenge. Angew Chem Int Ed Engl 2011, 50, 6, 1260–78.

[22] Lewinski N, Colvin V, Drezek R. Cytotoxicity of nanoparticles. Small 2008, 4, 1, 26–49.

[23] Monteiro-Riviere NA, Inman AO, Zhang LW. Limitations and relative utility of screening assays to assess engineered nanoparticle toxicity in a human cell line. Toxicol Appl Pharmacol 2009, 234, 2, 222–35.

[24] Kroll A, Pillukat MH, Hahn D, Schnekenburger J. Interference of engineered nanoparticles with *in vitro* toxicity assays. Arch Toxicol 2012, 86, 7, 1123–36.

[25] Ong KJ, MacCormack TJ, Clark RJ, Ede JD, Ortega VA, Felix LC, Dang MK, Ma G, Fenniri H, Veinot JG, Goss GG. Widespread nanoparticle-assay interference: Implications for nanotoxicity testing. PLoS One 2014, 9, 3, e90650.

[26] Andraos C, Yu IJ, Gulumian M. Interference: A much-neglected aspect in high-throughput screening of nanoparticles. Int J Toxicol 2020, 39, 5, 397–421.

[27] Principles and Methods to Assess the Risk of Immunotoxicity Associated with Exposure to Nanomaterials. Environmental Health Criteria Document 244. WHO-IPCS. (2019). (accessed at: https://www.who.int/publications/i/item/9789241572446). Last accessed May 3, 2021.

[28] Stoudmann N, Schmutz M, Hirsch C, Nowack B, Som C. Human hazard potential of nanocellulose: Quantitative insights from the literature. Nanotoxicology 2020, 14, 1241–57.

[29] Faria M, Björnmalm M, Thurecht KJ, Kent SJ, Parton RG, Kavallaris M, Johnston APR, Gooding JJ, Corrie SR, Boyd BJ, Thordarson P, Whittaker AK, Stevens MM, Prestidge CA, Porter CJH,

Parak WJ, Davis TP, Crampin EJ, Caruso F. Minimum information reporting in bio-nano experimental literature. Nat Nanotechnol 2018, 13, 9, 777–85.

[30] Faria M, Björnmalm M, Crampin EJ, Caruso F. A few clarifications on MIRIBEL. Nat Nanotechnol 2020, 15, 1, 2–3.

[31] Leong HS, Butler KS, Brinker CJ, Azzawi M, Conlan S, Dufés C, Owen A, Rannard S, Scott C, Chen C, Dobrovolskaia MA, Kozlov SV, Prina-Mello A, Schmid R, Wick P, Caputo F, Boisseau P, Crist RM, McNeil SE, Fadeel B, Tran L, Hansen SF, Hartmann NB, Clausen LPW, Skjolding LM, Baun A, Ågerstrand M, Gu Z, Lamprou DA, Hoskins C, Huang L, Song W, Cao H, Liu X, Jandt KD, Jiang W, Kim BYS, Wheeler KE, Chetwynd AJ, Lynch I, Moghimi SM, Nel A, Xia T, Weiss PS, Sarmento B, Das Neves J, Santos HA, Santos L, Mitragotri S, Little S, Peer D, Amiji MM, Alonso MJ, Petri-Fink A, Balog S, Lee A, Drasler B, Rothen-Rutishauser B, Wilhelm S, Acar H, Harrison RG, Mao C, Mukherjee P, Ramesh R, McNally LR, Busatto S, Wolfram J, Bergese P, Ferrari M, Fang RH, Zhang L, Zheng J, Peng C, Du B, Yu M, Charron DM, Zheng G, Pastore C. On the issue of transparency and reproducibility in nanomedicine. Nat Nanotechnol 2019, 14, 7, 629–35.

[32] Steinhäuser KG, Sayre PG. Reliability of methods and data for regulatory assessment of nanomaterial risks. NanoImpact 2017, 7, 66–74.

[33] Baun A, Sayre P, Steinhäuser KG, Rose J. Regulatory relevant and reliable methods and data for determining the environmental fate of manufactured nanomaterials. NanoImpact 2017, 8, 1–10.

[34] Gao X, Lowry GV. Progress towards standardized and validated characterizations for measuring physicochemical properties of manufactured nanomaterials relevant to nano health and safety risks. NanoImpact. 2018, 9, 14–30.

[35] Oberdörster G, Kuhlbusch TAJ. *In vivo* effects: methodologies and biokinetics of inhaled nanomaterials. NanoImpact. 2018, 10, 38–60.

[36] Miller G, Wickson F. Risk analysis of nanomaterials: Exposing nanotechnology's naked Emperor. Rev Policy Res 2015, 32, 4, 485–512.

[37] Stone V, Führ M, Feindt PH, Bouwmeester H, Linkov I, Sabella S, Murphy F, Bizer K, Tran L, Ågerstrand M, Fito C, Andersen T, Anderson D, Bergamaschi E, Cherrie JW, Cowan S, Dalemcourt JF, Faure M, Gabbert S, Gajewicz A, Fernandes TF, Hristozov D, Johnston HJ, Lansdown TC, Linder S, Marvin HJP, Mullins M, Purnhagen K, Puzyn T, Sanchez Jimenez A, Scott-Fordsmand JJ, Streftaris G, van Tongeren M, Voelcker NH, Voyiatzis G, Yannopoulos SN, Poortvliet PM. The essential elements of a risk governance framework for current and future nanotechnologies. Risk Anal 2018, 38, 7, 1321–31.

[38] Isigonis P, Afantitis A, Antunes D, Bartonova A, Beitollahi A, Bohmer N, Bouman E, Chaudhry Q, Cimpan MR, Cimpan E, Doak S, Dupin D, Fedrigo D, Fessard V, Gromelski M, Gutleb AC, Halappanavar S, Hoet P, Jeliazkova N, Jomini S, Lindner S, Linkov I, Longhin EM, Lynch I, Malsch I, Marcomini A, Mariussen E, de la Fuente JM, Melagraki G, Murphy F, Neaves M, Packroff R, Pfuhler S, Puzyn T, Rahman Q, Pran ER, Semenzin E, Serchi T, Steinbach C, Trump B, Vrcek IV, Warheit D, Wiesner MR, Willighagen E, Dusinska M. Risk governance of emerging technologies demonstrated in terms of its applicability to nanomaterials. Small 2020, 16, 36, e2003303.

[39] Maynard AD, Aitken RJ, Butz T, Colvin V, Donaldson K, Oberdörster G, Philbert MA, Ryan J, Seaton A, Stone V, Tinkle SS, Tran L, Walker NJ, Warheit DB. Safe handling of nanotechnology. Nature 2006, 444, 7117, 267–69.

[40] Maynard AD, Aitken RJ. 'Safe handling of nanotechnology' ten years on. Nat Nanotechnol 2016, 11, 12, 998–1000.

[41] Governance: The World Bank's Experience. World Bank, Washington, DC (1994). See: https://documents.worldbank.org/en/publication/documents-reports/documentdetail/711471468765285964/governance-the-world-banks-experience. Last accessed May 3, 2021.

[42] Introduction to the IRCG Risk Governance Framework. EPFL. International Risk Governance Center. (2017). (accessed at: https://infoscience.epfl.ch/record/233739). Last accessed May 3, 2021.

[43] Malsch I, Isigonis P, Dusinska M, Bouman EA. Embedding ethical impact assessment in nanosafety decision support. Small 2020, 16, 36, e2002901.

[44] An overview of concepts and terms used in the European Commission's definition of nanomaterial. JRC. (2019). See: https://ec.europa.eu/jrc/en/publication/overview-concepts-and-terms-used-european-commissions-definition-nanomaterial. Last accessed May 3, 2021.

[45] Guidance for Industry: Considering whether an FDA-regulated product involves the application of nanotechnology. US FDA. (2014). See: https://www.fda.gov/regulatory-informa tion/search-fda-guidance-documents/considering-whether-fda-regulated-product-involves-application-nanotechnology. Last accessed May 3, 2021.

[46] Giubilato E, Cazzagon V, Amorim MJB, Blosi M, Bouillard J, Bouwmeester H, Costa AL, Fadeel B, Fernandes TF, Fito C, Hauser M, Marcomini A, Nowack B, Pizzol L, Powell L, Prina-Mello A, Sarimveis H, Scott-Fordsmand JJ, Semenzin E, Stahlmecke B, Stone V, Vignes A, Wilkins T, Zabeo A, Tran L, Hristozov D. Risk management framework for nano-biomaterials used in medical devices and advanced therapy medicinal products. Materials (Basel) 2020, 13, 20, 4532.

[47] US EPA. 2017. Toxic Substances Control Act Information Gathering Rule on Nanomaterials. See: https://www.epa.gov/reviewing-new-chemicals-under-toxic-substances-control-act-tsca/control-nanoscale-materials-under. Last accessed May 3, 2021.

[48] COMMUNICATION FROM THE COMMISSION TO THE EUROPEAN PARLIAMENT, THE EUROPEAN COUNCIL, THE COUNCIL, THE EUROPEAN ECONOMIC AND SOCIAL COMMITTEE AND THE COMMITTEE OF THE REGIONS The European Green Deal COM/ 2019/640. See: https://ec.eu ropa.eu/info/strategy/priorities-2019-2024/european-green-deal_en. Last accessed May 3, 2021.

[49] COMMUNICATION FROM THE COMMISSION TO THE EUROPEAN PARLIAMENT, THE COUNCIL, THE EUROPEAN ECONOMIC AND SOCIAL COMMITTEE AND THE COMMITTEE OF THE REGIONS. Chemicals Strategy for Sustainability Towards a Toxic-Free Environment. COM/ 2020/667. See: https://ec.europa.eu/environment/strategy/chemicals-strategy_en. Last accessed May 3, 2021.

[50] Rasmussen K, González M, Kearns P, Sintes JR, Rossi F, Sayre P. Review of achievements of the OECD Working Party on Manufactured Nanomaterials' testing and assessment programme. From exploratory testing to test guidelines. Regul Toxicol Pharmacol 2016, 74, 147–60.

[51] Teunenbroek TV, Baker J, Dijkzeul A. Towards a more effective and efficient governance and regulation of nanomaterials. Part Fibre Toxicol 2017, 14, 54.

[52] Fadeel B, Kagan V, Krug H, Shvedova A, Svartengren M, Tran L, Wiklund L. There's plenty of room at the forum: Potential risks and safety assessment of engineered nanomaterials. Nanotoxicology 2007, 1, 2, 73–84.

[53] Yan L, Zhao F, Wang J, Zu Y, Gu Z, Zhao Y. A safe-by-design strategy towards safer nanomaterials in nanomedicines. Adv Mater 2019, 31, 45, e1805391.

[54] Salamanca-Buentello F, Persad DL, Court EB, Martin DK, Daar AS, Singer PA. Nanotechnology and the developing world. PLoS Med 2005, 2, 5, e97.

[55] Nanotechnologies, Ethics and Politics. ten Have, H. (Editor). UNESCO. (2007). (accessed at: https://digitallibrary.un.org/record/618003). Last accessed May 3, 2021.

[56] Understanding Public Perception of Nanomaterials and Their Safety in the EU. ECHA. (2020). See: https://echa.europa.eu/sv/-/what-do-eu-citizens-think-about-nanomaterials. Last accessed May 3, 2021.

[57] Larsson S, Jansson M, Boholm Å. Expert stakeholders' perception of nanotechnology: Risk, benefit, knowledge, and regulation. J Nanopart Res 2019, 21, 57.

[58] Besley JC, Kramer VL, Priest SH. Expert opinion on nanotechnology: Risks, benefits, and regulation. J Nanopart Res 2008, 10, 549–58.

[59] Foss Hansen S, Maynard A, Baun A, Tickner JA. Late lessons from early warnings for nanotechnology. Nat Nanotechnol 2008, 3, 8, 444–47.

[60] Friedersdorf LE, Bjorkland R, Klaper RD, Sayes CM, Wiesner MR. Fifteen years of nanoEHS research advances science and fosters a vibrant community. Nat Nanotechnol 2019, 14, 11, 996–98.

[61] Nymark P, Bakker M, Dekkers S, Franken R, Fransman W, García-Bilbao A, Greco D, Gulumian M, Hadrup N, Halappanavar S, Hongisto V, Hougaard KS, Jensen KA, Kohonen P, Koivisto AJ, Dal Maso M, Oosterwijk T, Poikkimäki M, Rodriguez-Llopis I, Stierum R, Sørli JB, Grafström R. Toward rigorous materials production: New approach methodologies have extensive potential to improve current safety assessment practices. Small 2020, 16, 6, e1904749.

[62] Feliu N, Fadeel B. Nanotoxicology: No small matter. Nanoscale 2010, 2, 12, 2514–20.

[63] Clausen LPW, Hansen SF. The ten decrees of nanomaterials regulations. Nat Nanotechnol 2018, 13, 9, 766–68.

[64] Hartung T. Toxicology for the twenty-first century. Nature 2009, 460, 7252, 208–12.

[65] Ingber DE. Is it time for Reviewer 3 to request human organ chip experiments instead of animal validation studies? Adv Sci 2020, 7, 2002030.

[66] Roco MC. The long view of nanotechnology development: The national nanotechnology initiative at 10 years. J Nanopart Res 2011, 13, 427–45.

[67] Fadeel B. The right stuff: On the future of nanotoxicology. Front Toxicol 2019. Nov 26. [Epub ahead of print].

[68] Trump BD, Cegan JC, Wells E, Keisler J, Linkov I. A critical juncture for synthetic biology: Lessons from nanotechnology could inform public discourse and further development of synthetic biology. EMBO Rep 2018, 19, 7, e46153.

Marcel Van de Voorde and Michael Vlerick

3 Nanotechnology: Ethical guidelines for a disruptive technology

Abstract: In this chapter, we discuss the potential societal impact of nanotechnology, highlighting the big impact of these tiny things, and reflect on the ethical issues it raises. Firstly, we discuss the harms nanotechnology might cause to health and environment and argue that the development of nanoproducts and applications should therefore be regulated. Secondly, we take on societal issues following from the development of nanotechnology, such as economic and social (in)justice, potential violations of privacy and autonomy, and unethical applications. In the face of these ethical issues, we argue that there is a need for transparency, interdisciplinary panels concerned with "nano-ethics" and institutions of democratic control. Finally, we offer a series of concrete recommendations in the face of this promising and disruptive technology.

Keywords: nanotechnology, risks, stakeholders, regulation, interdisciplinarity

3.1 Introduction

The idea that nanotechnology has the ability to drastically change the world is most probably the main reason why nanotechnology is often referred to as a "disruptive" technology. The number of areas that could be revolutionized by nanotechnology is mind-boggling. They include health care, information and communication technology, energy and transport, environment and climate change, and security and safety. Nanotechnology has the potential to revolutionize our everyday life with nano-foods, clothing, cosmetics, and cleaning products, and to solve major world problems such as global warming, environmental damage, access to clean water, public health, poverty, protection against harmful bacteria and viruses, and even against terrorism and crime.

On the other side, we cannot exclude that nanoproducts harm human health and the environment, as reported by certain toxicity studies. Moreover, the development of nanotechnology raises a host of societal and ethical issues. It could increase economic and social inequality, erode human privacy and autonomy, and be used for unethical purposes. In this chapter, we discuss the threats posed by nanotechnology and offer solutions to ensure that nanotechnology benefits all stakeholders. We end the chapter with a series of concrete recommendations.

Marcel Van de Voorde, University of Technology, Delft, The Netherlands
Michael Vlerick, Tilburg University, The Netherlands; University of Johannesburg (RSA), Johannesburg, South Africa

https://doi.org/10.1515/9783110669282-003

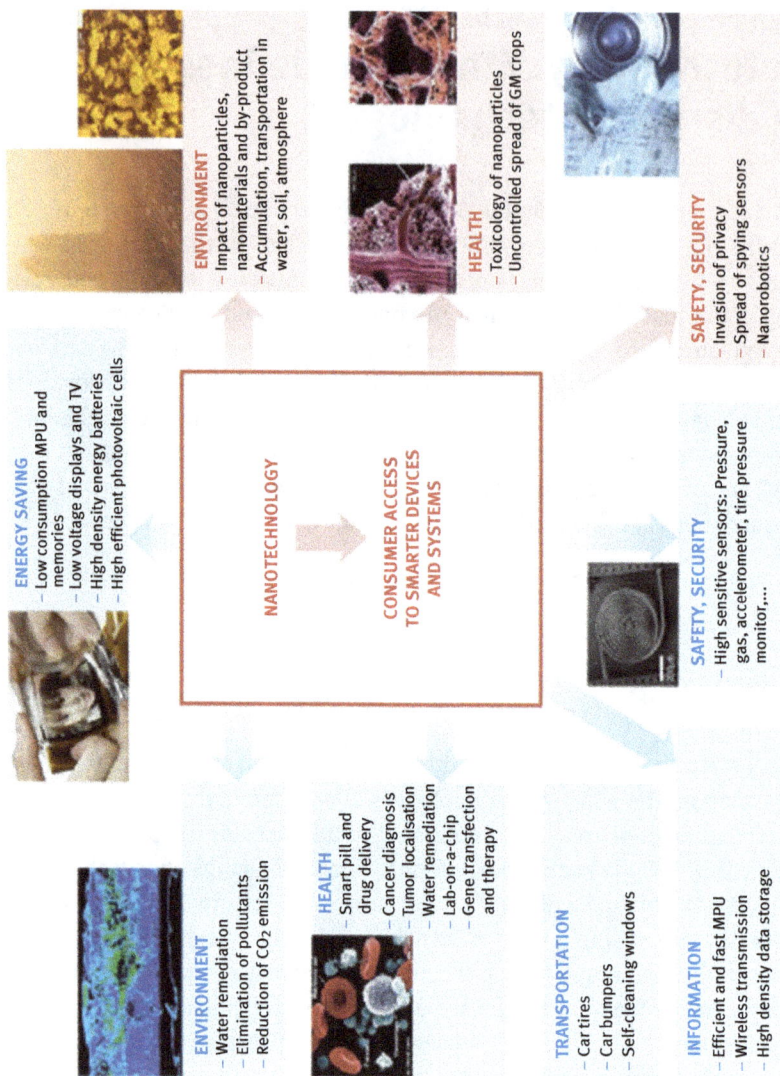

Figure 3.1: The benefits (left) and possible risks (right) of nanomaterials [1–6].

Benefits and risks of nanotechnology; some examples.

3.2 Potential harm to health and environment [7, 8]

3.2.1 The threat of uncertainties

Alongside the benefits of nanotechnologies, we must of course pay attention to the possible risks to human health and the environment, along with the societal and ethical issues it raises. In most nanotechnology applications, the nanoparticles or nanostructures are entirely inert and are constrained within a larger device. However, when the technology relies on nanoparticles, we must take into account the possible consequences of a release of these particles in our environment (and into our bodies).

Nanoparticles can enter the body through inhalation, ingestion, or direct contact with the skin. Nanoparticles may lead to toxicity or disease. They could cause inflammation, by interfering with the normal operation of body or organ chemistry, or even lead to the generation of cancers. Figure 3.2 shows the diseases that are associated with exposure to nanoparticles.

3.2.1.1 Entry by inhalation

Once inhaled, nanoparticles are either simply exhaled, or can be deposited anywhere in the respiratory tract from the nose, mouth, and larynx, down to the bronchi and alveoli of the lungs. This could lead to the exacerbation of asthma symptoms, cardiovascular adverse effects, and possibly carcinogenicity.

3.2.1.2 Entry by contact

Intact skin can effectively block the penetration of micro- and nano-objects. However, if the skin barrier is compromised by injury, sunburn, or skin disease, nanoparticles may diffuse through damaged or weakened skin and enter the bloodstream.

3.2.1.3 Entry by ingestion

Nanoparticles can be ingested directly (e.g., with food and drink) or indirectly (e.g., through the nose due to postnasal drip). Once in the digestive tract, the particles can be transported into the circulatory system.

At present, we only have limited understanding of the human health and safety risks associated with nanotechnology. Public health agencies are actively conducting research on the potential adverse health effects of unintended exposure to nanoparticles. At this point, we cannot exclude that the use of some nanoparticle products has unintended adverse effects. For example, silver nanoparticles used in socks as an

DISEASES ASSOCIATED TO NANOPARTICLE EXPOSURE

C. Buzea, I. Pacheco, & K. Robbie, Nanomaterials and nanoparticles: Sources and toxicity, Biointerphases 2 (2007) MR17-MR71

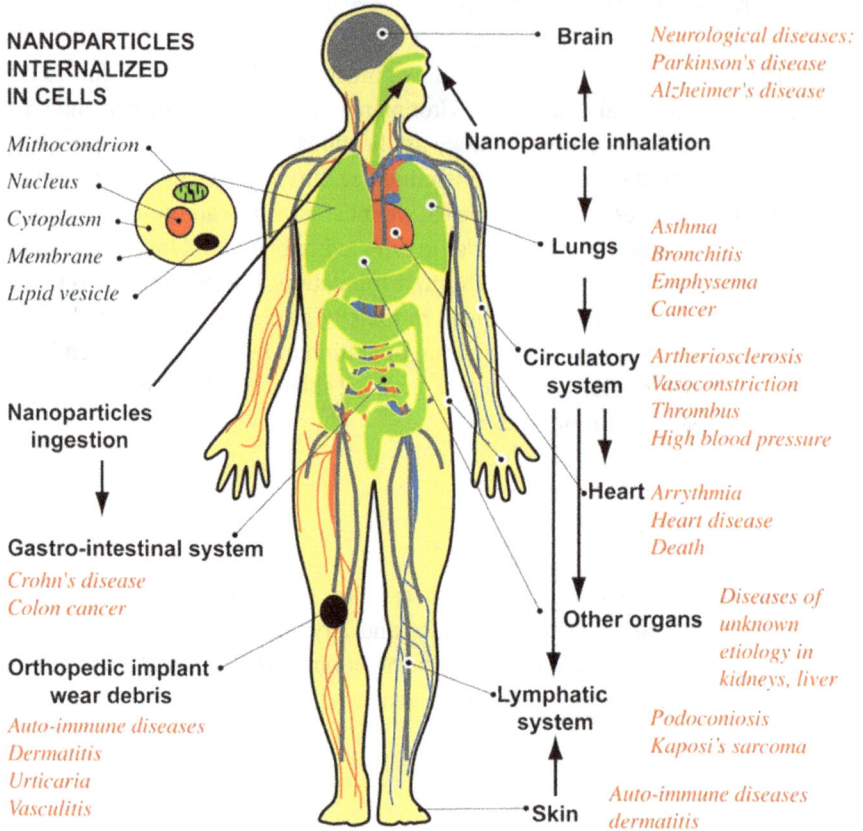

NANOPARTICLES INTERNALIZED IN CELLS

Mithocondrion
Nucleus
Cytoplasm
Membrane
Lipid vesicle

Nanoparticles ingestion

Gastro-intestinal system
Crohn's disease
Colon cancer

Orthopedic implant wear debris
Auto-immune diseases
Dermatitis
Urticaria
Vasculitis

Brain Neurological diseases: Parkinson's disease Alzheimer's disease

Nanoparticle inhalation

Lungs Asthma Bronchitis Emphysema Cancer

Circulatory system Artheriosclerosis Vasoconstriction Thrombus High blood pressure

Heart Arrythmia Heart disease Death

Other organs Diseases of unknown etiology in kidneys, liver

Lymphatic system Podoconiosis Kaposi's sarcoma

Skin Auto-immune diseases dermatitis

Figure 3.2: Main routes of nanoparticles entering the body and resulting disease possibilities (right part) [1].

antibacterial coating that reduces odors may be released when washed and then flushed into the wastewater stream. There they may destroy bacteria, which are critical components of natural ecosystems, farming, and waste treatment processes.

Given these potential harms, nanotechnology should be closely regulated by governments during the initial stages of its introduction into the marketplace. This is not an easy task. We must weigh off benefits against (mostly unknown) risks [10]. We must also be aware that it is virtually impossible to eliminate all risks.

3.2.2 The need for regulation

The growing manufacturing of nanotechnology products and their commercialization calls for the development of a regulatory framework to avoid harming people – both workers with nanomaterials and the consumers of nanoproducts – and the environment. This framework should be developed not only on a national scale by national governments but also on a supranational or even global scale by institutions such as the European Union and the United Nations [10]. Given the scientific complexity of nanotechnology, this should be done in close collaboration with scientists, toxicologists, and industrial players. [9, 10]

In the workplace, guidelines and procedures should be developed to protect exposed workers as much as possible. When it comes to everyday consumers, no substantial harm to their health has been detected, but we must remain vigilant. We cannot exclude with absolute certainty that people are not or will not be intoxicated by nanoparticles.

The regulation of nanotechnology should satisfy different criteria [4, 10]:

– We should strive for the harmonization and standardization of regulation on a global scale since nanotechnology has a global impact. First steps in this direction are being taken, with the European Union and the United States attempting to harmonize the regulation of nanomaterials.

– A sui generis regulation should be developed for all categories of nanomaterials. Nanomaterials can have significantly different characteristics than the bulk form of the same material. Many regulatory agencies have not yet devised specific regulations for the use of different nanosized versions of the same substance.

– Nanoproducts should be labeled: All nanomaterials in substances, mixtures, and so on should be clearly indicated in the labeling of the product. Today, this information is often lacking. The importance of (more) research can hardly be overstated. At present, we do not have sufficient understanding of nanomaterials. In order to have a better understanding of the risks involved, which in turn is necessary to draft appropriate regulation, we must have a better understanding of nanoparticles.

The European Union and the Organisation for Economic Co-operation and Development in collaboration with many national – for example, the US National Science Foundation – and international bodies – such as International Standardization Office in Geneva – have been working for years to develop guidelines for risk assessment of nanoproducts and technologies [4]. These efforts must be ramped up now that more nanoproducts are finding their way to the market.

Nanotechnology, however, does not only come with health and environmental risks, it also raises a host of societal issues. In the next section, we discuss these issues and suggest ways of addressing these.

3.3 Societal issues and ethical dilemmas [9, 11–13]

Nanotechnology does not only call for regulation because of the risks it poses to human health and environment. This disruptive technology also has the power to radically transform human societies. We must ask ourselves, what kind of world are we creating? As we will discuss below, while nanotechnology and its applications could have massive benefits and radically improve the human condition, they could also increase economic and social injustice, decrease privacy, and have a host of other morally undesirable effects.

Therefore, the development of nanoscience and nanotechnology should not be left to engineers and nanoscientists alone. It is of utmost importance that experts in the social sciences and humanities and the public at large are involved. Developments in nanotechnology should therefore be transparently communicated and closely monitored. For this purpose, scientists, engineers, and industrial players should work closely together with ethicists and social scientists and the public should be informed and consulted.

We must also realize that the ethical issues raised by developments in nanotechnology are not fixed. They will change over time with the development of new insights and applications. Moreover, global problems such as environmental problems and health issues (e.g., pandemics) also evolve over time and require new solutions and regulation. What is beyond any doubt, however, is that nanoscience and technology has the potential to revolutionize human life (and the lives of other species) and will therefore continue to raise important ethical questions. What is needed is an ongoing ethical debate involving all relevant experts and stakeholders (more on this below) [9, 10, 12].

In addition to protecting human health and the environment, we must be vigilant that nanotechnology does not exacerbate (and ideally reduces) economic and social injustice. We must also protect human dignity and autonomy in the face of nanotechnology-enabled privacy eroding technologies. Great power, as they say, comes with great responsibility, and nanotechnology has the potential to provide us with great power, so we must take our responsibility and develop this disruptive technology in an ethically desirable way. Nanotechnology can and should increase the well-being of all stakeholders.

3.3.1 Economic and social (in)justice [4, 6, 8]

In a morally perfect world, all stakeholders or possible beneficiaries would have equal access to nanotechnology applications. In the real world, it is very likely that at least initially nanotechnology will benefit wealthy nations and individuals (substantially) more than their poorer counterparts. Wealthy nations have the necessary infrastructure and can muster up the necessary investments to develop nanotechnology applications.

Developing nations do not, or at least not to the same extent, and risk missing out on these innovations.

In fact, third world countries have already been disappointed when it comes to nanotechnology. With the emergence of nanotechnology, 20 years ago in the United States, developing countries had great hopes that it would accelerate their economic development. Due to a lack of scientific and technological infrastructure, however, nanoscience and technology never took off, and the applications that were developed in industrialized countries were for the most part not made available to developing countries.

A major obstacle, in this context, is intellectual property rights or patent systems. Intellectual property rights threaten to make nanotechnology applications unafford- able and thus unavailable in developing countries. While the rationale of patents is to incentivize the development of new applications – by giving a time-limited monopoly on the production and commercialization of products developed in-house – it is un- clear whether this actually is the case when it comes to nanotechnology applications. Out of fear of missing moneymaking patents, competing firms are incentivized to file for many and broadly defined patents. Often these firms cannot or will not develop these patents themselves, but they hope to cash on the development of their patent by other developers. This in turn might make it less, not more, attractive to develop the said applications [14]. The current patenting system, we believe, should be revised. In addition to creating economic injustice by making useful applications unavailable in poorer regions, it seems to miss its primary target, which has to promote innovation.

In any case, intellectual property rights are likely to increase economic injustice between the developed regions and the developing regions of the world. The latter stand to miss out on many applications that could increase their wealth and well-being. This is especially problematic, given that nanotechnology applications could offer solu- tions to important problems of developing nations, such as the purification of water, energy production, efficient food production, and customized health solutions.

Finally, at the speculative end of things, nanotechnology – together with other cutting-edge technological developments such as Clustered Regularly Interspaced Short Palindromic Repeats (CRISPR) gene editing – has been linked to human en- hancement. By manipulating molecular structures, nanotechnology could in princi- ple be used by humans to enhance their appearance, physical strength, longevity, cognitive capacities, and others. Applications targeted at delivering such enhance- ments raise a long list of ethical issues and dilemmas.

Among those ethical issues is the realistic possibility that the commercialization of such applications would exacerbate social inequality. If access to these applications is limited to the wealthy (who can afford these "treatments"), this could create a huge gulf of unequal opportunities and outlook in life between the rich and the poor. In ad- dition to having less means, the latter would also find themselves at the losing end when it comes to their biological endowment. This in turn would make it very hard for them to aspire to a better future, since they could not compete in the job market with luckier cognitively enhanced people.

In short, with respect to the possibility that nanotechnology negatively influences economic and social inequality, we believe that the following moral imperatives apply:

1. Nanomaterials and nanotechnologies should, as much as is realistically possible, be made available to everyone in the world. This is more than a moral imperative for wealthy nations. A wealth gap between rich and poor nations negatively affects those rich nations because it leads to mass migration to wealthy regions, which can be profoundly destabilizing for those nations [15, 16].

2. Nanotechnologies should be developed to target specific problems faced by developing countries and made available to these countries in order to enhance the standard of living and the prospects of people in these regions.

3. The development of nanoproducts in industrialized countries should not have an adverse effect on wealth and economic prospects in developing countries. This could be the case if new nanoproducts developed in industrial countries outcompete the existing conventional products manufactured in developing countries. In this case, compensations should be made and those technologies should be made available to the developing countries in question (as pointed out above).

3.3.2 Privacy and other ethical issues [17, 18]

Nanotechnology could revolutionize health measurements by implanting microchips or biosensors in (human) organisms. This could lead to a marked improvement of human health (and to more efficient managing of chronic diseases such as diabetes). However, such technological developments may well be co-opted for surveillance and personal information gathering purposes. This raises a host of privacy issues.

Can we use personal data for scientific research? Could these data be shared with insurance companies? More extremely, we can imagine that all humans have a nanosensor implanted, which tracks their every movement and activity at all time. Michael Mehta [17] has coined the term "nano-panopticism" to describe a society in which nano-surveillance technology has created a "panopticon": a context in which everybody is "watched" and monitored at all times.

While the latter scenario is manifestly undesirable, we cannot simply dismiss any surveillance with the aid of nanosensors as immoral. People living a healthy lifestyle would benefit from having their health data accurately measured and sent to an insurance company which could in response lower their premium. Moreover, many people would probably agree that if we can detect the preparation of (and therefore prevent) heinous crimes and terrorist activities, we would be better off. But how far are we willing to go and what is the price we pay for increased security? These are hard questions that will become ever more poignant as the technology develops.

In addition to privacy issues, the development of nanotechnology raises other ethical issues. Given that nanotechnology lends itself to a very broad range of applications, it will come as no surprise that it can be used for ethically undesirable or at least questionable purposes. In particular, it could be used to design weapons and military apparel ranging from highly protective armor to miniature nuclear bombs. Ominously, a sizable part of the US funding of the development of nanotechnology, emanating from the National Nanotechnology Initiative, goes toward the development of such military applications.

Another ethical issue worth mentioning is animal suffering in nano-experiments. In order to test the safety of nanomedicines, materials, and products, experiments are performed on animals. Given the rapidly increasing production and commercialization of such medicines, materials, and products, animal testing is likely to go up. In an overview of such experiments, many were described as being "moderately to severely distressful to the animals" [19]. This raises an important ethical dilemma: how much – if any – animal suffering is justified for human benefits?

3.3.3 Solutions: interdisciplinary debate, transparency, and democratic control

In order to avoid these societal woes and ensure that nanotechnology enhances the human condition, we need to reflect on its desired development. As pointed out above, with the power to transform our world, we must constantly ask ourselves what world we want to create. The development of nanotechnology applications should not be left to nanoscientists and engineers, let alone to industrial players responding to economic incentives. Social scientists and philosophers have an indispensable role to play in guiding the development of these disruptive technologies.

3.3.3.1 Interdisciplinary debate [4, 9]

Technological innovation and ethics are age-old allies. New technologies raise a host of important ethical issues such as the potential harm and the economic or social injustice they may cause. These risks and potential woes, of course, are no reason to dismiss the development of new technologies. Technological innovations are the driver of the improvement of standards of living throughout history. We should not stop innovation – quite to the contrary! – but we should remain vigilant to the possible harm they may cause.

As pointed out, these ethical issues are prominent in the development of nanotechnology. This follows from the enormous potential impact of nanotechnology on the world we inhabit. Nanotechnology, some have argued, could hand us the power to radically reshape our world by modifying the very structure of natural materials.

While this is still very speculative, we cannot exclude that it will one day be a feasible endeavor. So how are we to reshape our world? Or rather, how are we to decide how we are going to use these disruptive technological innovations?

We believe this crucial question calls for an ongoing debate bringing together experts from a broad range of fields. In particular, experts in the exact and applied sciences and experts in the social sciences (economists, sociologists, psychologists, anthropologists, etc.) and humanities (philosophers of science, ethicists, etc.) should join forces. We need platforms and think tanks that bring together nano-scientists and engineers, on the one hand, and social scientists and philosophers, on the other hand. Their task is to give direction to development of future innovations and ensure that these developments benefit all stakeholders.

It is vital that these experts deliberate together. Too often today these two worlds remain isolated from each other. Nanoscientists and engineers work on advancing nanoscience and its applications and philosophers reflect on these developments from the sideline. This is problematic for two reasons. Firstly, the philosophical reflection often does not reach the scientists and engineers and so it misses its goal of steering the innovative developments in a desirable direction. Secondly, the philosophical reflection is somewhat disconnected and therefore poorly informed about what is actually happening in the lab. To remedy this, we should organize permanent interdisciplinary think tanks. Today, these are for the most part lacking. Finally, of course, in order to be effective, there must be proper channels through which these interdisciplinary deliberations are communicated to the policymakers (on national and international levels) who make up the regulation.

But in addition to experts in a variety of fields, the world citizens themselves – the ultimate stakeholders – should have their say. Democratic control – or rather input by the stakeholders – is important for two reasons. First, it gives the necessary legitimacy to the policy or regulation proposed in these think tanks and/or adopted by the governmental and supranational bodies (such as the EU and the UN – see above). Second, having the input of a diverse group of people is likely to lead to better policy [20, 21]. More on this important second point below.

For there to be efficient democratic control and input from the population, three conditions must be satisfied. Clear and objective information must reach the population. The population – or at least a representative sample of the population – must reflect on this information and on desirable policy. Finally, there must be proper channels so that their conclusions or views reach the policymakers.

3.3.3.2 Transparent information

The first condition (objective and clear information) requires a high degree of transparency from scientists, engineers, and other experts and industrial players involved. This is not always the case. The most important reason why information may be biased or withheld is a conflict of interest [10]. Such a conflict of interest may lead companies producing nanoproducts or applications to underplay the risks involved. The same goes for scientists and engineers who are on the payroll of these companies and policymakers who are funded or otherwise influenced by lobbying parties with a vested economic interest. In the past, parties with a conflict of interest have acted as "merchants of doubt" [22] to prevent precautionary measures against smoking, which led to pollution and environmental degradation.

We must therefore remain vigilant that industrial players and their spokespersons do not withhold or distort information about their products. Today, labeling information on nanoproducts is often missing. This is problematic. The consumer has a right to know what products they are consuming, together with an objective estimation of the risks this involves. This, of course, is no easy feat given the high degree of uncertainty that possible harm would occur (and calls – as pointed out above – for more research urgently). Nevertheless, the available information should be communicated in a transparent way.

3.3.3.3 Democratic control [10]

Democratic control can be exercised in different ways. While a lengthy discussion of democratic tools and institutions is beyond the scope of this chapter, we would like to bring a promising tool – that has been gaining momentum worldwide over the last decade – to attention: so-called citizen councils.

Traditionally, national democracies are "representative" or "indirect," which means that the decision-makers – the ones drafting policies and laws – are elected by (and represent) the people. The people exert control to the extent that they vote for those representatives or parties that propose policies that they (the people) believe are desirable. When they are disappointed in elected representatives, they can express this by not voting for them in the next election. This traditional democratic system has a lot of merit. It often leads to policy that is acceptable to the majority of the population. The condition of course for efficient democratic control is – as pointed out above – transparent information on policy proposals and the issues at stake.

Nevertheless, representative or indirect democracy has an import shortcoming. As Fishkin [23] has pointed out, it leads to "rational ignorance." When you have one vote in millions, you are not incentivized to take the trouble to inform yourself in-depth on the issues at stake and the policy proposals advanced by different politicians and/or parties to tackle these issues. This is especially relevant when it comes to complex matters such as policy regulating the development and commercialization of nanoproducts and applications.

A promising alternative democratic tool, in this context, is citizen councils [10, 23–25]. Such councils are composed of a representative sample of the population (often assembled by sortition). After being thoroughly informed by experts, they deliberate with the aim of reaching a high degree of consensus around certain policy proposals. Citizen councils have the added advantage that they can be organized on local, regional, national, supranational, and even global levels. Given the importance of coordinating national policy when it comes to nanotechnology (see above), this is an important feature. Ideally, policy and democratic control would emanate from the global level. After all, we all have equal stakes in the development of nanotechnology.

An often-voiced criticism of citizen councils and other forms of direct democracy is that laypeople are not equipped to deal with complex issues. According to this view, we should leave it up to experts. Social experiments with citizen deliberation, however, show that they often lead to thoughtful proposals even on complex issues [21]. In fact, research shows that a diverse lot of laypeople often outperforms a small group of experts. As Hong and Page [20] put it forcefully, "diversity trumps ability."

3.4 Conclusion

Mnyusiwalla et al. [9] claimed almost two decades ago that we must close the gap between the development of nanotechnology and the ethical reflection on its societal implications. Today, their call is more relevant than ever. Nanotechnology has the potential to revolutionize human life; hence, we must guide its development to ensure that it benefits all stakeholders. For this, we need sui generis regulation to protect human health and the environment – without robbing people of the many benefits these innovations bring [10] – and we need an ongoing debate between experts (from various fields) and between citizens guiding the policymakers.

In short, this debate should be:
- *Critical*: based on transparent information and expert estimations
- *Thoughtful*: leading to creative solutions
- *Inclusive*: including not only experts but also citizens and taking into account the interests of all stakeholders (both human and nonhuman, present and future generations)
- *Respectful*: admitting of divergent opinions and views, with the aim of reaching proposals that are widely accepted

Given nanotechnology's extraordinary economic and societal potential, it would be unethical to attempt to halt its development. It holds the promise to substantially improving human (and nonhuman) well-being. But this progress and innovation must be thoughtful. As is the case for many breakthroughs in technology, we cannot exclude harmful applications. We must therefore proceed with caution.

What is needed is an interdisciplinary field of "nano-ethics," together with transparent communication and democratic input. First steps have been taken, but more remains to be done. We cannot allow ethical reflection to lag behind technological development. It is our joint responsibility to make nanotechnology a force for good and create a better world.

We end with a series of following concrete recommendations:

1. Sui generis regulations of existing nanoproducts and applications should be drafted. These regulations should be internationally harmonized. More research should be done to have a better grasp of possible risks.
2. There needs to be transparent information on existing nanoproducts and applications. In particular, the labeling of nanoproducts should be accurate and clear.
3. The development of military applications of nanotechnology should be strictly limited to purposes of security and defense.
4. Interdisciplinary "nano-ethics" panels should be erected with the task to analyze present and future developments and formulate recommendations for the policymaker.
5. The next generation of scientists and engineers should be educated in the promises and perils of nanotechnology (and other promising disruptive technologies). A multidisciplinary training in "nano-ethics" at universities worldwide would contribute to this.
6. Institutions of democratic control of the development of nanoproducts and applications should be erected. Citizen councils are especially suited for this purpose.
7. There should be proper channels to connect the output of those interdisciplinary panels and the institutions of democratic control with the various national and supranational policymakers.
8. We must take particular care to protect economically disadvantaged states and people. They should not be the victim of developments in nanotechnology (either because they have no access to products and applications or because the development of these products and applications in developed countries hurts their economy). The development of nanotechnology should contribute to increasing social and economic justice.
9. The scientists and engineers at the cutting edge of research and development should be aware of their moral responsibility. They should communicate transparently about risks and benefits of existing products and about (the implications of) their latest findings and projects. This transparent information should be directed at other scientists, policymakers, and the public at large.

References

[1] Fonash, Van de Voorde M. Engineering, Medicine, and Science at the Nano-scale Study book for MSc and PhD students, WILEY Publ. Co, November 2018.
[2] Cornier J, Keck C, Van de Voorde M. Nano cosmetics. SPRINGER Publ. July 2019.

[3]   Van de Voorde M. Nanoscience and Nanotechnology: Advances and developments in Nano-sized Materials, DE GRUYTER Publ. Co, June 2018.
[4]   Van de Voorde M. Nanotechnology, Innovations & Applications.: Encyclopedia-15 Volumes WILEY Publ. Co in May 2017. Nano Health Care, Nano Information and Communication, Nano Industry.
[5]   Van de Voorde M, Fecht HJ, Werner M, Van de Voorde M. The Nano-Micro Interface, 2 Volumes, WILEY publ. Co, 2015.
[6]   Ngo, C, Van de Voorde M. Nanotechnology in a Nutshell, From Simple to Complex Systems. SPRINGER Publ. 2014.
[7]   Dosch H, Van de Voorde M. GENNESYS White Paper: A new European partnership between nanomaterials science and technology and synchrotron radiation and neutron facilities. Max Planck Gesellschaft, 2011.
[8]   Ruehle M, Mittemeyer H, Dosch & M. Van de Voorde. European White Book on Fundamental Research in Materials Sciences, Max Planck Institute, Stuttgart, 2005.
[9]   Mnyusiwalla A, Daar AS, Singer PA. Mind the gap: Science and ethics in nanotechnology. Nanotechnology 2003, 14, R9–R13.
[10] Vlerick M. Calibrating the balance: The ethics of regulating the production and use of nanotechnology applications. In: Van de Voorde M ed. Nanoethics. In Press.
[11] Eigler D, IBM Fellow, IBM Almaden Research Centre, US; Private communication Nanotechnology: The Challenge of a new Frontier.
[12] Sparrow R. The social impacts of nanotechnology: an ethical and political analysis. J Bioethical Inq 2009, 6, 1, 13–23, January.
[13] Bainbridge WS. Social and Ethical Implications of Nanotechnology. In: Bhushan B eds. Springer Handbook of Nanotechnology. Springer Handbooks. Berlin, Heidelberg, Springer, 2004, 1135–1151.
[14] Vaidhyanathan S. Nanotechnologies and the Law of Patents: A Collision Course. In: Hunt G, Mehta M eds. Nanotechnology: Risks, Ethics and Law. London, Earthscan, 2006, 225–36.
[15] Vlerick M. De tweede vervreemding: Het tijdperk van de wereldwijde samenwerking. Tielt, Belgium, Lannoo, 2019.
[16] Collier P. Exodus: How Migration is Changing Our World. 2013, Oxford, Oxford University Press.
[17] Mehta M. On Nano-panopticism: A Sociological Perspective. 2002.
[18] Gutierrez E Privacy implications of nanotechnology. Electronic Privacy Information Center. 2004.
[19] Sauer UG. 2009. Animal and non-animal experiments in nanotechnology – the results of a critical literature survey. ALTEX 2009, 26, 2, 109–28.
[20] Hong L, Page S. Groups of diverse problem solvers can outperform groups of high-ability problem solvers. Proc Natl Acad Sci U S A 2004, 101, 46, 16385–89.
[21] Landemore H. Deliberation, cognitive diversity, and democratic inclusiveness: An epistemic argument for the random selection of representatives. Synthese 2013, 190, 7, 1209–31.
[22] Oreskes N, Conway E. Merchants of Doubt: How a Handful of Scientists Obscured the Truth on Issues from Tobacco Smoke to Global Warming. 2010, New York, Bloomsbury Press.
[23] Fishkin J. When the People Speak: Deliberative Democracy and Public Consultation. 2009, Oxford, Oxford University Press.
[24] Vlerick M. Towards global cooperation: The case for a Deliberative Global Citizens' Assembly. Global policy 2020, 11, 3, 305–14.
[25] Steiner J. The Foundations of Deliberative Democracy: Empirical Research and Normative Implications. 2012, Cambridge, MA, Cambridge University Press.

Dave H. A. Blank

4 Defining the nanoethics frontier

Abstract: Nanotechnology is at the stage where industry-wide implementation of this technology is approaching, safety is guaranteed, and social acceptance is a fact. Future questions that arise, while speculative, are whether people will live longer, safer, more sustainably, and healthily? Will the quantum computer make its appearance? Do we no longer communicate only via electrons but mainly via photons? Are the materials so intelligent that they fully adapt to the circumstances or provide us with full renewable energy? Will the food production be such that the expected shortages belong to the past? Is our safety guaranteed by a conditioned environment? Or to mention an important goal: is cancer history? These are questions to which there are no ready-made answers. It is clear that the advancement of technology, such as nanotechnology, will play an essential role.

For a further development of nanotechnology, it is important to link scientific agendas with innovative challenges, to integrate a component of societal relevance, and to recognize the need for collaboration between science and industry, including the ethical aspects. This must include the entire secure chain, from science–spin-off–start-up–small and medium-sized enterprises–large companies–end user.

Keywords: NanoNextNL, NanoNed, nanotechnology, green deals, safe by design, safe innovation

4.1 Introduction

The Netherlands has in recent years invested severely in nanotechnology. It has already established itself in the field of nanotechnology at an early stage by initiating various (national) programs such as NanoNed and its successor NanoNextNL. As a result, the Netherlands has acquired a high level of knowledge and an excellent position in both the scientific and applied domains. These national initiatives have ensured that different disciplines, based on their own excellence and expertise, have started to work together. This has resulted in a clear added value in knowledge and valorization. The Netherlands is now one of the top countries in terms of scientific output, in both the number of articles and granted patents in the field of nanotechnology, and has a great international reputation with companies such as ASML, NXP, and DSM.

Dave H. A. Blank, MESA+ Institute for Nanotechnology, University of Twente, The Netherlands; SAXION University of Applied Sciences, The Netherlands

https://doi.org/10.1515/9783110669282-004

In the coming decade, nanotechnology will enter a new phase, in which, in addition to "traditional" applications (e.g., in nanoelectronics), many new ones will emerge. These are mainly related to public and environment and make a major contribution solving important social issues. Examples of relevant applications of nanotechnology are technologies for clean water, food and health, energy supply and energy saving, and nanomedicine (innovations in the field of medicine). Moreover, because nanotechnology will have an increasing impact on our society, researchers in the field of behavioral, social, nutritional, and health sciences have become full partners. In Figure 4.1, the societal challenges (missions) and which nanotechnology will play an important role are given. This matrix includes the connection with other key technologies as well. This is not comprehensive, but gives a good idea of the extent of the applications of nanotechnology.

missions / technologies	agriculture water, food	health & care	safety	energy & sustainability	along with key technology
nano-fluidics					
molecular and synthetic biology					biotechnology
Nano-sensors					
nano-photonics & devices					photonics
nano-analytical & production equipment					
(artificial) nano-materials					materials

Figure 4.1: Missions given by societal challenges versus nanotech technologies. Green means important technology. The connection with other key technologies is given in the right column. Taken from vision document Nano4Society 2020 (by NanoNextNL).

The possibilities offered by nanotechnology mean that, as with any innovation, one must carefully consider the impact on society and the possible risks. Within the national program NanoNextNL, it was decided to make the theme "risk analysis and technology assessment" as part of the program. Within this theme, it is investigated – in interaction with the other themes in nanotechnology – how we can deal with the latest developments safely, sustainably, and responsibly, in order to ensure that safety is guaranteed and that we consider the social debate. The latter is of great importance given the position and application that nanotechnology has obtained as a key technology for solving societal challenges. This relates with the four themes that Europe has chosen: health and care, energy and sustainability, agriculture, water, and food, and safety, and everything in line with the European green deal.

4.2 Nano inside

In the early years, nanotechnology has transitioned from an exotic phenomenon with potential applications in electronics as a driver of various applications in medical technology, energy materials, CMOS (complementary metal–oxide–semiconductor) transistors, quantum electronics, photonics, and power. These opportunities arose because we started to view the technological change differently. Distant concepts came closer and could become reality through nanotechnology. Viewing and manipulating at the nanoscale has led to new materials and insights into chemical processes. Food technology and the medical world are making productive use of this development. Due to the allocation of the national programs in the field of nanotechnology, in particular, the market grew. Infrastructure, on the one hand, and scientific breakthroughs, on the other hand, have resulted in a technology that is truly regarded as a key technology. It can even be said that many of the key technologies mentioned in several policy documents make use of nanotechnology. The term "nano inside" is therefore appropriate. The role of nano is not always visible, but essential to its impact. Nano is an essential building block of many applications and solutions for sustainable development. Nanotechnology is at the stage where industry-wide implementation of this technology is approaching, safety is guaranteed, and social acceptance is a fact. In Figure 4.2 the innovation scheme of nanotechnology is given. The figure shows the connections between knowledge, researchers, training and processes, and society and financials.

4.3 Nanotechnology as a key technology

Key technologies are strategically important because they allow the development of new goods and services. In addition, they provide the restructuring of industrial processes that are needed to modernize the industry, research and development, and safeguarding innovation base. Challenges such as affordable health care, sustainable energy, healthy food, clean (drinking) water, and safe information exchange for an increasingly connected but also aging population require innovations that are acceptable from an environmental and citizen perspective. Nanotechnology is considered one of the most important key technologies underlying innovations in the twenty-first century. It is indispensable to generate new knowledge, just as quantum technology and photonics are actually applications that arise and depend on the advancement of

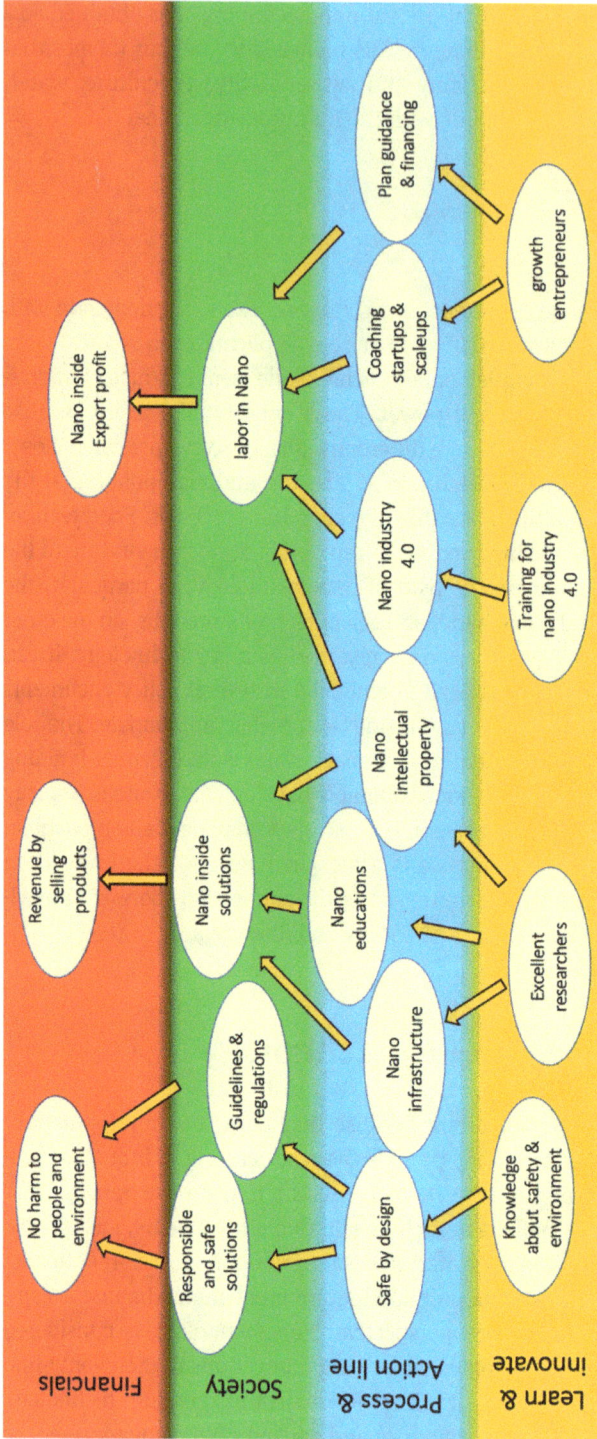

Figure 4.2: Innovation scheme of nanotechnology: from learning and innovation toward financials: crossing society and ethics. Taken from vision document Nano4Society 2020 (by NanoNextNL).

nanotechnology. This is also underlined by the recently awarded Nobel Prizes to nano-technology researchers.

4.4 Responsible nanotechnology

Risk governance is required to ensure that innovations meet the requirements and wishes of both government and society and to guarantee the safety of people and the environment. The Netherlands has an international leading position in the field of nanotechnology to achieve competent risk management. Developed concepts such as green deals, safe by design, and safe innovation approach are building blocks for developing risk management.

At the moment, however, there is still a lack of structure and instruments to promote technological development, societal challenges, and risk management in conjunction. In the forthcoming research into successful nanotechnology applications, it is necessary to pay attention to developments in risk management and social and ethic acceptance in order to arrive at an efficient and targeted innovation process. Fortunately, we can nowadays use standardization and internationally accepted guidelines (in both Organization for Economic Co-operation and Development and International Organization for Standardization), bringing together society readiness levels (SRLs) and regulatory readiness levels in one system with technology readiness levels (TRLs).

A new toolbox is needed for the realization of social innovation: models, strategies, processes, and tools that we refer to as key methodologies or key enabling methodologies, where, from a technological economic perspective, the development of the technology is placed along the bar of the TRL. This technology can be compared with the bar of SRL from a societal perspective.

4.5 Ethical component of nanotechnology

Nanotechnologies offer society a great potential of economic and technological opportunities, but technological innovations also pose a challenge to safeguard safety for people and the environment. This is a consequence of the difference in pace in which the innovations and the so-called risk governance are developed. This difference in tempo and the need to build a bridge between these worlds was recognized as one of the first in the Dutch NanoNextNL program. NanoNextNL included the earlier mentioned specific theme "risk analysis and technology assessment" (RATA) that was aimed at identifying and, where possible, addressing attention to nano-specific safety aspects from the early stages of innovation.

The ethical component of the nanotechnology research programs was initially part of the RATA. The development of technology and its embedding in society became increasingly complex, and the emergence of new forms of technology assessment can be seen as one of the answers. It is important within this kind of program that, in addition to the scientific content and its applications, the way in which new science and technology was imbedded in society was studied. One of the most important lessons we learned is the timely provision of tools and approaches to make the coevolution of technology and society more reflective.

While ethics was mainly linked to major themes, such as nuclear energy or the human genome project, in nanotechnology it also intervenes in the development of new applications, such as DNA determination and recognition, detection of diseases, functional materials by adding nanoparticles, but also nanotechnology in food, and others. Ethics and risk analysis have become an important part of innovation development. This means that various relevant actuators, such as scientists, policymakers, technology developers, insurance companies to non-governmental organizations and civil society groups, must increasingly collaborate and recognize the interdependence between research activities, scientific fields, funding opportunities, and social visions. Public dialogues are essential here.

The fact that nanotechnology is increasingly emerging as a key technology that can make important contributions to a number of essential themes makes this ethical approach even more crucial. On the one hand, innovation continues to be seen as progress and competitiveness, on the other hand, interaction with users and society is important for successful embedding in society.

A solution for such a dialogue has been found in the Dutch research program NanoNextNL, which, in addition to generic nanotechnological topics, also envisaged the applications. After all, it was a program funded by the Ministry of Economic Affairs. However, this matrix had an important ethical component, classified in RATA. In order to stimulate and facilitate dialogue, technology assessment programs have been developed that address the ethical aspects of nanotechnology. Participation in such programs was open to both PhD students and employees from science and industry. In addition, each doctoral candidate must investigate and write an assay in their thesis about the relationship between their scientific content and ethical issues, of course where applicable. This raised attention to understanding the dynamics of nanotechnology development and its embedding in society. This approach was unique for such a large research program that aims to be relevant and effective. The evaluation of the program showed that such an approach should be included at the beginning of the research program and not afterward. Here, the comparison with safe by design, which will be discussed later, is noticeable. Ethics clearly fulfills the broad interpretation of "safe."

4.6 Nanotechnology in social challenges

It is clear that nanotechnology has matured. It may fade into the background, but it stays essential. This key technology definitely plays an important role in the development and resolution of social challenges and problems. To give insight on what nanotechnology can signify, four themes are mentioned with its substantive contribution, its challenges, and its (economic) perspective, respectively – subjects in which nanotechnology sometimes plays a minor role, sometimes a major one, but nano inside everywhere.

4.6.1 Health and health care

Solutions that are made possible by nanotechnology will give people a better quality of life and will be able to take care of themselves for longer through tailor-made care. Another five quality-adjusted life years is a feasible ambition with this key technology. The lab is expected to perform more in the people instead of going to the lab, and will monitor nanotech more in the home environment. Medication and therapy can be performed by microfluidics, and nanotechnology-based lab-on-a-chip solutions are more adapted to the individual (personalized medicine). Nanotechnology-based DNA analysis systems are used to analyze the whole genome, gut flora, cancer cells, and much more. Organ-on-a-chip is expected to accelerate drug development and reduce animal testing. Nanotechnology improves drug delivery and contrast agents for imaging techniques. Nanomechanically tuned materials and structural properties reduce implant rejection processes, enable portable medical measurement and monitoring instruments, and are patient friendly.

Nanotechnology is essential to gather that information: ultrafast DNA analysis, ultra-sharp medical images (positron emission tomography–magnetic resonance imaging, computed tomography, etc.) obtained with nanoparticles, nanosensors such as sensor pills, organs on the chip of your own cells, and personalized and targeted administration of medicines. Figure 4.3 shows a platform for organ-on-a-chip measurements. With such a platform, over 150 processes and organ tissues can be observed, for example, if they are exposed to medicine. These techniques make possible a major cost reduction in health care: with a digital twin, many diseases can be prevented, citizens can lead a much better and longer independent life, while with organs on chip the costs and development time of medicines will greatly decrease.

Due to the presence of high-tech SME (small and medium-sized enterprises) and a clear market demand for solutions, the chance of impact in the short and medium term is high. The time to market is mainly determined by regulation and financing.

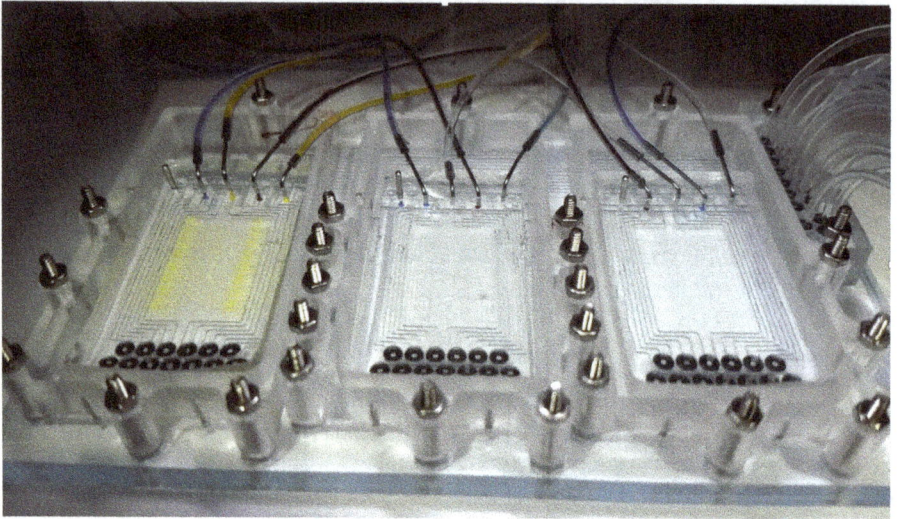

Figure 4.3: Organ-on-a-chip platform with 3 × 64 micro-organs from BIOS group, University of Twente.

4.6.1.1 Challenges in health and health care

There are several solutions to future challenges in health care where nano will play an important role. Examples are the developed organ-on-a-chip systems that mimic different interacting organ systems for better understanding of diseases, testing drugs, foods, cosmetics, chemicals, and others. Organ on a chip will play an important role in the development of personalized medication and treatments. Furthermore, these systems will lead to reduction in animal testing and possibly to a faster and safer route to make products available (shorter time to market). By developing new techniques in fabricating (artificial) nanoparticles, these particles become available for cell-specific drug delivery and active drug targeting, for example, imaging techniques.

There will be an ongoing development for the application of nanosensors combined with self-management through (home) health monitoring by means of wearables and insideables. In addition, the application of (functional) nanosystems will control the behavior of tumors that reprogram the microenvironment of the tumor and thus inhibit growth.

4.6.1.2 Economic perspective

The resulting ecosystem has led to a large number of high-tech SMEs with scaling-up potential and a good basic infrastructure that make the chance of an economic impact very high. The mostly medical nature of research and products means that there is a great demand for thorough research, validation, and risk management. These are necessary to convert an innovation into a marketable and approved product. Timeline of 15 years of research and development is a rule rather than the exception. Investments in the ecosystem already created are therefore necessary to leverage the already existing investments to benefit from it. The highly specialized market can also be developed in the short term (0–7 years) for medical research with instruments and equipment made available. Through collaboration with prominent individuals, the medical centers within the program are expected to have good market access.

With sufficient investments, the export potential for the Netherlands is predictable and will make the sector an even more important source of scale-ups and export earnings. Many markets and market segments in which nanotechnology is the key to growth are growing much faster than the average market. Market growth of more than 15% is not an exception.

4.6.2 Safety and security

Safety will become an increasingly important point of attention in the future. How do we get a grip on the advancing digitization and how do we prevent unencrypted data from falling into the hands of others? How can we protect ourselves against terrorist biochemical attacks in the future? As citizens, do we have a grip on the fact that the multitude of medical data that are generated is increasing? Can we protect our own DNA code? What about food safety? In many cases, future nanotechnologies offer the solution. Nanophotonic data encryption will play an important role, as do nanobiosensors that can rapidly identify a virus or microorganisms. We may obtain a fast "nano inside" handheld DNA test or new nanocoatings that protect surfaces against bacteria.

Nanoforensic research is used for the development of nanosensors, which is applicable in DNA research; and nano-based sensors are used for the detection of minute quantities of relevant substances in traces.

Nano for food safety: for consumer health and food security, and for adequate food and waste prevention.

Nano for safe clothing: for the development of protective clothing against aggressive substances and weather influences that meet today's health and safety standards.

Nano and fraud in advanced methods: for identification and authentication of products.

Nano-based solutions for tracking and tracing products (track and trace) and technology for protecting sensitive communications, for example, through quantum communications.

Other applications are nano-enabled quantum security (secure Internet). Nano for safety in traffic and living and working environment. Optimal use of new materials to make processes more efficient or to combine functions. Developing methods and tools to measure unwanted or harmful substances.

4.6.2.1 Challenges in safety and security

As a key technology, nanotechnology will contribute to various social issues. Developments within nanotechnology are often essential to take important steps. If the innovation lines are to be successful, knowledge institutions, the business community, and civil society organizations must work together on solutions. These innovation lines not only fall under the aforementioned four missions but also overlap or depend on mutual success. The challenges that apply to nanotechnology in relation to safety and protection are a good example and cover many aspects within health, food, and consumer.

Nanotechnology for safety equipment: protective clothing against aggressive substances and weather conditions that meet health and safety standards is one of these aspects. Furthermore, determination of nano-indicators measures whether food is still safe to consume and thus to reduce unnecessary waste. Another important development is nanodevices that support or are even based on quantum computing and encryption. With this latest development, nanophotonics plays a significant role. Applications include data storage for DNA coding and forensic research.

4.6.2.2 Economic perspective

The security sector offers plenty of opportunities for new companies. The sector will monitor developments in all technologies that affect the safety of humans and nature and will seriously monitor and assess their impact. Responses to these (potentially threatening) developments will also have to come from high-tech innovations. There appears to be a willingness to invest in technology, provided it is applied in a result-oriented manner.

4.6.3 Energy and sustainability

How do we ensure a sustainable energy supply in the future? Can we increase efficiency with the help of nanotechnology, say 50%? Can we reduce the capacity and weight of batteries by a factor of 5 or 10, so that electric flying may also become possible? Or should we move toward radically new nanoquantum concepts such as "spin batteries" that can in principle be charged and discharged indefinitely with an almost unlimited energy content? Closer to home is innovative nano-electrochemical breakthroughs to make hydrogen from sunlight or electricity. A completely different area is that of brain-inspired computing, where the unimaginable energy efficiency of the human brain serves as an example for completely new innovative computing concepts that allow us to follow the trend of breaking through more and more energy-consuming data centers.

Lithium-ion batteries are the most popular rechargeable batteries today as they have become the primary power source for many applications such as portable electronics, power tools, and hybrid/all electric vehicles. Although enormous efforts have been made to investigate the electrochemical performance of many lithium-based materials, today's rechargeable batteries exhibit energy density, long life, and safety that are still far below theoretical capabilities. None of the current rechargeable batteries can fully meet all the challenging requirements for our current energy storage.

New research directions aimed at understanding ionic diffusion and electron transport and the regulation of reversible electrochemical reactions are crucial for developing batteries with greatly improved energy storage and longer life. Nanotechnology and advanced materials, including innovative material-making and -shaping techniques, are the most important technologies as they enable intervention at the critical nanolength scale.

In addition, research is being done into the use of nanotechnology for the reusability of materials, more sustainable use of energy, and the development of alternatives to plastics and insulation materials.

4.6.3.1 Challenges in energy and sustainability

One of the most exciting opportunities for nanotechnology is the combination of properties of electrons, photons, ions, spins, phonons, and others. Because many of these phenomena work specific at the nanoscale, an integration with sensors and devices improves sensitivity of testing and applicability in health and intensive agriculture, and provides energy-efficient communication and energy storage.

Due to the exponential use of energy in communication and storage, it is important to look at alternatives to the current energy-intensive computer architecture by developing energy-efficient information and communication technology devices and circuits, for example, based on neuromorphic techniques and on efficient functioning of the human brain. Figure 4.4 shows an example of a network of nanoparticles. These particles are connected with thiols. With such network, computer calculations can be carried out. The idea is based on brain activity and is energy efficient compared to the standard CMOS.

In addition, we can use nano-based batteries that work on ionic diffusion and electron transport, realizing photovoltaic cells with a higher efficiency than conventional cells based on stable, durable, non-toxic, and recyclable materials. In any case, we have to work on nanomaterials that are reusable and sustainable in use or an environmental alternative to plastic.

a

b

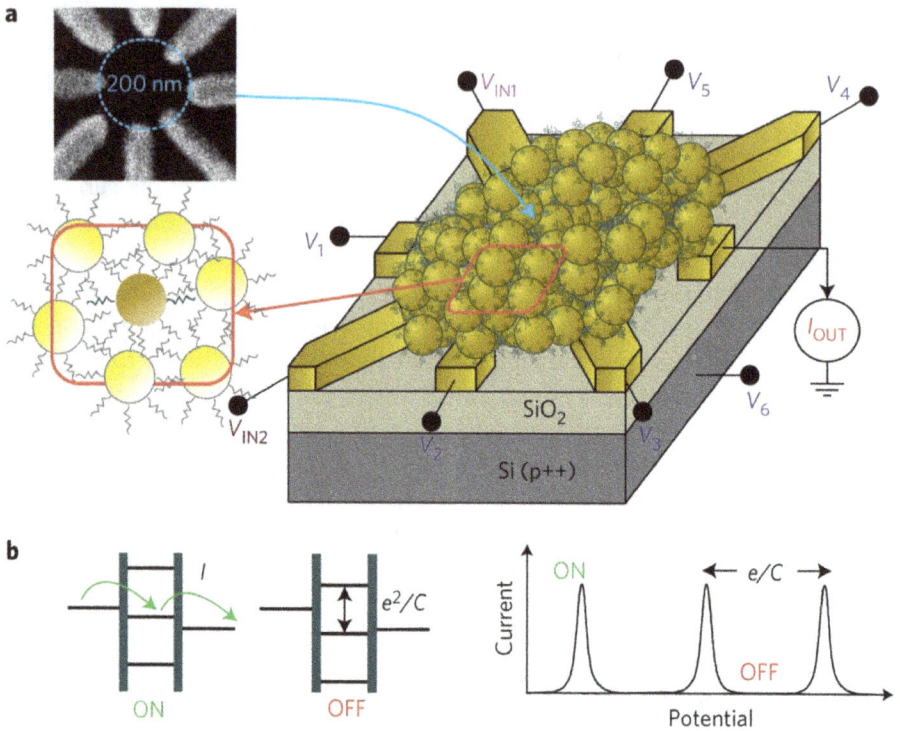

Figure 4.4: (a) Disordered network of ~20 nm Au NPs separated by ~1 nm 1-octanethiols in between eight Ti/Au electrodes, shown in the scanning electron micrograph (top inset). Two time-varying input-voltage signals V_{IN1} and V_{IN2} and six static control voltages V_1–V_6 give rise to an output current I_{OUT}. (b) The Au NPs act as SETs at low temperatures. The potential diagrams illustrate the single-electron tunneling (ON state) and the Coulomb blockade (OFF state). The conductance state of a single NP can be switched periodically between ON and OFF by varying its electrostatic potential. Data were taken from S. Bose et al. *Nature Nanotech* **10**, 1048–1052 (2015).

4.6.3.2 Economic perspective

In the field of sustainability, people will have to adapt to new standards. The climate agreement provides guidelines that cannot be achieved without the innovative contribution that nanotechnology offers us.

Using energy more consciously and efficiently shows that other forms of energy flows are needed. The sector will work on standards that have a positive influence on people and nature. The entrepreneurs and ministries in this sector are joining forces for this, and the willingness to invest in high-tech solutions is great. The technological maturity of some innovations is already fairly high and the current applications are on the way. However, the size, complexity, and dynamics of the field also require more in-depth research to achieve the desired results. Because the program is embedded in an ecosystem of knowledge institutions and companies (from large to small), the flow of results runs smoothly. In addition, it is linked to a number of major initiatives, and some of which are already underway so that the results can be optimally utilized.

4.6.4 Agriculture, water, and food

Good food and clean water are essential for people's health and well-being. However, it must be tailored to the body's needs and individual's preferences, and must be produced sustainably. Nanotechnology is crucial in this. Nanosensors complete the set of parameters used in primary production and post-harvest processing that need to be monitored to optimize processes. Process innovations based on nanotechnology, such as membrane technology in separation processes, lead to more sustainable and circular production. Nanostructured ingredients, nutrients, and pesticides ensure a better release and therefore less waste and side effects. All this ultimately leads to optimal food quality and safety, less food waste, and sufficient and pure water for the consumer with the smallest possible ecological footprint.

Food is a nanostructured material and its processing in the body also takes place on a nanoscale. That is why the possibilities of nanotechnology are used to realize better product properties and/or new functionalities ("food/body interaction") that fit into a preventive care concept.

As a key enabling technology, nanotechnology will almost never be immediately visible in the end result, but it is essential for that. The program aims to use and, if necessary, expand the potential of nanotechnology to provide a sustainable way to meet our needs for high-quality food and clean water, thus providing a healthy and enjoyable life, including for future generations.

4.6.4.1 Challenges in agriculture, water, and food

An obvious challenge is further development of nanosensors for sustainable food production and monitoring of food products. In addition, the interaction between food and body with regard to efficient absorption of food, allergies, and digestive problems takes place at the nanolevel. Optimization of ingredients, nutrients, and pesticides must ensure the quality and safety of food and water. Furthermore, membrane technology with nano-pores for water filtering and separation processes will lead to a more sustainable and circular production. In Figure 4.5, a sensor based on functionalized piezo cantilevers is shown. By using different thiols, this device can detect different molecules, originated from cancer cells or chemical reactions. This so-called nano-nose is extremely sensitive due to its nano-dimensions.

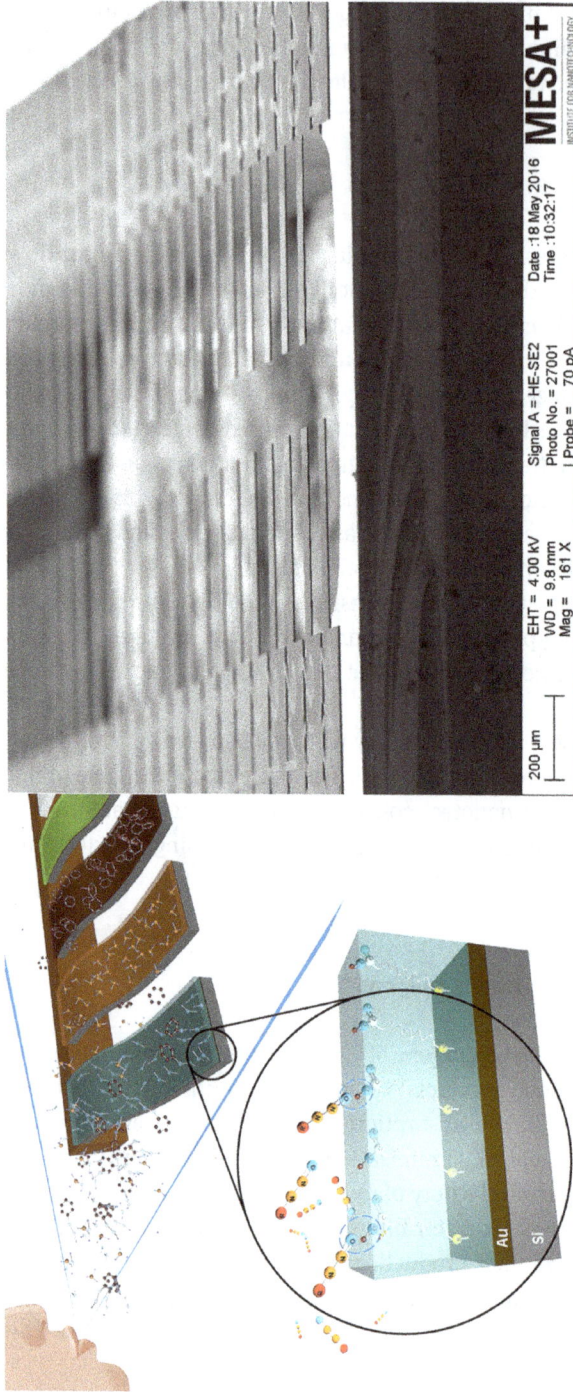

Figure 4.5: (a) Principle of nano-nose based on functionalized piezo cantilevers. (b) Realization of piezo cantilevers for nano-nose with nanoelectromechanical technology from R.J.A. Steenwelle (University of Twente).

4.6.4.2 Economic perspective

The agricultural and food sectors are aware of the fact that something must change radically due to the burden on nature and the environment. The entrepreneurs in these sectors are all looking for opportunities to improve that, and many have placed their hopes on high-tech applications to provide lighting. The willingness to invest in these sectors is great.

4.7 Safe by design

As mentioned before, it is important to take ethical issues into account for new developments and challenges that nanotechnology holds. In the early days of nanotechnology, research on exploring nanotechnology and risk and technology assessment were disconnected. Scientific breakthroughs were followed by applications, without realizing that such an application was not trustworthy or socially accepted. This was due to the fact that the research took place in two different communities that hardly had any mutual contact with each other. For example, it could happen that spin-offs that exploit the latest developments in nanotechnology turned out to be not or less successful than one would expect at first sight. In fact, this situation was rather contrafactive than reinforced new developments. Real breakthroughs can be achieved with the realization that safety and technology assessment had to be included immediately at starting new developments, both within research and in setting up business.

This "safe by design" has become an important part of nanotechnology. Within Safe by Design (SbD), various aspects must be considered. First to mention is regulations. Test methods and strategies must be harmonized. It is too costly and time-consuming to test each nanomaterial individually, especially when you consider that many materials can come in countless forms. In addition, nanomaterials are becoming increasingly complex. This complexity and the new functionalities offer unprecedented possibilities, but make safety testing even more difficult. Figure 4.6 shows an artist's impression of a new material designed through stacking layers with different properties. The stack of layers could have properties different from the individual layers. Furthermore, the interface between the layers can have additional features, for example, an interface between two nonconducting materials can be a good conductor. The multilayer in Figure 4.6 is made by pulsed laser deposition. The electron energy loss spectroscopy and transmission electron microscopy measurements show the accurate stacking of the individual layers using this deposition technique.

In addition, testing methods and strategies have not been harmonized at the international level, leading to incomparable results. Second is data management. It is very important that the outcomes of research results are unambiguous and transparent. Given the enormous increase in new materials and often complex properties, an

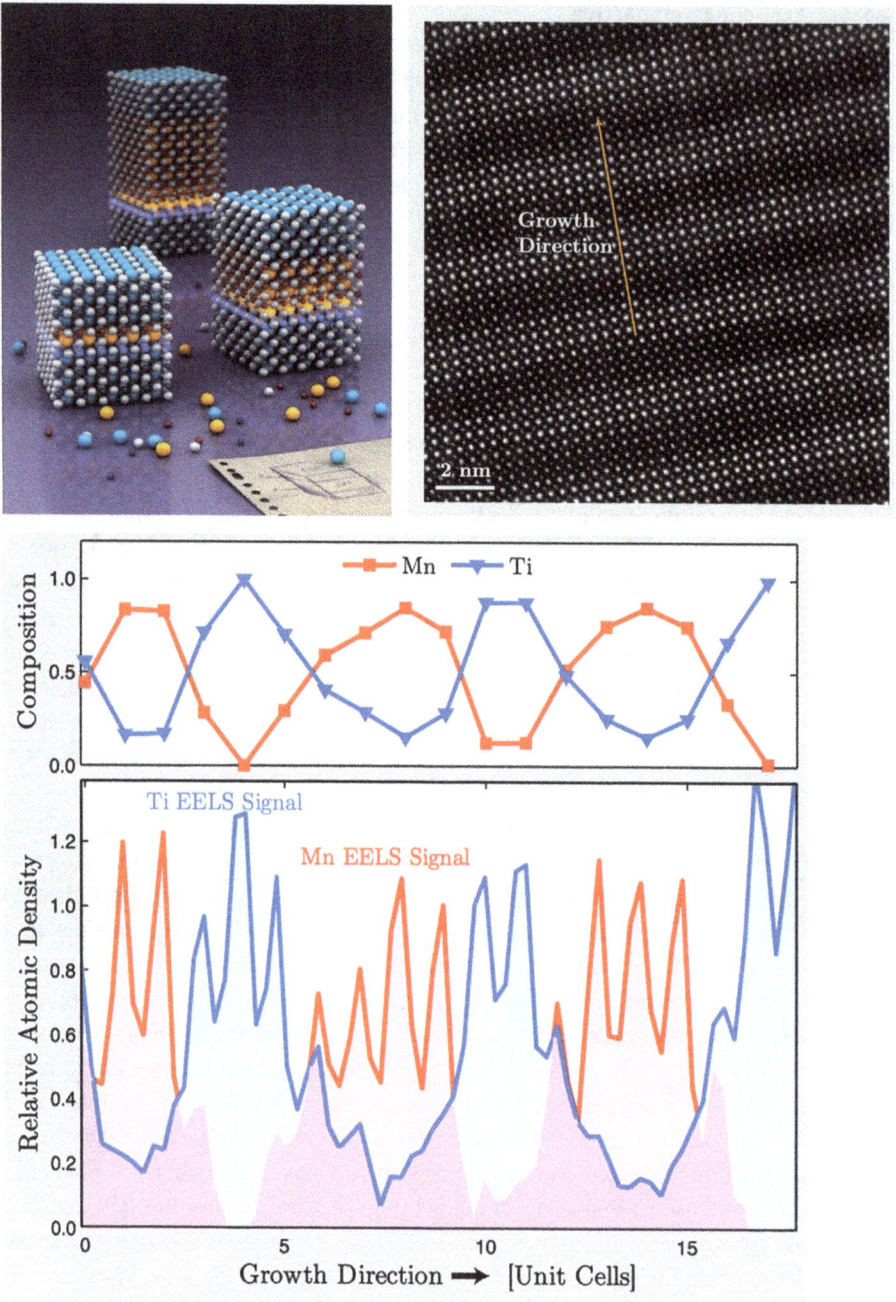

Figure 4.6: (a) An artist's impression of functional structures by atom for atom deposition from J. Huijben (University of Twente). (b) Deposition of multilayers of complex oxides of Ti and Mn by pulsed laser deposition. (Left) Transmission electron microscopy and (right) electron energy loss spectroscopy (EELS) measurements from M. Huijben et al. (University of Twente).

overview of information is time-consuming and produces large data files. The EU can play an important role in this. This can be done, for example, by introducing and enforcing an obligation to share the results of nano-safety research as a condition for funding project partners. A sustainable system for managing data on the environment, health, and safety is essential.

It is recommended to adjust the existing regulations, also by considering new and unknown risks and how to deal with them. Anticipate the potential risks from the start of the design process. Consider safe by design as a natural starting point, not because it is regulated by law but because it must be ethically responsible.

This contribution was made possible with the cooperation of the NanoNextNL foundation, in particular, the board members A. Sips, R. van't Oever, and A. van den Berg. Furthermore, fruitful discussions with partners in the NanoNextNL community are much appreciated.

Karolina Jurczyk, Urs Braegger and Mieczyslaw Jurczyk

5 Application of nanotechnology for dental implants

Abstract: Nanotechnology as a new area of research becomes one of the most favorable technologies and one that will change the application of materials in different fields. Over the past, nanomaterials have become very popular in dental applications. These materials can exhibit enhanced mechanical, biological, and chemical properties compared with their conventional microcrystalline counterparts. This chapter describes the basic concept of nanomaterials, recent innovations in nanomaterials, and their applications in restorative dentistry. Many new nanostructured Ti-based alloys have been developed and successfully tested in vivo and in clinical trials. Available products in the market, however, are sparse. Improvement of both short- and long-term tissue integration of nano-titanium implants can be achieved by modification of the surface roughness at the nanoscale level for increasing protein adsorption and cell adhesion, by biomimetic calcium phosphate coatings for enhancing osteoconduction, and by the addition of biological drugs for accelerating the bone healing process in the peri-implant area. Additionally, the ongoing research on Ti-based nanomaterials for tooth replacement is presented. Finally, an overview of the emerging nanotechnology trends in dental implants is discussed with its ethical implications on society.

Keywords: nanotechnology, dental implants, titanium, ethics

5.1 Introduction

In 1959, Richard Feynman, a Physics Nobel Laureate, presented his famous theory on nanostructure materials production [1]. He stated: "The principles of physics, as far as I can see, do not speak against the possibility of maneuvering things atom by atom." Feynman proceeded to describe a building with atomic precision and outlined a pathway involving a series of increasingly smaller machines. Nanostructures represent key building blocks for nanoscale science and technology. Recently, nanotechnology has led to a remarkable convergence of disparate fields including medicine [2–5].

Karolina Jurczyk, Urs Braegger, Department of Reconstructive Dentistry and Gerodontology, School of Dental Medicine, University of Bern, Bern, Switzerland, karolajur@gmail.com
Mieczyslaw Jurczyk, Institute of Materials Science and Engineering, Poznan University of Technology, Poznan, Poland

https://doi.org/10.1515/9783110669282-005

During the past years, interest in the study of biomaterials with nanostructure has been increasing at an accelerating rate, stimulated by recent advances in materials synthesis and characterization techniques and the realization that these materials exhibit many interesting and unexpected properties with many potential medical applications [3, 6, 7]. Nanotechnology provides tools and technology platforms for the investigation and transformation of biological systems, and biology offers inspiration models and bio-assembled components to nanotechnology [8]. For example, the London Centre for Nanotechnology offers a wide range of bionanotechnology and health-care research programs: bionanoparticles, bionanosensors, biocompatible nanomaterials, advanced medical imaging, technologies for diagnosis, self-assembled biostructures, degenerative disease studies, molecular simulation, lab on a chip and screening, drug screening technologies, and molecular simulation.

The application of new materials such as biomaterials and implants increases steadily. However not all replacement systems have provided trouble-free service. In dental implants, the rate of success is 98.6% after 5 years post-loading [9]. None of the failed implants presented the regular surface composition and depth profile of the TiO_2 overlayer; foreign elements (Ca, Na, P, Si, Cl, Zn, Pb, and Al) were observed on some implants [10]. Fibrosis, lymphocytic and plasmacytic infiltrates, and granulomatous lesions were detected in the surrounding tissues. An appropriate surface modification would prevent their transfer into nearby tissues. The absence of debris particle generation is crucial for the prevention of implant malfunction. The determination of the mechanisms of debris generation and modification of implant surface bulk structure and properties is one of the main aims of current research projects.

This chapter focuses on the current status and the potential future applications of nanotechnology in dental implantology. Additionally, the ongoing research on Ti-based nanomaterials for tooth replacement is presented. Finally, an overview of nanotechnology and its ethical implications in dental medicine is summarized. The discussion includes a range of different types of available nanoproducts and their potential in dentistry.

5.2 Mimetizing the stomatognathic system

To discuss the therapeutic or reconstructive issues of the stomatognathic system needs a good knowledge of the morphology and function of all elements [11–13]. For this purpose, we usually use materials, which would be capable of reconstructing tissues that were disabled, due to the aging process, trauma, or disease. Biomedical engineering has rapidly developed over the past years, creating analogs of the tissues but is still far from perfection [14]. Improving the quality of life is a rewarding goal for scientists and engineers. One of the fields of science is based on understanding the relationship between the human system and the environment. Indeed, those areas of research and new technologies in the field of rehabilitation improve the quality of life for disabled as well as abled patients.

Most of the global population, sooner or later, needs a repair or reconstruction within the craniofacial region [15]. It may be caused by a minor pathological process, such as dental caries occurring in a tooth, or an oncological process leading to resection of the tissues. Most of the human tissues are hybrids, such as the tooth composed of enamel, dentine, cementum, and pulp tissue. Therefore, material engineering is focused on developing single tissues, as well as hybrid organs and interfaces. Understanding the interaction between molecules of the extracellular matrix and attached cells to the materials is essential for proper biomaterial design. Applications of biomaterials need experimental verifications in different laboratory models to prove their functional effectiveness [4]. In addition to the physical, chemical, mechanical, and biological performance, the clinical performance of the material is also reviewed through different trials.

Obtaining the same chemical structures as those of the body tissues, however, does not solve the problem that the artificially obtained material is non-vital. For example, new structures of materials incorporating sensory systems and signal processing are proposed for the next generation of prostheses. They will help better tailor the response in stiffness and damping equipment or provide spectral properties closer to those of natural materials.

In the rehabilitation of oral tissues, the fundamental role is played by osseointegration, which means a direct connection between the bone and implant without interposed soft tissue layers [5]. However, a 100% connection of the bone to the implant material does not occur. Therefore, a more suitable approach to understanding this term would be based on stability and not histological criteria: "A process whereby clinically asymptomatic rigid fixation of alloplastic materials is achieved, and maintained, in bone during functional loading" [15–17].

Although modern implants have improved substantially over the last 50 years, the general concept has remained unchanged: replacing a missing tooth with an inert non-biological material. It can be a metal, ceramic, or a combination of both. The rate of technological improvements in implants has reached a plateau and new developments will require major changes to the basic approach. Recent researches focused on developing new materials, designs, and the introduction of new surfaces, although not all those tendencies were without pitfalls. For example, the ceramic-coated metal implants turned out to be not so successful in the clinical trial due to the brittle coating, which tends to flake away.

A great number of investigations focusing on the new material designs in implantology should be considered with great caution. In vitro studies bring into our knowledge many interesting findings; however, we must remember that those trials are conducted on glass disk with an artificial environment without the influence of hormones, nervous system, and blood flow. It so happens that in vivo studies not always confirm the in vitro results. Therefore, it is essential to finish all the in vivo models to initiate the clinical trials.

Within the stomatognathic system, we can distinguish two environments. One of them is composed of hard and soft tissues as well as body fluids. From the biophysical point of view, the tissue environment has the characteristics of an electrochemical membrane. Moreover, it possesses membranes and membrane potentials, which are responsible for ion-selective effect and metabolism. Body fluids, on the contrary, are characterized by ion conductivity. Into this environment, due to a loss of tissue, bio-materials are introduced to repair and to bring back the function of the system. Mostly, these are metal implants or screws or bone materials used in maxillofacial surgery, implantology, and prosthetics. The osseointegrated interface depends on the physical and biophysical characteristics of the surrounding environment and the physicochemical properties of the surface of the implant.

The second environment is composed of the oral cavity, to which materials are introduced by restorative dentistry, prosthetics, and orthodontics. Those implants have a hybrid composite structure, where the core is composed of a metal, an electric conductor. However, the surface of the implant is coated by a polymer or bioceramics, which has the characteristics of an insulator or a semiconductor. Moreover, the whole structure of the biomaterial forms a complex system, which is an essential issue for the development of corrosion or biodegradation of implants. Some of the prosthetic reconstructions are localized in both environments, such as implant-supported fixed and mobile prosthesis, which through abutment connections are attached to the osseous implants.

Referring to the complexity of the system, biochemical, biomechanical, and bioelectronic factors should be considered, which are undoubtedly connected to the ongoing metabolic, bacteriological, and immunological processes within the system.

Microflora of the oral cavity is specific: containing unique molecular, structural, and microbial characteristics, each of the body sites harbors a normal microflora, therefore, forming an ecosystem [11]. It consists of species that are responsible for protecting the surrounding tissues against the pathogenic bacteria. In a physiological state, all bacteria remain in equilibrium. The change in a pathological state occurs when the influence of the negative factors, such as antibiotic administration, fermentation of food, and dysfunction of the immunological system, is stronger than the protective ability of the system. The physical and chemical nature of the oral cavity is not uniform and changes depending on the localization. For example, anaerobes can be found in the gingival sulcus, whereas aerobes exist on supragingival surfaces, due to the presence of oxygen. Also, the presence of food changes the oxygen concentration, pH, the metabolites, which in consequence also has its impact on the bacteria species existing in the oral cavity. The environment in dental implant sulcus is considered similar to that of dental ones, such as neutral pH, anaerobiosis, and rich nutrition. The environment may support the anaerobic microflora growth in the crevices around implants, particularly at the interface between histocompatible artificial material and mucosal epithelium. Moreover, the type and shape of the implant, connection type, abutment, and suprastructure material, and the type of prosthetic suprastructure also affect the peri-implant soft and hard tissues. By adhering to the abutment implant surfaces, microbiota may trigger inflammation in the tissue around the implants.

The biocompatibility of biomaterials depends on many factors, mainly on composition, location, and interaction with surrounding tissues within the oral cavity [18]. Due to differences in the composition of materials, we encounter different biological responses. However, the reaction of the host body depends on whether the implanted material releases any of its components, which may be toxic, mutagenic, or immunogenic to the tissues. Not without any influence is also the location of the implant. Some materials that may be toxic in contact with oral mucosa are biocompatible with the hard tissues and vice versa [18]. We have to take into consideration the fact that the stomatognathic system is variable, taking, for instance, the pH, the effects of body fluids, which also have their effect on the biocompatibility of materials. The morphology of the biomaterial surface should be designed so, or to elicit the growth of cells, or to prevent the attachment of other cells or retention of plaque.

5.3 Nanotechnology in dental implants

Prostheses on osseointegrated dental implants have become a standard of care in the management of edentulous and partially edentulous patients. Titanium and titanium alloys have been the preferred materials in the production of fixed substitutes for roots of lost teeth [19, 20]. Titanium appears in two different allotropic forms. At low temperatures, it has a closed packed hexagonal crystal structure (space group: P63/mmc–hcp), which is known as α, whereas above 882 ± 2 °C, it has a body-centered cubic structure (space group: Im-3m–bcc) termed β. The alloying elements, such as N, O, and Al, tend to stabilize the α phase, whereas elements V, Cr, Nb, and Mo stabilize the β phase, at room temperature.

Zirconia implants are new to the implant arena. Zirconia is the crystal form of the transitional metal zirconium, and zirconia implants are often marketed as "metal free." A dental implant is essentially a titanium or zirconia screw or cylinder, between 4 and 16 mm long, which is inserted into a prepared bony socket in the jaw and acts as a replacement root for the missing tooth. A special attachment called the abutment is fitted to the top of the implant and forms the external connection to the replacement tooth (crown) or teeth (bridge or denture). Once an implant has been placed in the jaw, it needs to heal before the crowns can be added. During this healing, which can take between 2 and 6 months, the surface of the implant fuses with the surrounding bone, a process known as osseointegration. There are also one-piece implants, where the abutment and the implant are part of a single piece.

A satisfactory clinical outcome depends on the capability of the implant to bear loads, which is obtained by primary stability immediately following implantation. However, long-term outcomes are the result of solid osseointegration of the implant into the host bone. The characteristics of the implant surface itself are of critical importance for the progression toward osseointegration [21]. Cell biological as well as biomechanical properties of the biomaterial play a crucial role in the initial formation of the bone–implant system. An open-porous structure and adequate pore sizes are the determinants for bone ingrowth [22, 23].

The mechanical properties of the implants acting as a scaffold for bone ingrowth should be adjusted to the mechanical properties of the surrounding tissue. Several factors have been demonstrated to have an influence on bone ingrowth into porous implant surfaces, such as the porous structure (pore size, pore shape, porosity, and interconnecting pore size) of the implant, duration of implantation, biocompatibility, implant stiffness, and micromotion between the implant and adjacent bone [24–28].

Developments in material engineering resulted in the fabrication of porous scaffolds that mimic the architecture and mechanical properties of the natural bone. The porous structure functions as a framework for bone cell ingrowth into the pores and therefore leading to the integration with the host tissue known as osseointegration [29]. Appropriate mechanical properties, in particular, a low elastic modulus similar to the bone may minimize the stress-shielding problem.

The preparation of porous metals can be obtained by several methods [28]. Aiming at bone ingrowth, long-term stability, and load-bearing capacity, porous titanium should represent the following features: (i) high porosity and interconnected pore structure for sufficient space, enabling attachment and proliferation of new bone tissue and the transport of body fluids; (ii) a critical pore size usually ranges within 150–500 mm; and (iii) appropriate mechanical properties are adjusted to the surrounding bone tissue for load bearing and transforming.

In the past, efforts have been made to enhance the topography of implant surfaces to accelerate the healing process. Titanium surfaces with microscopic scale roughness have been proposed as an alternative to more conventional implant surfaces produced by machining or titanium plasma spraying. Various techniques such as sandblasting, acid etching, and combinations of both have been applied to obtain micro-rough surfaces used nowadays (Table 5.1).

Current research focuses on improving the mechanical performance and biocompatibility of metal-based systems through changes in alloy composition, microstructure, and surface treatments [30–40]. In the case of titanium, a lot of attention is paid to enhance the biocompatibility of commercial purity grades to avoid potential biotoxicity of alloying elements, especially in dental implants [41–43]. Nanostructured materials can exhibit enhanced mechanical, biological, and chemical properties compared with their conventional counterparts [40, 43].

Table 5.1: Present dental Ti-type implants and new Ti-type materials with nanostructure under investigation for dental implant applications.

Present α-Ti-type implants	Bulk nanostructured α-Ti-type implants	Bulk α-Ti-type nanocomposites	Bulk β-Ti-type alloys with nanostructure	Implants with nanosurface
Machined	SPD	Ti–HA	$Ti_{23}Mo$	Ti–1Ag
TPS	ECAP	Ti-45S5 bioglass	$Ti_{23}Mo$–45S5 bioglass	Ti–HA
SLA®	ECAP with TMP	Ti–SiO$_2$	$Ti_{31}Mo_5HA$–Ag or CeO$_2$	
SLActive®	Nanoimplants®	Ti–bioglass–Ag	Ti–Zr–Nb	
Roxolid®				
TiUnite®				
Inicell® Thommen				

TPS, titanium plasma spray layers; SLA®, sandblasted, large grit, acid etched; SLActive®, SLA® chemically modified; Roxolid®, a metal alloy composed of 15% zirconium and 85% titanium; TiUnite®, Nobel Biocare's proprietary titanium oxide dental implant surface; Inicell® Thommen with APLIQUIQ® system; SPD, severe plastic deformation; ECAP, equal channel angular pressing process; TMP, thermomechanical processing; Nanoimplants® Timplant Ltd; HA, hydroxyapatite.

Improvement of the physicochemical and mechanical performance of Ti-based implant materials can be achieved through microstructure control, and the top-down approaches are known as severe plastic deformation and mechanical alloying (MA) techniques [44]. Recent studies showed clearly that nanostructuring of titanium can considerably improve not only the mechanical properties but also the biocompatibility [45, 46]. Moreover, this approach also has the benefit of enhancing the biological response to the commercially pure titanium surface (cp-Ti) [46].

An alternative method for changing the properties of Ti and Ti-based alloys is the production of a composite, which will combine the favorable mechanical properties of titanium and the excellent biocompatibility and bioactivity of ceramics [43, 46–49]. The main ceramics used in medicine are hydroxyapatite (HA, $Ca_{10}(PO_4)_6(OH)_2$) and 45S5 bioglass (44.8% SiO_2, 24.9% Na_2O, 24.5% CaO, 5.8% P_2O_5) [50]. The ceramic coating on titanium improves surface bioactivity, but often flakes off as a result of poor ceramic/metal interface bonding, which may cause an early or late failure. However, the nanocomposite materials containing metal and bioceramic as a reinforced phase are promising alternatives compared to conventional materials, because they can potentially be designed to match the properties of bone tissue to enhance bone healing.

The manufacturing of Ti-based nanomaterials and their upscaling for industrial use is still challenging [40, 51–56]. New types of bulk three-dimensional (3D) porous Ti-based nanocomposite biomaterials with a preferred size of porous and 3D capillary-porous coatings were developed. Ongoing research aims at producing new generations of titanium–bioceramic nanocomposites by constructing porous structures with a strictly specified chemical and phase composition, porosity, and surface morphology, which will promote good adhesion of the substrate and show increased hardness, high resistance to biological corrosion, and good biocompatibility [31, 33, 34, 43, 46]. Materials with nanoscaled grains would offer new structural and functional properties for innovative products in dental applications.

5.2.1 Titanium nanocomposite implants

Many researchers, focusing on the synthesis of nanoscale Ti-based biocomposites, have achieved better mechanical and corrosion properties of nanomaterials compared with microcrystalline titanium [46, 48, 49]. For the Ti–HA nanocomposites, the Vickers hardness increased (Ti–20 wt% HA nanocomposites (1,030 $HV_{0.2}$)) and was four times higher compared with pure microcrystalline Ti metal (250 $HV_{0.2}$) [46]. No significant difference in corrosion resistance among Ti–3 wt% HA ($i_c = 9.06 \times 10^{-8}$ A/cm^2, $E_c = -0.34$ V) and Ti–20 wt% HA ($i_c = 8.5 \times 10^{-8}$ A/cm^2, $E_c = -0.55$ V) was noted although there was a significant difference in porosity. A titanium–45S5 bioglass nanocomposite with a unique microstructure, higher hardness, and better corrosion resistance was produced using MA and a powder metallurgy process [48, 49]. Microhardness tests showed that the obtained material exhibited Vickers microhardness as high as 770 $HV_{0.2}$ for Ti–20 wt% 45S5 bioglass, which was more than three times higher than that of conventional microcrystalline titanium. Young's modulus of the obtained Ti–10 wt% 45S5 bioglass composite was measured to be 110 GPa, whereas for the microcrystalline titanium the value was 150 GPa. The corrosion resistance tests proved that Ti–10 wt% of 45S5 bioglass nanocomposites ($i_c = 1.20 \times 10^{-7}$ A/cm^2, $E_c = -0.42$ V vs. SCE) were more corrosion resistant than microcrystalline titanium.

Nanograined materials are characterized by large surface energy due to the very high number of atoms on the surface. This property can explain their entirely different behavior compared with the micron-sized grains. Osteoblasts adhere to the surface with a roughness in the nanometer range. However, not only the roughness but also the composition and the surface energy affect the initial contact and spreading of cells [56, 57].

In vitro biocompatibility tests were performed with titanium–10 wt% 45S5 bioglass nanocomposites [48, 49]. The morphology of the cell cultures obtained on the tested nanocomposite was similar to those obtained on microcrystalline titanium. On the other hand, on porous scaffolds, the cells adhered with their entire surface to the insert penetrating the porous structure, while on the polished surface, more spherical cells with a smaller surface of adhesion were noticed. The study has demonstrated that titanium–10 wt% 45S5 bioglass scaffold nanocomposite is a promising biomaterial for dental tissue engineering.

The surface roughness R_a of commercial Ti dental implants produced by machine is in the range of 0.08–1.3 mm [31]. Recent studies show that surface roughness influences cell spreading and proliferation but not cell attachment of human osteoblast-like cells [25, 58]. Therefore, it seems reasonable to search for modifications of surface roughness, which would provide a more suitable microenvironment for early osteoblast response to implant materials.

The design of titanium–10 wt% 45S5 bioglass–1.5 wt% Ag composite foams with 70% porosity and pore diameter of about 0.3–1.1 mm is one of the most promising approaches to achieving optimal antibacterial activities against *Staphylococcus aureus* and *Staphylococcus mutans* [59, 60]. The Ti–bioglass–Ag composite showed the highest antibacterial activity against *S. aureus* and *S. mutans*. In both cases, the biofilm formation was reduced by more than 90% in comparison to microcrystalline titanium. According to the previously published research, when Ti–45S5 bioglass–Ag material can be immersed in the body fluid, the metallic Ag particles on the surface of the Ti–45S5 bioglass–Ag could react with the body fluid and release ionized Ag into the surrounding fluid [61]. A steady and prolonged release of the silver biocide in a concentration level (0.1×10^{-9}) is capable of rendering antibacterial efficacy [48]. Silver particles had the highest antibacterial effect with minimal inhibition concentration of 4.86 ± 2.71 µg/mL and minimal bactericidal concentration of 6.25 µg/mL, respectively [62].

5.2.2 Ti β-type alloys and composite implants

Titanium β-type alloys attract attention as biomaterials for dental applications [42, 43, 63]. Nanostructured β-type $Ti_{23}Mo$–x wt% 45S5 27 bioglass ($x = 0$, 3, and 10) composites were synthesized [63]. The crystallization of the amorphous material upon annealing led to the formation of a nanostructured β-type $Ti_{23}Mo$ alloy with a grain size of approximately 40 nm. With the increase of the 45S5 bioglass content in $Ti_{23}Mo$ nanocomposite, an increase of the α-phase was noticeable. The electrochemical treatment in phosphoric acid electrolyte resulted in a porous surface, followed by a bioactive ceramic Ca-P deposition. Implants, due to the corrosive environment of the tissue and body fluids, may undergo unexpected local corrosion attacks, leading to a release of the corrosion products into the tissue with a toxic effect. In vitro test cultures of normal human osteoblast (NHOst) cells showed very good cell proliferation, colonization, and multilayering. The key factors for the success of implant integration with the surrounding hard tissues are the surface topography and the chemical composition. Therefore, the biofunctionalized nanocrystalline $Ti_{23}Mo$ 3 bioglass composite may represent an important step forward in the development of such a structure, which will support the primary retention and initial healing of the implant.

Recently, the crystal structure, microstructure, and mechanical and corrosion properties of bulk $Ti_{31}Mo_xHA$ composites ($x = 0$, 2.5, 5, and 10 wt%) were investigated [43]. All these composites have elastic modulus lower than cp micro-crystalline α-Ti, and their hardness is two times higher. The ultrafine $Ti_{31}Mo_5HA$ composite was more corrosion resistant in Ringer's solution than the bulk $Ti_{31}Mo$ alloy. Surface wettability measurements revealed the higher surface hydrophilicity of the bulk ultrafine-grained $Ti_{31}Mo_{10}HA$ sample in comparison to the microcrystalline Ti sample. $Ti_{31}Mo_5HA$ composites with the addition of 1 wt% Ag, 2 wt% Ta_2O_5, or 2 wt% CeO_2 were synthesized, too. The antibacterial activity of $Ti_{31}Mo_5HA$ composite contain-ing silver (Ag), tantalum (V) oxide (Ta_2O_5), or cerium (IV) oxide (CeO_2) against *S. aureus* was studied. In vitro bacterial adhesion, the study indicated a significantly reduced number of *S. aureus* on the bulk ultrafine-grained $Ti_{31}Mo_5HA–Ag$ (Ce_2O_3) plate surfaces in comparison to microcrystalline Ti plate surface. Ultrafine-grained $Ti_{31}Mo_5HA–Ag$ or CeO_2 biomaterials can be considered to be the future generation of medical implants.

Independently, the technical viability of preparing Ti–Zr–Nb alloys by high-energy ball-milling in a SPEX 8000 mill has been studied [42]. These materials were prepared by the combination of MA and powder metallurgy approach with cold powder compac-tion and sintering. Changes in the crystal structure as a function of the milling time were investigated using X-ray diffraction. Our study has shown that MA supported by cold pressing and sintering at the temperature below α→β transus (600 °C) can be applied to synthesize single-phase, ultrafine-grained, bulk Ti(β)-type $Ti_{30}Zr_{17}Nb$, $Ti_{23}Zr_{25}Nb$, $Ti_{30}Zr_{26}Nb$, $Ti_{22}Zr_{34}Nb$, and $Ti_{30}Zr_{34}Nb$ alloys. Alloys with lower content of Zr and Nb need higher sintering temperatures to have them fully recrystallized. The properties of developed materials are also engrossed in terms of their biomed-ical use with Young's modulus, which is significantly lower than that of pure titanium.

5.2.3 Implants with nanosurface

Nanoscale modification can alter the chemistry and/or topography of the implant surface [25, 32, 38, 54, 64]. There are many different methods to impart nanoscale features to the implant surface (Table 5.2) [25]. Many studies focused on how to im-prove the bone/implant interface. Two different approaches were discussed to ob-tain the enhanced interface: (i) chemically by incorporating inorganic phases, such as calcium phosphate (Ca-P), into the TiO_2 layer, and (ii) physically by modifying the architecture of the surface topography [24, 29, 32, 36, 58, 65, 66]. Both methods aimed at stimulating bone formation and therefore shortening the healing time and in consequence patient's rehabilitation time.

At the micrometer level, the concept of changing the surface topography is that a rough surface having a higher developed area than the smooth surface increases bone anchorage and reinforces the biomechanical interlocking of the bone with the implant, at least up to a certain level of roughness [67]. At the nanometer level, however, the roughness increases the surface energy, and thus improves matrix

Table 5.2: Overview of methods used for creating nanofeatures on cp-Ti implants.

Method	Modified layer	Objective
Physical		
Plasma spray	Coatings such as HA, bioglass, boron surface-modified layer	Improve wear resistance, corrosion resistance, and biological properties
Physical vapor deposition	TiN, TiC, TiCN, thin-film HA coating	Improve wear resistance, corrosion resistance, and biological properties
Ion implantation and deposition	Surface modification layer, thin film	Surface composition modification – improve wear resistance, corrosion resistance, and biological properties
Chemical		
Sol–gel	Thin films of calcium phosphate, TiO$_2$, silica	Improve wear resistance, corrosion resistance, and biological properties
Chemical vapor deposition	TiN, TiC, TiCN, thin-film HA coating	Improve wear resistance, corrosion resistance, and biological properties
Anodic oxidation	TiO$_2$ layers with specific surface topographies	Improve wear resistance, corrosion resistance, and biological properties
Cathodic deposition	Ca-P deposition	Improve corrosion resistance and biological properties
Chemical treatments such as acidic, alkaline, and hydrogen peroxide treatments	Oxide layers, sodium titanate gel	Remove oxides and contaminations, and improve biocompatibility
Biochemical methods	Coating deposition	Induce specific cell and tissue reaction
Mechanical		
Severe plastic deformation	Bulk ultra-fine grain metals/alloys	Improve mechanical and biological properties
Mechanical alloying and powder metallurgy	Alloys with nanostructure and nanocomposites	Improve mechanical properties, corrosion resistance, and biological properties

protein adsorption, bone cell migration, and proliferation, and finally osseointegration [40, 68].

Numerous reports are demonstrating that the surface roughness of titanium implants affects the rate of osseointegration and the biomechanical fixation [69]. The biological properties of titanium depend on its surface oxide film [24]. The modification of the surface morphology and properties of titanium dental implants can be obtained by several mechanical and chemical treatments. One possible method of improving dental implant biocompatibility is to increase surface roughness and decrease the contact angle. Surface profiles in the nanometer range play an important role in the adsorption of proteins, adhesion of osteoblastic cells, and thus the rate of osseointegration [70].

Improvement of the interfacial properties between the surrounding tissues and the existing implants, for example, Ti and Ti-based alloy is still in the interest of many researchers. The electrochemical technique, a more simple and fast method, can be used as a potential alternative for producing porous Ti-based metals for medical implants. The good corrosion resistance of the titanium is provided by the passive titanium oxide film on the surface, which has the thickness of a few nanometers. This layer is important for good biocompatibility. Its thickness can be increased up to the micrometer range by anodic oxidation and is dependent on the electrochemical etching conditions, for example, current density, voltage, and electrolyte composition [32]. In the electrochemical etching of titanium, electrolytes containing H_3PO_4, CH_3COOH, and H_2SO_4 are used. In the Ti anodization, the dissolution is enhanced by HF- or NH_4F-containing electrolytes, which results in pore or nanotube formation. The current density, in this case, is much higher than in electrolyte without HF or NH_4F [65]. Fluoride ions form soluble $[TiF_6]^{2-}$ complexes resulting in the dissolution of the titanium oxides, limiting the thickness of the porous layer. Porous implant layers have a lower density than the respective bulk structures; therefore, good mechanical strength is ensured by the characteristics of the latter.

It has been shown that a surface with appropriate chemical composition and topography after combined electrochemical anodic and cathodic surface treatment supports osteoblast adhesion and proliferation on the Ti–6Zr–4Nb sintered nanocrystalline alloy [32]. The porous surface of Ti–6Zr–4Nb sintered nanocrystalline material was produced by anodic oxidation in 1 M H_3PO_4 + 2% HF electrolyte at 10 V for 30 min. Then onto the formed porous surface, the Ca-P layer was deposited, using cathodic potential of −5 V kept for 60 min in 0.042 M $Ca(NO_3)_2$ + 0.025 M $(NH_4)_2HPO_4$ + 0.1 M HCl electrolyte. The deposited Ca-P layer is anchored in the pores. In vitro test cultures of NHOst cells resulted in very good cell proliferation, colonization, and multilayering.

Recently, much attention has been paid to the TiO_2 nanotube preparation on dense titanium implants using anodization [71]. Porous titanium scaffolds with a porosity of 70% with pore sizes in the range of 200–300 μm were prepared using the space holder method using titanium and ammonium hydrogen carbonate particles. Finally, the bioactive anatase nanotubes with the size of approximately 100 nm were successfully fabricated on the titanium scaffold by anodization and heat treatment, which improved the biocompatibility obviously, as assessed by apatite formation ability.

5.2.4 Biomimetic porous scaffolds

Surface biomodification has been applied to titanium to improve its poor surface activity [32, 70]. Moreover, new techniques have been invented for the production of biomimetic porous titanium scaffolds for bone substitution [32, 64]. As it is well known, for osteoconductive function, the pore size distribution within the range of 200–500 μm is essential and can be provided by the space holder sintering method, which adjusts the pore shape and the porosity. The deposition of even coatings on porous implants can also be achieved by the biomimetic process, due to its non-line-of-sight characteristics. The process was based on the heterogeneous nucleation of Ca-P from simulated body fluid (SBF), in which titanium implants were soaked directly in the SBF solution and a Ca-P layer was coated on the surface.

Another effective method of improving bioactive properties is HA coating on the surface of Ti implants [71]. Synthetic HA is known as a bioactive and biocompatible material and bonds directly to the bone without the formation of connective tissue on the interface. Apatite is the main component of bone crystal, and also preferentially adsorbs proteins that serve as growth factors.

Adjusting the mechanical properties of porous titanium with a 3D interconnected pore structure can be obtained by selecting the proper porosity enabling bone ingrowth. Additionally, the 3D pore structure can function as a transport channel for body fluids within the pore network. The surface of the TiUnite® implant (Nobel Biocare/Sweden) is a highly crystalline, phosphate-enriched titanium oxide characterized by open pores in the low micrometer range [72]. In comparison to machined implant surfaces, this surface has repeatedly proven to elicit a more enhanced bone response. Furthermore, the TiUnite® surface maintains primary stability better than the machined surfaces and shortens the healing time needed to accomplish secondary stability. Shibuya et al. determined the success of Brånemark System TiUnite® implants placed in partially or completely edentulous jaws restored with fixed or removable prostheses [73]. The TiUnite® implant employed in Shibuya's study had an overall success rate of 96.56%, 6 years after implantation. On the other hand, SLActive® has been developed to optimize early implant stability and to reduce the risk during the critical early treatment [74–76]. The in vitro and in vivo studies of the SLActive surface demonstrate a stronger cell and bone tissue response than for the predecessor, the SLA surface, produced by the same company. Immediate and early loading with Straumann SLActive® implants yields excellent survival rates (98% and 97% after 1 year; 97% and 96% after 3 years, respectively).

5.3 Ethical issues

It is clear that nanotechnology could fundamentally influence humans' health and will pose new ethical questions [19, 77–81]. Despite the many benefits of nanotechnology, there are potential risks and ethical issues involved in its implementation. All biomaterials, after the initial research and developmental phase, undergo extensive preclinical in vitro testing for mechanical, toxicity, and immunological properties. Although animal studies give us an insight into what to expect from tested materials, there have been serious adverse reactions in human subjects exposed to a dose of nanomedicine 500 times lower than the recorded toxic dose in the animal model. Strict regulations have been introduced to control the use of novel nanomaterials in humans; however, great caution and ethical dilemma rests whether being faced with a wide range of therapeutical options, whether to choose from some having a long track of supporting data of clinical trials, or whether turn to novel materials with promising, appealing properties but only short-term clinical studies.

For example, nanoparticles can penetrate the blood–brain barrier. But there are differing views on the harmfulness of this process. On the one side, there is a risk of nanoparticle accumulation in body fluids causing complications [82]; on the other hand, it is known that nanocolloids do not settle, implicating no change in their concentration in air and no harm in contamination of the environment. Due to the lack of strict ethical regulations in nanoscientific activities rises the need to define conditions for responsible research in that field.

The nanotechnology stakeholders must be informed about the short-term and long-term benefits, limitations, and risks of nanotechnology to support sustainable, ethical, and economic nanotechnological development [81, 83]. Nanotechnology, like its predecessor technologies, will have an impact on all areas of health care. It will lead to fully automated diagnostics of patients, which in consequence will be more time efficient, excluding human errors or needs for further physical exams performed by medical staff. With nanomedicines, the average human life span will likely increase, leading to an increase in the number of elderly persons requiring medical attention and with the result of increased health expenditures.

Social concern for the nanotechnology stakeholders falls under four objectives: (1) education on local and global forces and issues that affect people and societies, (2) guidelines for local/global societies for appropriate use of technology, (3) risk and failure alerts, and (4) committee for problem solving in a technological world [81, 84]. Developments in nanotechnology present numerous challenges and risks in health and environmental areas. For that reason, a deeper understanding of the science is required, including risk/benefit analysis and ethical considerations during the development process.

5.4 Future trends in dental medicine

Nanotechnology is foreseen to change medicine and dentistry fundamentally. It already forms the basis of novel methods for the diagnosis and prevention of diseases. It will be useful in treatment procedures tailored to the patient's profile and in drug delivery and gene therapy. However, being still in the development phase of research, their clinical use is limited due to safety and cost-effectiveness issues, and further clinical exams are required to obtain unbiased results.

The development of many new nanostructured Ti-based alloys has been noticed over the years, and in vivo tests as well as clinical trials have shown their success. Nevertheless, products on the market are sparse. Enhancement of both short- and long-term tissue integration of nano-titanium implants can be accomplished by increasing protein adsorption and cell adhesion through surface roughness modification at the nanoscale level, or by improving osteoconductivity through biomimetic Ca-P coatings, or by facilitating the bone healing process in the bone–implant are through adjunction of biological drugs.

The development of new strategies can be achieved by a better understanding of the interactions between proteins, cells, tissues, and implant material. The ideal implant will be based on a surface with a controlled and custom-made topography and chemistry. The local release of bone stimulating drugs may help in advanced and complex clinical situations with poor bone quality and quantity. Reduction of osseointegration time for immediate loading cases will allow patients to have a shorter and safer rehabilitation treatment.

The environmental impact of nanoparticles and their effects such as toxicity, persistence, and bioaccumulation still need to be examined [85, 86], as well as the exposure pathways to the manufactured particles in treated patients are topics of future research.

Adequate adhesion and initiation of the cell subdivision process is the desired property of the implant's surface. However, these processes should be stable to avoid uncontrolled growth rates, which could lead to the escape of cells from the surveillant agents thus exceeding barriers of carcinogenesis. It still seems very important to develop such titanium-type alloys for the production of dental implants, which will demonstrate a reduced susceptibility to bacterial colonization and will not elicit pathogenic effects. It seems that the nanocrystalline Ti-type implants may be tailored to the production of structures that support the process of continuous adaptation to the implant by the host organism.

Current trends in clinical dental implant therapy include the use of endosseous dental implant surfaces embellished with nanoscale topographies. Nowadays, there is still little evidence of the long-term benefits of nanofeatures, as the promising results achieved in vitro and animals have still to be confirmed in humans. Currently, a large range of different types of nanoproducts is available and their potential ethical implications on society should be discussed as well.

References

[1] Feynman RP. There's plenty of room at the bottom. Eng Sci 1960, 23, 22–26.

[2] Baker Jr JR. What is nanomedicine? Nanomed Nanotechnol Biol Med 2005, 1, 243.

[3] Bugunia-Kubik K, Susisaga M. From molecular biology to nanotechnology and nanomedicine. J Biosystems 2002, 65, 123–38.

[4] Bayne SC. Dental restorations for oral rehabilitation – testing of laboratory properties versus clinical performance for clinical decision making. J Oral Rehab 2007, 34, 921–32.

[5] Lindhe J, Karring T, Lang NP. Clinical Periodontology and Implant Dentistry. Oxford, Blackwell Munksgaard, 2003.

[6] Gleiter H. Nanostructured materials: State of the art and perspectives. Nanostruct Mat 1995, 6, 3–14.

[7] Roco MC. International strategy for nanotechnology research and development. J Nanoparticle Res 2001, 3, 353–60.

[8] Horton MA, Khan A. Medical nanotechnology in the UK: A perspective from the London Centre for Nanotechnology. Nanomed Nanotechnol Biol Med 2006, 2, 42–48.

[9] Beschnidt SM, Cacaci C, Dedeoglu K, Hildebrand D, Hulla H, Iglhaut G, Krennmair G, Schlee M, Sipos P, Stricker A, Ackermann KL. Implant success and survival rates in daily dental practice: 5-year results of a non-interventional study using CAMLOG SCREW-LINE implants with or without platform-switching abutments. Int J Implant Dent 2018, 4, 33–46.

[10] Arys A, Philippart C, Dourov N, He Y, Le QT, Pireaux JJ. Analysis of titanium dental implants after failure of osseointegration: Combined histological, electron microscopy and X-ray photoelectron spectroscopy approach. J Biomed Mater Res 1998, 43, 300–12.

[11] Turp JC, Greene CS, Strub JB. Dental occlusion: A critical reflection on past, present and future concepts. J Oral Rehabil 2008, 35, 446–53.

[12] Farantalz GJ, Beckler IM, Gremilion H, Pink F. The effectiveness of equilibration in the improvement of signs and symptoms in the stomatognathic system. Intern J Periodontics Restorative Dent 1998, 18, 595–99.

[13] Tylman SD, Malone WF. Tylman's Theory and Practice of Fixed Prosthodontics. St. Louis, Mosby Company, 1978.

[14] Scheller EL, Krebsbach PH, Kohn DH. Tissue engineering: State of the art in oral rehabilitation. J Oral Rehabil 2009, 36, 368–89.

[15] Ferreira CF, Magini RS, Sharpe PT. Biological tooth replacement and repaid. J Oral Rehab 2007, 34, 933–39.

[16] Zarb GA, Albrektsson T. Osseointegration: A requiem for the periodontal ligament? Int J Periodontics Restorative Dent 1991, 11, 88–91.

[17] Tannock GW. Normal Microflora: An Introduction to Microbes Inhabiting the Human Body. London, Chapman & Hall, 1995.

[18] Craig RG, Ward ML. Restorative Dental Materials. 10th, St. Louis, Mosby, 1997.

[19] Long M, Rack HJ. Titanium alloys in total joint replacement – A materials science perspective. Biomaterials 1998, 19, 1621–38.

[20] Wang K. The use of titanium for medical applications in the USA. Mater Sci Eng A 1996, 213, 134–37.

[21] Albrektsson T, Brånemark P-I, Hansson HA, Lindstrom J. Osseointegrated titanium implants. Requirements for ensuring a long lasting, direct bone-to-implant anchorage in man. Acta OrthopScand 1981, 52, 155–70.

[22] Li JP, Wijn JR, van Blitterswijk CA, de Groot K. Comparison of porous Ti6Al4V made by sponge replication and directly 3D fiber deposition and cancellous bone. Key Eng Mater 2007, 330–332, 999–1002.

[23] Spoerke ED, Murray NG, Li H, Brinson LC, Dunand DC, Stupp SI. A bioactive titanium foam scaffold for bone repair. Acta Biomater 2005, 1, 523–33.
[24] Le Gu´ehennec L, Soueidan A, Layrolle P, Amouriq Y. Surface treatments of titanium dental implants for rapid osseointegration – Review. Dental Mater 2007, 23, 844–54.
[25] Mendonça G, Mendonça, DBS, Araga͠o FJL, Cooper LF. Advancing dental implant surface technology – From micron to nanotopography – Review. Biomaterials 2008, 29, 3822–35.
[26] Cachinho SCP, Correia RN. Titanium scaffolds for osteointegration: mechanical, in vitro and corrosion behaviour. J Mater Sci Mater Med 2008, 19, 451–57.
[27] Ehrenfest DMD, Coelho PG, Kang BS, Sul´YT, Albrektsson T. Classification of osseointegrated implant surfaces: materials, chemistry and topography – Review. Trends Biotechnol 2010, 28, 198–206.
[28] Singh R, Lee P, Dashwood R, Lindley T. Titanium foams for biomedical applications. Review Mater Technol 2010, 25, 127–36.
[29] Brånemark PI. Osseointegration and its experimental background. J Prosthet Dent 1983, 50, 399–410.
[30] Jurczyk MU, Jurczyk K, Niespodziana K, Miklaszewski A, Jurczyk M. Titanium-SiO$_2$ nanocomposites and their scaffolds for dental applications. Mater Charac 2013, 77, 99–108.
[31] Jurczyk MU, Jurczyk K, Miklaszewski A, Jurczyk M. Nanostructured titanium -45S5 Bioglass scaffold composites for medical applications. Mater Design 2011, 32, 4882–89.
[32] Jakubowicz J, Adamek G, Jurczyk MU, Jurczyk M. 3D topography study of the biofunctionalized nanocrystalline Ti-6Zr-4Nb/Ca-P. Mater Charac 2012, 70, 55–62.
[33] Miklaszewski A, Jurczyk MU, Kaczmarek M, Paszel-Jaworska A, Romaniuk A, Lipinska N, Zurawski J, Urbaniak P, Jurczyk M. Nanoscale size effect in in situ titanium "titanium" based composites with cell viability and cytocompatibility studies. Mater Sci Eng C 2017, 73, 525–36.
[34] Pye AD, Lockhart DEA, Dawson MP, Murray AA, Smith AJ. A review of dental implants and infection. J Hosp Infect 2009, 72, 104–10.
[35] Hu K, Yang XJ, Cai YL, Cui ZD, Wei Q. Preparation of bone-like composite "composite" coating using a modified simulated body fluid with high Ca and P concentrations. Surf Coat Technol 2006, 201, 1902–06.
[36] Choi J, Bogdanski D, Koller M, Esenwein SA, Muller D, Muhr G, Epple M. Calcium phosphate coating of nickel–titanium "titanium" shape-memory alloys. Coating procedure and adherence of leukocytes and platelets. Biomaterials 2003, 24, 3689–96.
[37] Estrin Y, Kasper C, Diederichs S, Lapovok R. Accelerated growth of preosteoblastic cells on ultrafine grained titanium "titanium". J Biomed Mater Res A 2008, 90, 1239–42.
[38] Jakubowicz J, Adamek G. Preparation and properties of mechanically alloyed and electrochemically etched porous Ti-6Al-4V. Electrochem Commun 2009, 1, 1772–75.
[39] Niespodziana K, Jurczyk K, Jakubowicz J, Jurczyk M. Fabrication and properties of titanium "titanium" – hydroxyapatite nanocomposites. Mater Chem Phys 2010, 123, 160–65.
[40] Webster TJ, Ejiofor JU. Increased osteoblast adhesion on nanophase metals: Ti, Ti6Al4V, and CoCrMo. Biomaterials 2004, 25, 4731–39.
[41] Matusiewicz H. Potential release of in vivo trace metals from metallic medical implants in the human body: From ions to nanoparticles – A systematic analytical review. Acta Biomater 2014, 10, 2379–403.
[42] Marczewski M, Miklaszewski A, Maeder X, Jurczyk M. Crystal structure evolution, microstructure formation and properties of mechanically alloyed ultrafine-grained Ti-Zr-Nb alloys at 36≤Ti≤70 (at %). Materials 2020, 13, 587.
[43] Sochacka P, Miklaszewski A, Jurczyk M, Pecyna P, Ratajczak M, Gajecka M, Jurczyk MU. Effect of hydroxyapatite and Ag, Ta2O5 or CeO2 addition on the properties of ultrafine-grained Ti31Mo alloy. J Alloys Compds 2020, 823, 153749.

[44] Jurczyk M ed. Bionanomaterials for Dental Applications. Singapore, Pan Stanford Publishing, 2012.

[45] Park JW, Kim YJ, Park CH, Lee DH, Ko YG, Jang JH, Lee CS. Enhanced osteoblast response to an equal channel angular pressing-processed pure titanium substrate with microrough surface topography. Acta Biomat 2009, 5, 3272–80.

[46] Kaczmarek M, Jurczyk MU, Rubis B, Banaszak A, Kolecka A, Paszel A, Jurczyk K, Murias M, Sikora J, Jurczyk M. In vitro biocompatibility of Ti-45S5 Bioglass nanocomposites and their scaffolds. J Biomed Mater Res A 2014, 102, 1316–24.

[47] Tulinski M, Jurczyk M. Nanostructured nickel-free austenitic stainless steel composites with different content of hydroxyapatite. Appl Surf Sci 2012, 260, 80–83.

[48] Jurczyk K, Niespodziana K, Jurczyk MU, Jurczyk M. Synthesis and characterization of titanium -45S5 Bioglass nanocomposites. Mater Design 2011, 32, 2554–60.

[49] Jurczyk K, Niespodziana K, Jurczyk MU, Jakubowicz J, Jurczyk M. Titanium-10 wt% 45S5 Bioglass nanocomposite for biomedical applications. Mater Chem Phys 2011, 131, 540–46.

[50] Cao WP, Hench L. Bioactive materials. Ceramics Int 1996, 22, 493–507.

[51] Van de Voorde M, Tulinski M, Jurczyk M. Engineered Nanomaterials: A Discussion of the Major Categories of Nanomaterials, In: Mansfield E, Kaiser D, Fujita D, Van de Voorde M eds. Chapter 3 in Metrology and Standardization of Nanomaterials: Protocols and Industrial Innovations. Wiley-VCH, 2017, 49–73.

[52] Tulinski M, Jurczyk M. Nanomaterials Synthesis Methods, Chapter 4 in Metrology and Standardization of Nanomaterials: Protocols and Industrial Innovations". Eds Mansfield E, Kaiser D, Fujita D, Van de Voorde M Wiley-VCH, 2017, 75–98.

[53] Khurshid Z, Zafar M, Qasim S, Shahab S, Naseem M, Abu Reqaiba A. Advances in nanotechnology for restorative dentistry – review. Materials 2015, 8, 717–31.

[54] Zhang L, Webster TJ. Nanotechnology and nanomaterials: Promises for improved tissue regeneration. Nano Today 2009, 4, 66–80.

[55] Ozak ST, Ozkan P. Nanotechnology and dentistry. Eur J Dent 2013, 7, 145–51.

[56] Ward BC, Webster TJ. Increased functions of osteoblasts on nanophase metals. Mat Sci Eng C 2007, 27, 575–78.

[57] Webster TJ, Siegel RW, Bizios R. Nanoceramic surface roughness enhances osteoblast and osteoclast functions for improved orthopaedic/dental implant efficacy. Scr Mater 2001, 44, 1639–42.

[58] Albrektsson T, Wennerberg A. Oral implant surfaces: Part 2 – review focusing on clinical knowledge of different surfaces. Int J Prosthodont 2004, 17, 544–64.

[59] Jurczyk K, Miklaszewski A, Niespodziana K, Kubicka M, Jurczyk MU, Jurczyk M. Synthesis and properties of Ag-doped titanium -10 wt.% 45S5 Bioglass nanostructured scaffolds. Acta Metall Sin (Engl Lett) 2015, 28, 467–76.

[60] Jurczyk K, Kubicka MM, Ratajczak M, Jurczyk MU, Niespodziana K, Nowak DM, Gajecka M, Jurczyk M. Antibacterial activity of nanostructured Ti-45S5 Bioglass-Ag composite against Streptococcus mutans and Staphylococcus aureus. Trans Nonferrous Met Soc China 2016, 26, 118–25.

[61] Kumar R, Münstedt H. Silver ion release from antimicrobial polyamide/silver composites. Biomaterials 2005, 26, 2081–88.

[62] Hernández-Sierra JF, Ruiz F, Pena DC, Martinezgutiérrez F, Martinez AE, Guillén Ade J, Tapia-Pérez H, Castañón GM. The antimicrobial sensitivity of streptococcus mutans to nanoparticles of silver, zinc oxide, and gold. Nanomedicine 2008, 4, 237–40.

[63] Jurczyk K, Miklaszewski A, Jurczyk MU, Jurczyk M. Development of betta type Ti23Mo-45S5 Bioglass nanocomposites for dental applications. Materials 2015, 8, 8032–46.

[64] Jakubowicz J. Formation of porous TiO_x biomaterials in H_3PO_4 electrolytes. Electrochem Commun 2008, 10, 735–39.
[65] Bonfante, EA, Witek L, Tovar N, Suzuki M, Marin C, Granato R, Coelho PG. Physicochemical characterization and in vivo evaluation of amorphous and partially crystalline calcium phosphate coatings fabricated on Ti-6Al-4V implants by the plasma spray method. Int J Biomaterials 2012, 2012, 603826. Last access date: 20.05.2021. https://doi.org/10.1155/2012/603826.
[66] Albrektsson T, Wennerberg A. Oral implant surfaces: Part 1 – review focusing on topographic and chemical properties of different surfaces and in vivo responses to them. Int J Prosthodont 2004, 17, 536–43.
[67] Wennerberg A, Albrektsson T. On implant surfaces: a review of current knowledge and opinions. Int J Oral Maxillofac Implants 2010, 25, 63–74.
[68] Elias CN, Rocha FA, Nascimento AL, Coelho PG. Influence of implant shape, surface morphology, surgical technique and bone quality on the primary stability of dental implants. J Mechanical Behavior Biomed Mater 2012, 16, 169–80.
[69] Wennerberg A, Hallgren C, Johansson C, Danelli S. A histomorphometric evaluation of screw-shaped implants each prepared with two surface roughnesses. Clin Oral Implants Res 1998, 9, 11–19.
[70] Brett PM, Harle J, Salih V, Mihoc R, Olsen I, Jones FH, Tonetti M. Roughness response genes in osteoblasts. Bone 2004, 35, 124–33.
[71] Nouri A, Hodgson PD, Wen C. Biomimetic Porous Titanium Scaffolds for Orthopedic and Dental Applications. In: Mukherjee A ed. Biomimetics Learning from Nature. InTech, 2010.
[72] Rocci A, Martignoni M, Gottlow J. Immediate loading of Brånemark System TiUnite and machined-surface implants in the posterior mandible: A randomized open-ended clinical trial. Clin Implant Dent Res 2003, 5, Suppl 1, 57–63.
[73] Shibuya Y, Kobayashi M, Takeuchi J, Asai T, Murata M, Umeda M, Komori T. Analysis of 472 Brånemark system TiUnite implants: A retrospective study. Kobe J Med Sci 2009, 55, E73-E81.
[74] Buser D, Broggini N, Wieland M, Schenk RK, Denzer AJ, Cochran DL, Hoffmann B, Lussi A, Steinemann SG. Enhanced bone apposition to a chemically modified SLA titanium surface. J Dent Res 2004, 83, 529–33.
[75] Zöllner A, Ganeles J, Korostoff J, Guerra F, Krafft T, Brägger U. Immediate and early non-occlusal loading of Straumann implants with a chemically modified surface (SLActive) in the posterior mandible and maxilla: interim results from a prospective multicenter randomized-controlled study. Clin Oral Implants Res 2008, 19, 5, 442–50.
[76] Nicolau P, Korostoff J, Ganeles J. et al., Immediate and early loading of chemically modified implants in posterior jaws: 3-year results from a prospective randomized multicenter study. Clin Implant Dent Relat Res 2013, 15, 4, 600–12.
[77] Jhaver HM, Balaji. Nanotechnology: The future of dentistry. J Indian Prosthodontic Soc 2005, 5, 15–17.
[78] Williams D. The risks of nanotechnology. Med Devices Technol 2005, 16, 9–10.
[79] Corley E, Scheufele DA, Hu Q. Of risks and regulations: How leading US nanoscientists form policy stances about nanotechnology. J Nanopart Res 2010, 11, 1573–85.
[80] Nanoethics. The Ethical and Social Implications of Nanotechnology. Allhoff F, Lin P, Moor J, Weckert J Eds, Hoboken, John Wiley & Sons, 2007. Last access date: 20.05.2021. https://doi.org/10.1002/anie.200785559.
[81] Khan A. Ethical and social implications of nanotechnology, QScience Proceedings (Engineering Leaders Conference 2014) 2015, 57. Last access date: 28.05.2021. http://dx.doi.org/10.5339/qproc.2015.elc2014.57.

[82] White GB. 2009, Missing the boat on nanoethics. Am J Bioethics 2009, 9, 18–19.

[83] Markova B. The main ethical issues with nanotechnology in the future context. Organ Manage 2015, 84, 1942.

[84] Thiele F, Mehlich J. Nanoparticles for medical purposes – Technical, medical and ethical aspects: research report. Bad Neuenahr-Ahrweiler: Europäische Akademie, 1, 2012.

[85] Hoet PHM, Bruske-Hohlfeld I, Salata OV. Nanoparticles – known and unknown health risks, – Review. J Nanobiotechnol 2004, 2, 12–27.

[86] Tsuda A, Gehr P. Nanoparticles in the lung-Environmental Exposure and Drug Delivery. Boca Raton, FL, CRC Press, 2015.

Daisuke Fujita

6 Nanomaterial characterization for nanoethics

Abstract: In this chapter, the current status of various key technologies for nanoscale characterization and analysis that are indispensable for dealing with nanoethics issues is discussed. Such advanced analytical methodologies include not only the precise shape and size measurements but also the in-depth characterization of various properties of nanomaterials in the bulk states and at the interfaces and surfaces. Especially, nanoscale physicochemical characterization technologies for manufactured and engineered nanomaterials that can be useful for nanoethics assessments have been focused on. Starting from the introduction of the history and status of nano-characterization, major microscopies or spectroscopic imaging tools with the nanoscale spatial resolution are introduced. Interesting applications of such nano-characterization and nano-analysis methods to nanomaterials research are introduced. Besides the major nano-characterization players such as scanning tunneling microscopy and atomic force microscopy, emerging techniques such as helium ion microscopy and tip-enhanced Raman spectroscopy are also introduced. Since some of the nano-characterization techniques are semi-quantitative, the establishment of quantitative methodologies and the standardization of nano-characterization technologies are the next steps toward reliable assessments for nanoethics. Another important direction based on the industrial application needs for engineered nanomaterials is to develop dynamic observation techniques involving operando spectroscopy for physicochemical properties at the nanoscale.

Keywords: nanotechnology, atomic force microscopy, engineered nanomaterial, manufactured nanomaterial, incidental nanomaterial, nanoethics

6.1 Introduction

The term "nanotechnology" has become one of the most circulated ones that have been created in the last few decades. The official definition adopted by the International Organization for Standardization (ISO) technical committee (TC) 229 in charge of nanotechnologies standardization is "the application of scientific knowledge to control and utilize matter in the nanoscale, where properties and phenomena related to size or structure can emerge" [1]. The unit prefix "nano" (symbol n) means one-billionth or 10^{-9}, which is used primarily with physics, chemistry, and precise

Daisuke Fujita, Research Center for Advanced Measurement and Characterization, National Institute for Materials Science, Japan

https://doi.org/10.1515/9783110669282-006

engineering. Actually the use of the prefix "nano" was adopted and standardized by the SI (*Le Système International d'Unités*) and ISO in 1960, where "nano" was originated from the Greek word for "dwarf" meaning a person of abnormally small stature. According to the ISO/TS80004-1:2015, "nanoscale" is defined as the "length range approximately from 1 nm to 100 nm." It should be noted that properties that are not extrapolations from larger sizes are predominantly exhibited in the nanoscale.

Among various nanotechnology products, novel nanomaterials are playing key roles in industrial and societal innovations. The term "nanomaterial" can be defined as a material with any external dimension in the nanoscale or having an internal structure or surface structure in the nanoscale. Based on the nanoscopic structures, nanomaterials can be classified into two categories, that is, nano-objects and nanostructured materials, as shown in Figure 6.1. A nano-object is defined as a discrete piece of a material with one, two, or three external dimensions in the nanoscale. Zero-dimensional fullerenes, one-dimensional carbon nanotubes, two-dimensional (2D) graphenes, and three-dimensional (3D) nanoparticles are typical examples of nano-objects. A nanostructured material is a material having internal nanostructure or surface nanostructure. Quantum well superlattices, nanopore materials, and carbon nanowalls are typical examples of nanostructured materials.

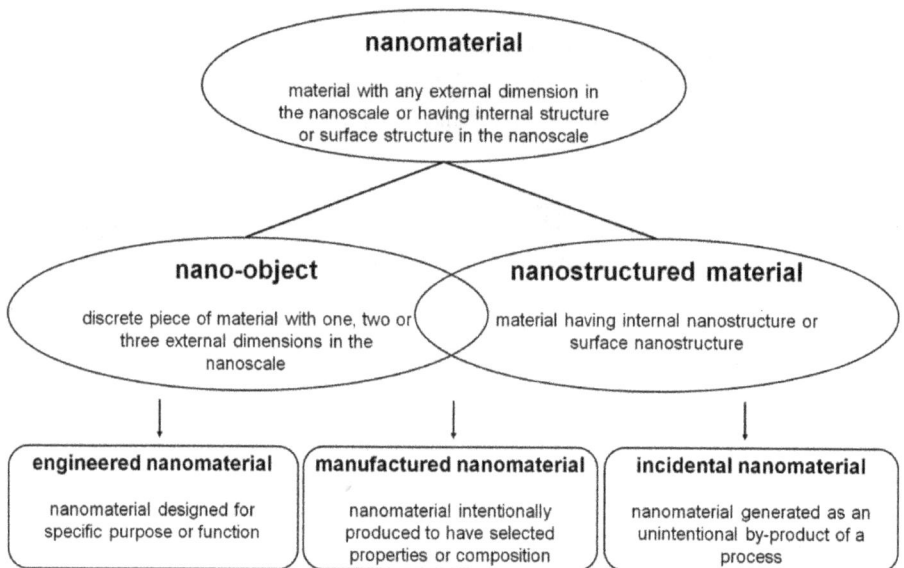

Figure 6.1: Categorization of nanomaterials based on ISO/TS80004-1:2015 [1].

It is also possible to categorize nanomaterials into three types: engineered, manufactured, and incidental nanomaterials, based on their attributes of production or generation as shown in Figure 6.1. Engineered nanomaterials (ENMs) are nanomaterials designed for specific purpose or function. Manufactured nanomaterials (MNMs) are nanomaterials that are intentionally produced to have selected properties or compositions. These two terms look remarkably similar and used almost interchangeably. Incidental nanomaterials are nanomaterials generated as unintentional by-products of processes. It should be noted that the incidental nanomaterials have been continuously produced and distributed worldwide since the Industrial Revolution two-and-a-half century ago [2].

The European Union (EU) adopted a definition of a nanomaterial in 2011 (recommendation on the definition of a nanomaterial (2011/696/EU)) [3]. According to the recommendation, "nanomaterial" means a natural, incidental, or manufactured material containing particles, in an unbound state or as an aggregate or as an agglomerate and where, for 50% or more of the particles in the number size distribution, one or more external dimensions is in the size range of 1–100 nm. Considering the significant industrial importance, even though derogated from the above definition, fullerenes, graphene flakes, and single-wall carbon nanotubes, and so on with one or more external dimensions below 1 nm should be considered as nanomaterials.

Expecting the rapid and global commercialization of nanotechnology and various nanomaterials, the so-called nanoethics has been emerging soon after the initiation of the National Nanotechnology Initiative proposed by the US President Clinton in 2001 [4]. Nanoethics is an interdisciplinary field of science, which aims to promote critical and ethical reflection in nanotechnology and related science and engineering fields.

It is well known that Clinton's initiative is highly inspired by Feynman's famous talk entitled "There's Plenty of Room at the Bottom" at an annual meeting of the American Physical Society at Caltech in 1959 [5]. One of the genius suggestions of Feynman's talk is that he pointed out the importance of the characterization on the nanoscale, by saying "What you should do in order for us to make more rapid progress is to make the electron microscopes 100 times better." Among various nanotechnology products, ENMs and MNMs have gained prominence in industrial advancement, especially related to medical, biological, and engineering applications. To implement the reliable risk assessment against increasing ENM and MNM applications, we must overcome the inherent difficulty in characterizing such nanomaterials [6]. It clearly means that the technologies to quantitatively measure the 3D size and shape of nanomaterials and to characterize the physicochemical properties of nanomaterial surfaces at the nanoscale shall be critical to the advancement of nanotechnology with the least negative effects on our sustainable society [7, 8].

In this chapter, we introduce the current status of such key technologies of nano-characterization that is necessary to deal with nanoethics issues. It includes not only the precise shape and size measurements but also the characterization of various physicochemical properties of nanomaterials in the bulk and at the surface.

6.2 Historical overview of nano-characterization

It should be noted that the first demonstration of atomic-resolution imaging of materials was achieved in 1955 by using a field ion microscope (FIM), which was invented by Müller in 1951 [9]. Such an epoch-making innovation that occurred in the 1950s might have been a potential trigger for Feynman's inspiration on nanotechnology. A following remarkable idea of a synthetic superlattice was proposed by Esaki in 1969 [10]. The semiconductor superlattices are artificial low-dimensional potentials fabricated with nanometer periodicities by human beings, where novel and useful quantum-mechanical effects have been emerging even up to now. Early in the 1980s, an innovative surface microscope named scanning tunneling microscope (STM) was invented by Binnig and Rohrer [11]. The operation principle of STM is based on a quantum-mechanical phenomenon called "electron tunneling." The first atomically resolved images of clean Si(111) surfaces were obtained by STM [12]. Soon after the invention of STM, atomic force microscopy (AFM) was invented by Binnig et al. [13]. In 1989, the first demonstration of atom manipulation using an STM tip was made by Eigler, who made it possible to manipulate individual Xe atoms on a surface and to write the smallest letters "IBM" in human history [14]. In the 1980s, a new era for the realization of true nanotechnology has opened, when humankind has obtained an amazing tool for observation, characterization, and manipulation of individual atoms for the first time.

After the prediction by Feynman, the spatial resolution of electron microscopes had been improved gradually up to the 1990s. One of the biggest obstacles had been the blurring of images caused by lens aberrations. In 1998, Heider, Rose, and Urban demonstrated the solution to this issue for a medium-voltage electron microscope which gives a stunning enhancement of image quality [15]. The breakthrough of aberration-corrected electron optics for transmission electron microscopes (TEMs) has enabled subatomic precision and made TEMs one of the most important tools for nanotechnology.

6.3 Categorization of nanoscale characterization

Since nanoethics aims to promote ethical reflection in the nanotechnology field, it includes proper characterization of the potential negative effects such as nanotoxicity of nano-objects and nanostructured materials. Corresponding to this issue, guidance on the physicochemical characterization of engineered nanoscale materials is summarized as one of the technical reports on the nanotechnologies standardization of ISO/TC 229 [16]. A 2D diagram of spatial resolution and physicochemical properties of nanomaterials to be measured by using nanoscale and microscale characterization tools is shown in Figure 6.2. Physicochemical properties of nanomaterials can be categorized

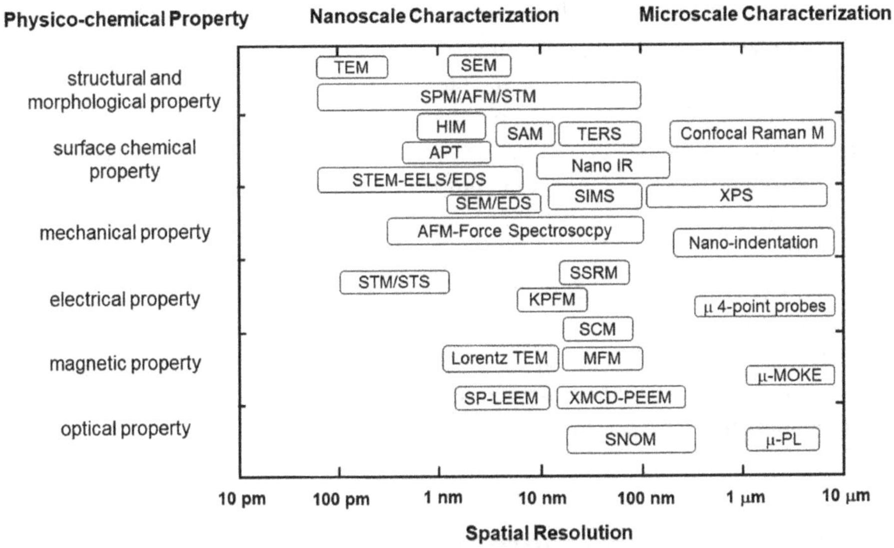

Physico-chemical Property	Nanoscale Characterization	Microscale Characterization

Figure 6.2: Categorization of physicochemical characterization tools of nanomaterials.

as structural and morphological property, surface chemical property, mechanical property, electrical property, magnetic property, and optical property. The nanoscale characterization tools have typical spatial resolution ranging from subnanometers to approximately 100 nm. For the detailed analysis of nano-objects, advanced characterization tools with atomic spatial resolution such as STM, AFM, and TEM are employed. On the other hand, the microscale characterization tools have typical spatial resolution ranging from submicrometers to approximately 10 μm.

The structural and morphological property means the size, size distribution, shape, and surface morphology of nanomaterials. Besides TEM, STM, and AFM and related scanning probe microscopy (SPM), conventional scanning electron microscopy (SEM) and newly emerging helium ion microscopy (HIM) can be used at the nanoscale for this purpose [17]. The surface and subsurface chemical property of nanomaterials such as chemical compositions and chemical states can be characterized at the nanoscale and better by scanning transmission electron microscopy (STEM) combined with electron energy loss spectroscopy (EELS) and energy-dispersive spectroscopy (EDS), SEM combined with EDS, scanning Auger microscopy (SAM), secondary ion mass spectrometry (SIMS), atom probe tomography (APT), tip-enhanced Raman spectroscopy (TERS), nanoscale infrared spectroscopy (nano IR), and so on [18, 19]. Microscale characterization tools for surface chemistry of nanomaterials are X-ray photoelectron spectroscopy (XPS), confocal Raman microscopy (CRM), and so on [20]. Mechanical properties such as hardness, strength, and elastic modulus can be characterized by AFM combined with force spectroscopy at the nanoscale, and nano-indentation techniques at the micron scale [21]. Electric properties of nanomaterials, such as electronic states, electric potentials, electric conductivities, carrier concentration at the nanoscale, can be characterized by scanning tunneling spectroscopy (STS) with STM, Kelvin probe force microscopy (KPFM), scanning spread resistance microscopy (SSRM), and scanning capacitance microscopy (SCM), respectively [22–24]. Microscale characterization of the surface conductivity of 2D nanomaterials such as graphene can be done by micro-four-point probes [25]. Magnetic properties of nanomaterials can be characterized at the nanoscale by Lorentz TEM, spin-polarized low-energy electron microscopy (SP-LEEM), X-ray magnetic circular dichroism (XMCD) with photoelectron emission microscopy (PEEM), and magnetic force microscopy (MFM) [26, 27]. Microscale magnetic properties of nanomaterials can be characterized by the micro-magneto-optic Kerr effect (μ-MOKE) and so on. Optical properties of nanomaterials can be characterized by scanning near-field optical microscopy (SNOM) at the nanoscale and by micro-photoluminescence (μ-PL) at the microscale [28].

6.4 Structural and morphological nano-characterization

6.4.1 Electron microscopy (TEM/SEM)

Since the wavelength of energetic electrons can be very much shorter than that of visible light, electron microscopes have a much higher resolution than a visible light microscope. Nowadays, electron microscopy is most widely used for the detailed analysis of various materials and biological objects since it can reveal the structure and property of nano-objects at the subatomic resolution. Due to the advantage of shorter wavelengths with higher accelerating voltages, high-voltage electron microscopes (HVEM) with the acceleration voltage of 1–3 MeV have been developed since 1960s. Those HVEMs have been applied for relatively thick specimens of biological and medical applications, radiation damage studies related to the development of nuclear reactor materials, and so on. Recent breakthroughs in spherical-aberration (C_s) and chromatic-aberration (C_c) corrections have contributed to the significant improvement of resolution. Especially the development of C_s correctors and their impressive applications to advanced materials research by Heider, Rose, and Urban have revolutionized the electron microscopy field [29]. Nowadays, advanced TEMs can achieve subatomic resolution of better than 50 pm using a highly coherent focused electron probe with a spherical-aberration correction [30].

TEM is the original form of electron microscope. The electron beam is focused by electrostatic and electromagnetic lenses and transmitted through the specimen. Because the high-resolution electron microscope (HREM) with the correction of spherical aberration can determine the positions of atomic rows in real space, HREM has been one of the most powerful tools for nano-characterization. SEM produces images of sample's surfaces with a focused electron beam that is scanned over the specimen. When the incident electron beam interacts with the specimen, emission of low-energy secondary electrons occurs, which provides signals for imaging the surface topography of specimen. Generally, the best image resolution of SEM is about 1 nm, one order poorer than that of TEM.

Recent trend of electron microscopy is operando observation with environmental control, which can produce images of sufficient quality and resolution with the samples being wet or contained in low vacuum, gas atmosphere or light irradiation with controlled temperatures. Environment TEM or SEM can facilitate not only imaging biological samples that are unstable in the high vacuum of conventional electron microscopes but also real-time observation of dynamic process such as catalytic reaction, phase transition, oxidation, reduction, photo-catalysis, charging and discharging of secondary batteries, and so on.

6.4.2 Atomic force microscopy (AFM)

Since AFM does not require electric conductivity, it is the most popular microscopy for surface topography imaging of nanomaterials. Feedback signal sources of AFM are short- and long-range forces interacting between a probe tip and a sample surface. The interacting force varies as the tip–surface distance changes as shown in Figure 6.3(a). As the tip approaches to the sample, long-range attractive interaction force (*van der Waals* force) between tip and sample atoms increases gradually. The attractive interaction force increases until the tip and sample atoms are so close that their wave functions start to repel each other at near the contact. The repulsion force progressively weakens the attractive force as the separation continues to decrease. The total force goes through zero and finally becomes positive (repulsive).

Various types of force sensors have been proposed. Among them, the most dominant one is a cantilever type, based on the quantitative measurement of the deflection induced by interacting forces. A probe tip with a nanoscale curvature radius is mounted at the end of a cantilever. Various quantities have been utilized for the detection of a minute bending of a cantilever, such as electron tunneling, optical reflection, interferometry, capacitance, and piezo-resistance [31–34]. The most popular one is optical reflection, where a fine laser beam reflected at the backside of a cantilever is detected by a position-sensitive photon detector. A conventional AFM is schematically illustrated in Figure 6.3(b).

Normal or lateral force acting on the tip is measured by vertical and/or torsional deflection of the cantilever. Either deflection causes a change in the optical path of a reflected beam, which is quantitatively detectable by a position-sensitive four-segment photodiode detector. For example, the signal of $Z = (A + C) - (B + D)$ is proportional to the normal force, whereas that of $X = (A + B) - (C + D)$ is related to the lateral force. By using a feedback signal like Z, gap distance is maintained constant. By scanning a

(a)

dynamic mode

intermittent contact mode

static mode

contact mode

Force [nN]

0

Tip-surface distance [nm]

noncontact mode

repulsive force   **attractive force**

(b)

Position sensitive
detector

Laser diode

Mirror

A   B

C   D

Feedback
electronics

Reflected beam

Cantilever

Laser beam

Probe tip

AFM images

z

Sample

Scanner for
xyz motion

x   y

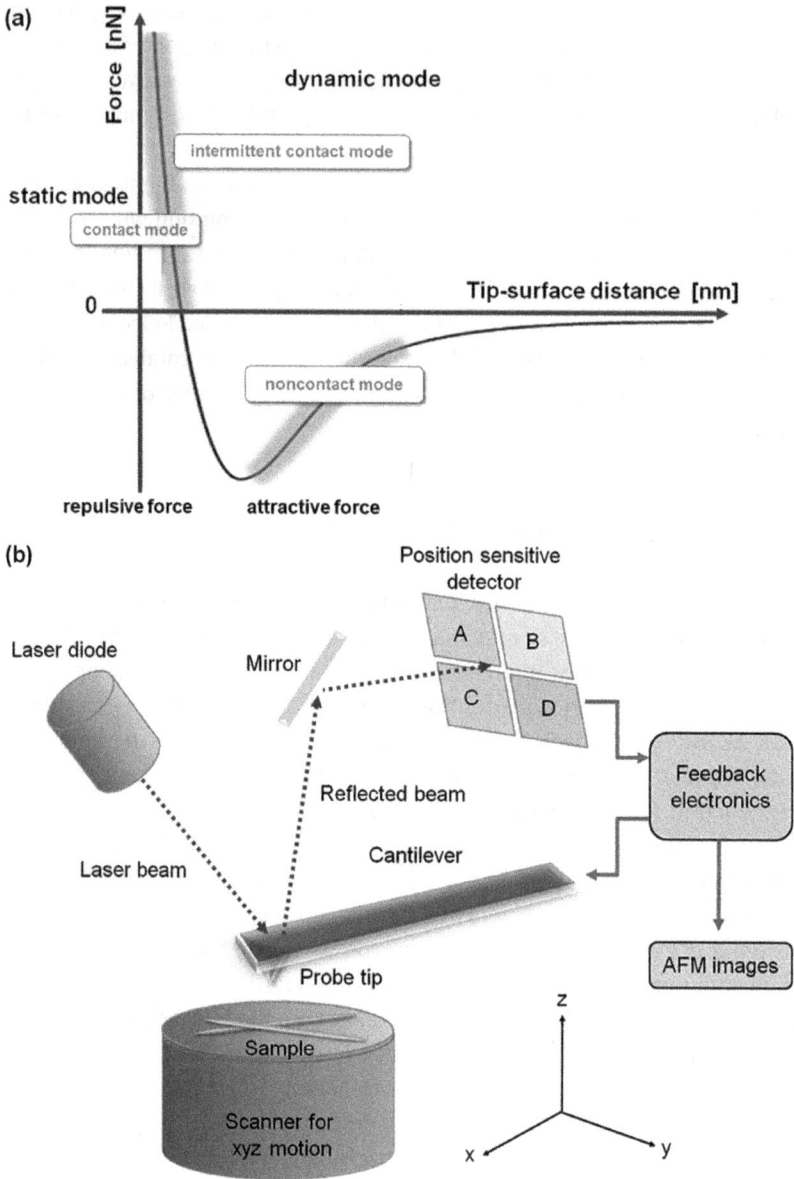

Figure 6.3: (a) Relationship between tip–surface distance and interaction force between a tip and a sample surface. (b) An operating illustration of the optical reflection-type AFM. A laser beam reflected from the cantilever's rear side is detected with a position-sensitive detector.

sample or a tip, a variety of tip–surface interaction is imaged by AFM, depending on the gap distance.

AFM operation modes can be categorized into static and dynamic modes. Static AFM produces a topographic image of a surface measured at a constant normal force. It is mostly operated in contact with the surface where the tip experiences repulsive force with a typical value of 10^{-9} N. Consequently, static AFM is generally called contact AFM, where a force between a tip and a sample causes the cantilever to deflect in accordance with Hooke's law. Thus, the normal force is calculated by using the normal spring constant of the cantilever. In principle, the spatial resolution of contact AFM is limited by the contact area. When the cantilever tip is scanned perpendicular to its length in the contact mode, a nanoscopic friction in the contact area produces a lateral force on the tip, which causes a torsional deflection. The measurement technique of lateral force imaging is called lateral force microscopy or friction force microscopy [35].

In the dynamic mode, the vibrational properties of a cantilever related to tip–sample interaction forces are measured while the cantilever is externally oscillated at or close to its resonance frequency. The dynamic mode can be categorized into frequency modulation (FM) and amplitude modulation (AM). In FM mode, the shift in resonance frequency caused by tip–sample interaction is used as a feedback signal, which is also called non-contact AFM (NCAFM). FM-mode NCAFM was the first reliable technique to demonstrate true atomic resolution on Si(111) surfaces in ultrahigh vacuum (UHV) [36]. Recently, submolecular resolving power by FM-mode NCAFM has been developed to reveal the 3D chemical structures of both planar and non-planar molecules [37].

In AM mode, change in the amplitude or phase of an oscillation excited at a fixed frequency close to the resonance frequency is used as the feedback signal. AM-AFM can be operated either in non-contact or intermittent contact regime. In the intermittent contact mode, a sinusoidal oscillation of the tip is excited at a little below the resonance frequency. The intensity of the oscillation amplitude is used for the feedback source signal. AM mode can also be used in the non-contact regime where the cantilever is oscillated at a frequency slightly above its resonance frequency. Attractive *van der Waals* forces act to decrease the resonance frequency, which also decrease the oscillation amplitude.

6.4.3 Helium ion microscopy (HIM)

HIM is a relatively new type of microscopes where a finely focused beam of helium ions is scanned over the sample surface. Based on the FIM technology and the development of gas field ion source (GFIS), an innovative microscope for surface observation and modification at the nanoscale was developed. Focused helium ion beam of high brightness, small energy spread, and small de Broglie wavelength is produced from GFIS by the field ionization of helium gas at a three-atom triangle, so-called trimer, formed on the refractory metal tip. The ion source is single atomic size because the helium ion beam is extracted from a single atom of the trimer through a fine aperture as shown in Figure 6.4. HIM is attractive for high-resolution secondary electron imaging, versatile Rutherford backscattered ion imaging, and high-resolution nanoscale modification of nanomaterials. The obtainable information of nanomaterials includes morphology, materials composition, crystal orientation, and surface potential. HIM is more powerful than conventional SEM and focused ion beam (FIB) techniques for some aspects such as spatial resolution, adaptation to insulating biomaterials, and depth of focus [38].

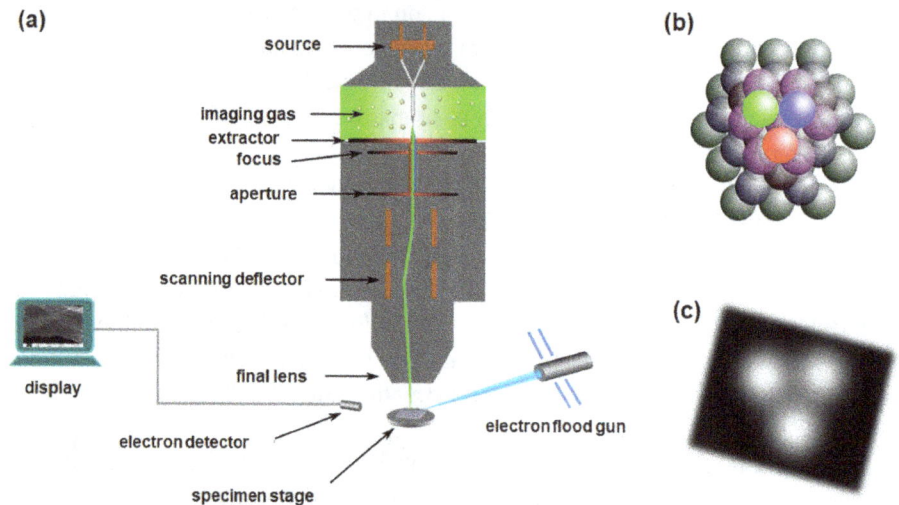

Figure 6.4: Schematic representation of principle and design of scanning helium ion microscope with a single-atom-sized ion source by using field ionization.

HIM provides not only a very high-resolution imaging of sub-nanometer level but also a very precise nanofabrication on various materials' surfaces. Typical applications of HIM in nanotechnology field is shown in Figure 6.5. High spatial resolution imaging of HIM with a large depth of focus is very suitable for critical dimension analysis of semiconductor nanodevices as well as for imaging insulating biological nanomaterials. Especially for the nanofabrication applications, HIM is advanced in surface lithography and pattering at the nanoscale. Since the spot size of the HIM is much smaller than that of conventional Ga-ion source FIB, the resolution of lithography is much higher than that of FIB system. Relatively much smaller etching rate of HIM is suitable for the fabrication of nanoribbon by precise nanolithography of a single-layer graphene nanosheet. With a gas injection system, various materials can be deposited on surfaces with designed patterns. For example, a metallic Pt nanowire with an extremely high aspect ratio can be formed on an AFM probe tip with the desired angle.

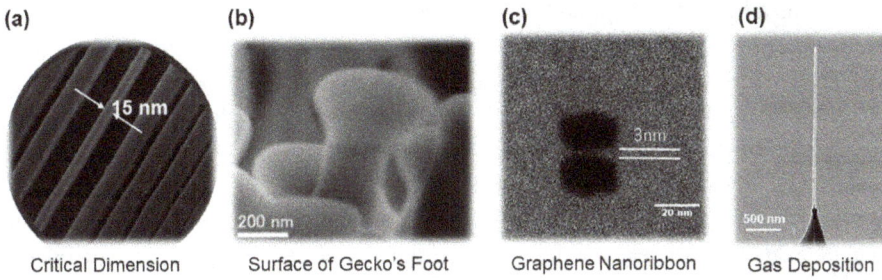

| (a) | (b) | (c) | (d) |

Critical Dimension   Surface of Gecko's Foot   Graphene Nanoribbon   Gas Deposition

Figure 6.5: Typical applications of HIM in nanomaterials: (a) critical dimension analysis of nanoscale devices; (b) imaging of insulating gecko's foot; (c) nanoribbon fabricated by precise nanolithography of a single-layer graphene; and (d) nanofabrication of a metallic Pt nanowire on AFM tip by ion-beam-induced gas deposition.

6.5 Surface physical nano-characterization

6.5.1 Scanning tunneling microscopy and spectroscopy (STM/STS)

STM is based on the concept of the quantum mechanical tunneling of electrons. When a conductive sharp tip is brought to the proximity of a conductive sample's surface, a bias voltage applied between two electrodes to allow electrons to tunnel through the insulating gap between them. STM image can be acquired in constant current mode or constant height mode. In the constant height operation, the current image can be related to charge density. In the constant current operation, the height image represents a constant charge density contour of the sample surface. Typical constant current STM image of single-layer graphene with atomic resolution is shown in Figure 6.6.

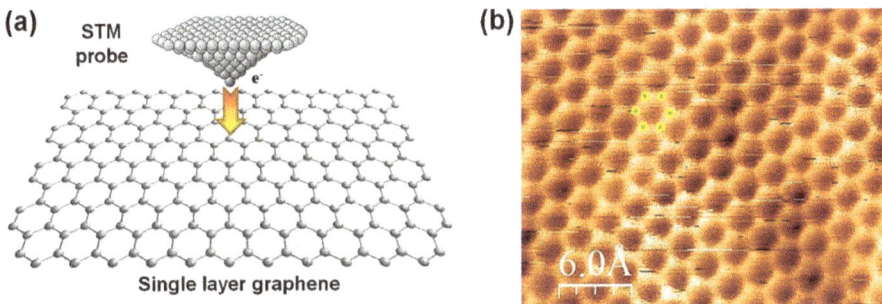

Figure 6.6: Atomic resolution imaging on a single-layer graphene by STM.

STS is one of the spectroscopic measurement modes of STM, which provides information on the electronic structure of a nanomaterial's surface by probing tunneling current I as a function of applied bias voltage V. For small bias, the Local Density Of States (LDOS) of probed surface can be obtained by the derivative dI/dV of site-specific I–V tunneling spectroscopy. Combined with UHV, atomically clean surface of single crystalline surfaces can be prepared and maintained for a reasonably long duration for the STS measurements.

STM has another type of versatility as a powerful nanofabrication tool, performing functions including atom manipulation, local oxidation, nanoscale lithography, and local deposition by the electric-field-induced transfer of tip atoms [39]. Thus, STM has great potential as a highly versatile tool for nanofabrication and nano-characterization.

6.5.2 Scanning probe microscopy (SPM)

Various AFM-based SPMs have been developed to characterize the functionalities of nanomaterials at the nanoscale. For example, using a magnetized probe brought close to a magnetic sample, the probe tip interacts with the magnetic stray fields near the surface. The local magnetostatic interaction results in the vertical deflection of the cantilever, which is utilized for magnetic force imaging as it scans across the sample surface. This operation mode is called MFM, which was introduced in 1987, shortly after the invention of the AFM [40]. Since the MFM does not need for special sample preparation or environmental conditions, MFM has become a popular technique immediately for the nanoscale characterization of magnetic nanomaterials, especially for high-density magnetic recording. Typical resolution of ~30 nm is attainable. Recent developments in MFM are focused on quantification, improvement of resolution, and in situ operation under external field application [41].

Another important force to be sensed is electrostatic force caused by coulomb interaction between a tip and surface. Electrostatic force microscopy (EFM) is a dynamic non-contact mode of AFM, where the distribution of electrostatic force on a sample surface is imaged by using a conductive probe. KPFM is a variant of EFM for the direct measurement of contact potential difference between the tip and sample surface with nanometer resolution [42]. Since the surface potential or work function relates to various electric properties, KPFM has diverse applications in functional nanomaterials.

6.6 Surface chemical nano-characterization

6.6.1 Atom probe tomography (APT)

Atom probe FIM (AP-FIM) is closely related to FIM, the first microscope capable of atomic resolution imaging from a sharp tip, which was developed by Müller in the 1950s. AP-FIM was introduced by Müller and Panitz in the 1960s, which is a destructive microscopy with elemental identification at an atomic level [43]. Removing surface atoms by field evaporation from a sample surface, the evaporated ions by application of electric pulses can be detected, identified, and imaged through magnification imaging method coupled with time-of-flight (TOF) mass spectrometry. Through controlled layer-by-layer field evaporation of material and computational methods, reconstruction of a 3D view of the nanomaterials is possible. Thus, this technique is called APT or 3D atom probe. Commercialization of laser-assisted APT in the 2000s has expanded the probed nanomaterials from highly conductive metals to poor conduction materials such as semiconductors, ceramics, and even insulating materials like biomaterials.

6.6.2 Scanning Auger microscopy (SAM)

SAM is a combination of the techniques of SEM and Auger electron spectroscopy (AES). In this technique, the sample surface is scanned by a fine focused electron beam with a typical energy of 3–20 keV, which results in the emission of secondary, backscattered, and Auger electrons. Since the kinetic energy of Auger electrons is a characteristic of the elements present on the sample surface, SAM enables imaging of the chemical species in the near-surface layer of samples. The characteristic energies of Auger electrons are such that only the electrons from the outer 2–10 atomic layers can escape, which makes SAM an extremely surface-sensitive microscopy. Kinetic energy of Auger electrons due to Auger transition between the three energy levels W, X and Y is given as $E_{WXY} = E_W - E_X - E_Y - \Phi_A$, where Φ_A represents the work function of the electron energy analyzer as shown in Figure 6.7(a). The intensity of Auger peaks as a function of the position of electron beam provides the so-called Auger map of the surface chemical species at the nanoscale. Typical SAM instrument uses a cylindrical mirror energy analyzer with a coaxial electron gun as shown in Figure 6.7(b).

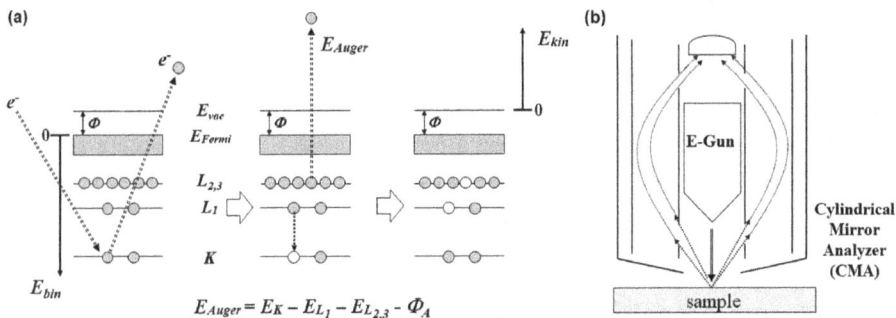

Figure 6.7: (a) Energy-level diagram representation of Auger process. (b) Schematic representation of scanning Auger microscope with a cylindrical mirror analyzer.

Using a field-emission electron gun, the spatial resolution of as small as 10 nm is achievable. Typical detection limits are 0.1–1% of a monolayer, or 10^{12}–10^{13} atoms/cm^2, depending on the signal-to-noise ratio. With Ar$^+$ ion bombardment, the surface layers can be removed in a controlled manner, and analysis can be carried out on new layers exposed after each sputtering cycle. This is known as depth profiling and provides the 3D chemical analysis of elements as a function of depth at the nanoscale. Quantitative identification of the layer number of graphene nanosheet using SAM is possible. Due to the short inelastic mean free paths for relatively low-energy electrons, quantitative identification of atomic-scale thickness is demonstrated with a lateral resolution of nanometer scale.

6.6.3 Secondary ion mass spectrometry (SIMS)

SIMS is a technique used to analyze the chemical composition of solid surfaces and thin films by sputtering the surface of the specimen with a focused primary ion beam and collecting and analyzing ejected secondary ions. The secondary ions are characteristics of the composition of the analyzed area. The mass/charge ratios of the secondary ions are measured with a mass spectrometer to determine the elemental, isotopic, or molecular composition of the surface. SIMS is the most sensitive spectroscopy for surface chemical analysis. Elemental detection limits range from parts per million to parts per billion. Due to the large matrix effects, SIMS is generally considered to be a qualitative technique. Quantitative chemical analysis by SIMS is only possible with the use of certified reference materials. SIMS is one of the most powerful techniques for dopant characterization with detection limits in the range of 10^{14}–10^{15} cm^{-3}. There are two types of SIMS methods: static SIMS and dynamic SIMS. Static SIMS involves surface atomic monolayer analysis, or surface molecular analysis under static limit, usually with a pulsed ion beam and a TOF mass spectrometer. Dynamic SIMS involves in bulk analysis using a Direct Current (DC) primary ion beam and a magnetic sector or quadrupole mass spectrometer. Nano-SIMS is a new generation of SIMS combining high lateral and spectral resolution with high sensitivity. The coaxial lens allows the primary beam to be focused to sub-50 nm for Cs$^+$ primary ions. Using a nanoscale FIB for SIMS imaging and the ability of depth profiling open the possibility of 3D imaging at the nanoscale, especially for the applications in biological systems [44].

6.6.4 X-ray photoelectron spectroscopy (XPS)

XPS is a type of photoelectron spectroscopy such as ultraviolet photoelectron spectroscopy and a surface-sensitive quantitative spectroscopic technique based on the photoelectric effect. XPS with soft X-rays is particularly surface sensitive due to relatively short attenuation depth (typically less than a few nanometers) through which photoelectrons can travel without energy loss via inelastic scattering [45]. Since XPS can identify not only chemical composition but also overall chemical states including electronic structure and local density of states of material surfaces, it is also called electron spectroscopy for chemical analysis (ESCA). By irradiating the sample surfaces with a beam of X-rays and measuring the energy of the generated photoelectrons, the constituent elements of the sample and their electronic states at the surface regions can be analyzed. When the surface of a nanomaterial is irradiated with X-rays of about several kiloelectron volts, the electrons in the atomic orbitals of the constituent elements absorb the photon energy (hv) and are knocked out as photoelectrons as shown in Figure 6.8(a). The emitted photoelectron has a kinetic energy $E_{kin} = hv - E_{bin} - \Phi$, where Φ is the work function for the specific surface of the material. If the incident X-ray photon energy is known, the value of the binding energy E_{bin} of core electrons can be obtained. Since the binding energy of the electrons that make up a substance is unique to each element, quantitative chemical composition analysis is possible. Subtle shifts in electron binding energy are called chemical shifts and are derived from interactions with surrounding atoms. Since it reflects the chemical state and electronic state (oxidation number, etc.) of the element, it is possible to investigate the detailed analysis of surface chemical state of nanomaterials, which are highly related to functionality and safety of the nanomaterials.

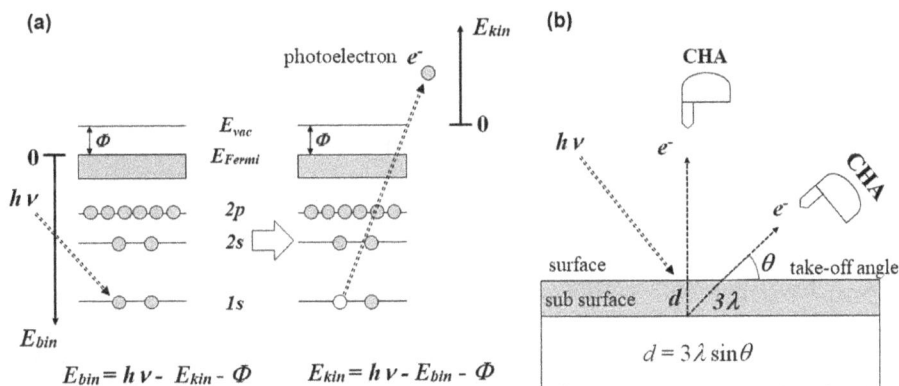

Figure 6.8: (a) Energy-level representation of XPS process. (b) Schematic representation of surface sensitivity enhancement by variation of the take-off angle of photoelectrons.

XPS requires UHV conditions to maintain the clean surfaces of specimens. As a photoelectron energy analyzer for XPS, concentric hemispherical analyzer is widely used. Surface sensitivity of XPS can be significantly enhanced by lowering the angle θ of photoelectron exit relative to the sample surface (the "take-off angle") as shown in Figure 6.8(b) [46]. If the attenuation length λ of the photoelectrons is known, then the effective vertical depth sampled d is given by $d = 3\lambda\sin\theta$. Ninety-five percent of the detected photoelectron intensity is derived from the subsurface region designated by the vertical depth.

By using finely focused beams of X-ray photons from synchrotron light sources, nano-beam spectrum imaging technologies such as "3D nano-ESCA" for 3D analysis of chemical states have been developed and spatial resolution of 70 nm has been reported [47]. However, typical spatial resolution obtainable with commercial XPS instruments for conventional laboratory use is approximately 10 μm, so normally XPS is a micron-scale characterization tool. Thus, conventional XPS can provide mainly population-averaged information of surface chemical properties of nanomaterials.

However, even conventional XPS can be a powerful characterization tool for the detailed characterization of electronic states of nano-objects if the size of the nano-objects is nearly uniform. It is well known that nano-objects such as nanoclusters exhibit quantum mechanical behaviors and physicochemical properties different from their bulk states. For example, although gold in bulk is known as almost inert as a catalyst, single-digit nanoscale gold nanoparticles or nanoclusters supported on titania exhibit a high catalytic ability for CO oxidation [48]. One of the interesting nanoscale effects of metallic nanoparticles is the so-called Coulomb blockade [49]. The charging energy $\Delta E = e^2/2C$, where C is a capacitance, required for the addition of an extra electron onto the nanoparticle may well exceed the thermal energy $k_B T$ and the current through the nanoparticle is blocked. The smaller the capacitance C, the larger the charging energy ΔE. The strong interaction of gold with the terminal sulfur atoms of dithiol molecules of a self-assembled monolayer on Au(111) effectively suppresses the penetration of deposited Au atoms through the dithiol layer and results in the formation of homogeneous Au nanoclusters. The small self-capacitance and uniform size distribution of the Au nanoclusters make it possible to observe the Coulomb blockade in the Au 4f core-level shifts in XPS at room temperature as shown in Figure 6.9 [50].

6.6.5 Raman microscopy

Raman microscopy consists of CRM and TERS. Both are applicable to various ENMs, including those of biological and medical uses. CRM is a microscopic type of Raman spectroscopy that allows 3D chemical imaging based on the use of optical confocality using a laser beam. Confocality means that an illuminated sample spot and a pinhole aperture within the beam path both share the same focal point. This confocal principle can be applied to spectrum imaging of Raman spectroscopy, thus enhancing lateral and depth resolution while also enabling 3D depth profiling. Modern state-of-the-art CRM can achieve a spatial resolution down to 200 nm laterally and 500 nm vertically using visible light excitation, which provides a useful tool for 3D physicochemical analysis of nanomaterials.

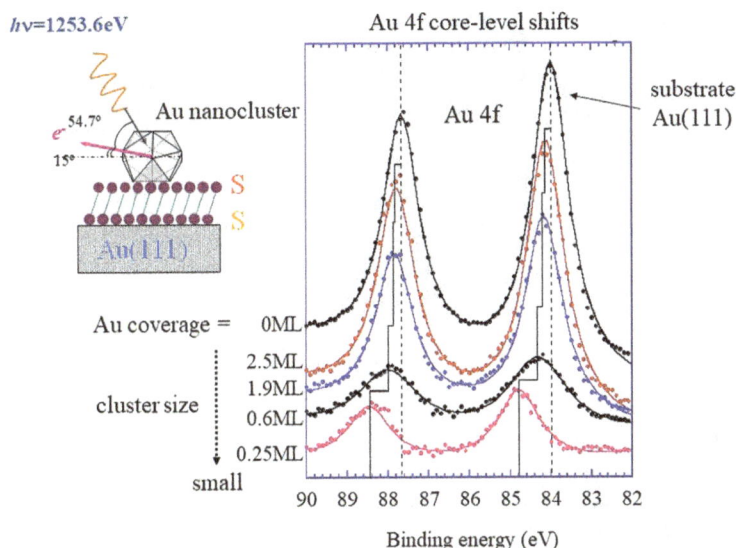

Figure 6.9: Coverage dependence of XPS Au 4f core-level shifts of Au nanoclusters on the dithiol self-assembled monolayer on Au(111). The smaller the coverage, the smaller the size of Au nanoclusters. The observed core-level shifts originate from the single-electron charging energy of Au nanoclusters.

Among recently emerging ENMs, 2D materials such as graphene are most successfully developing due to their remarkable electronic, optical, and mechanical properties, resulting in a variety of industrial applications. Raman spectroscopy is a versatile tool to identify and characterize the chemical and physical properties of 2D materials. Especially in the case of graphene, CRM is so powerful for the chemical state characterization that most scientific papers concerning graphene contain Raman data. Thus, CRM is widely used to provide a fast, non-destructive, and microscopic means of determining layer number for graphene films [51]. The typical Raman spectrum obtained from the single-layer graphene is shown in Figure 6.10. G band results from an in-plane vibrational mode involving sp^2-hybridized carbon atoms. D band is known as the defect band and is very weak in high-quality graphene. G' or 2D band originates from the double-resonance Raman process.

TERS is an emerging surface chemical nano-characterization tool that combines the chemical sensitivity of surface-enhanced Raman scattering with a high spatial resolution of AFM [52]. Using a conductive AFM tip or so-called TERS tip, the significant enhancement in the Raman signal intensity for nano-objects such as molecular adsorbates on a surface occurs due to the enhancement in the electric field between the probe and surface. Normally, TERS does not require UHV or low temperature (LT) but is operable at room temperature and in ambient conditions. Recently, TERS has achieved a single-molecule analysis with intramolecular spatial resolution using a state-of-the-art LT-UHV TERS system [53].

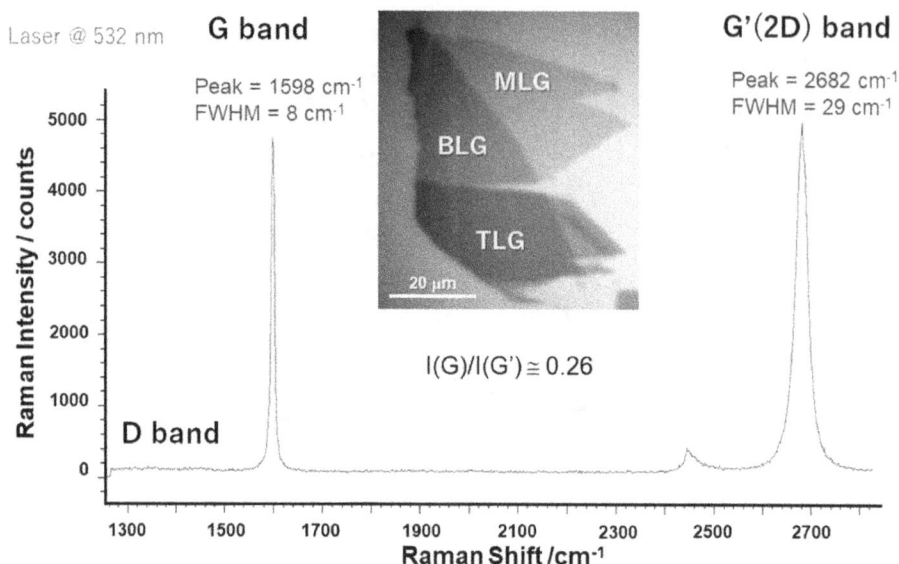

Figure 6.10: Typical Raman spectrum obtained from defect-free single-layer graphene on SiO$_2$/Si wafer, collected with 532 nm laser excitation.

6.7 Summary

Nanoscale physicochemical characterization technologies for MNM and ENM that can be useful for nanoethics assessments have been reviewed. Starting from the introduction of the history and status of nano-characterization, major microscopies or spectroscopic imaging tools with the nanoscale spatial resolution are introduced. Interesting applications of such nano-characterization tools to nanomaterials research are covered. Besides the major players such as STM and AFM, emerging techniques such as HIM and TERS are also introduced. Since some of the nano-characterization techniques are semiquantitative, the establishment of quantitative methodologies and the standardization of nano-characterization technologies are the next steps toward reliable assessments for nanoethics. Besides, one more important direction based on the industrial application needs is to develop in situ observation techniques involving operando spectroscopy for physicochemical properties at the nanoscale.

References

[1]   ISO/TS 80004-1:2015. Nanotechnologies – Vocabulary – Part 1: Core terms. ISO2015.
[2]   Hochella Jr. MF, Mogk DW, Ranville J. et al., Natural, incidental, and engineered nanomaterials and their impacts on the earth system. Science 2019, 363, eaau8299.

[3]   European commission (2011). Recommendation on the definition of a nanomaterial. 2011/ 696/EU.
[4]   Ball P. Ethics at the nanoscale. Nat Mater 2003, 2, 299.
[5]   Feynman RP. There's plenty of room at the bottom. J Microelectromech Syst 1992, 1, 60.
[6]   Xiarchos I, Moroznis AK, Kavouras P, Charitidis CA. Nanocharacterization, materials modeling, and research integrity as enablers of sound risk assessment: Designing responsible nanotechnology. Small 2020, 16, 2001590.
[7]   Xu MS, Fujita D, Kajiwara S. et al., Contribution of physicochemical characteristics of nano-oxides to cytotoxicity. Biomaterials 2010, 31, 8022.
[8]   Xu MS, Li J, Iwai H. et al., Formation of nano-bio-complex as nanomaterials dispersed in a biological solution for understanding nanobiological interactions. Sci Rep 2012, 2, 406.
[9]   Müller EW, Bahadur K. Field ionization of gases at a metal surface and the resolution of the field ion microscope. Phys Rev 1956, 102, 624.
[10]   Esaki L, Superlattice TR. Negative differential conductivity in semiconductors. IBM J Res Dev 1970, 14, 61.
[11]   Binnig G, Rohrer H. Scanning tunneling microscopy. Helv Phys Acta 1982, 55, 726.
[12]   Binnig G, Rohrer H, Gerber C, Weibel E. 7 × 7 reconstruction on Si(111) resolved in real space. Phys Rev Lett 1983, 50, 120.
[13]   Binnig G, Quate CF, Gerber C. Atomic force microscope. Phys Rev Lett 1986, 56, 930.
[14]   Eigler DM, Schweizer EK. Positioning single atoms with a scanning tunnelling microscope. Nature 1990, 344, 524.
[15]   Haider M, Uhlemann S, Schwan E, Rose H, Kabius B, Urban K. Electron microscopy image enhanced. Nature 1998, 392, 768.
[16]   ISO/TR 13014:2012. Nanotechnologies – Guideline on Physico-Chemical Characterization of Engineered Nanoscale Materials for Toxicologic Assessment. ISO 2012.
[17]   Guo HX, Itoh H, Wang CM, Zhang H, Fujita D. Focal depth measurement of scanning helium ion microscope. Appl Phys Lett 2014, 105, 023105.
[18]   Xu MS, Fujita D, Gao JH, Hanagata N. Auger electron spectroscopy: A rational method for determining thickness of graphene films. ACS Nano 2010, 4, 2937.
[19]   Masuda H, Ishida N, Ogata Y, Ito D, Fujita D. In situ visualization of Li concentration in all-solid-state lithium ion batteries using time-of-flight secondary ion mass spectrometry. J Power Sources 2018, 400, 527.
[20]   Wang HX, Zhang H, Da B, Shiga M, Kitazawa H, Fujita D. Informatics-aided Raman microscopy for nanometric 3D stress characterization. J Phys Chem C 2018, 122, 7187.
[21]   Wang HX, Zhang H, Tang DM, Goto K, Watanabe I, Kitazawa H, Kawai M, Mamiya H, Fujita D. Stress dependence of indentation modulus for carbon fiber in polymer composite. Sci Technol Adv Mat 2019, 20, 412.
[22]   Ishida N, Jo M, Mano T, Sakuma Y, Noda T, Fujita D. Direct visualization of the N impurity state in dilute GaNAs using scanning tunneling microscopy. Nanoscale 2015, 7, 16773.
[23]   Cai ML, Ishida N, Li X. et al., Control of electrical potential distribution for high-performance perovskite solar cells. Joule 2018, 2, 296.
[24]   Matsumura K, Fujita T, Itoh H, Fujita D. Characterization of carrier concentration in CIGS solar cells by scanning capacitance microscopy. Meas Sci Technol 2014, 25, 044020.
[25]   Hasegawa S, Shiraki I, Tanabe F. et al., Electrical conduction through surface superstructures measured by microscopic four-point probes. Surf Rev Lett 2003, 10, 963.
[26]   Yu XZ, Kanazawa N, Onose Y. et al., Near room-temperature formation of a skyrmion crystal in thin-films of the helimagnet FeGe. Nat Mater 2011, 10, 106.
[27]   Kudo K, Suzuki M, Kojima K. et al., Simulations of magnetic domain patterns on the surface of Co/Ni multilayers. Surf Interface Anal 2014, 46, 1174.

[28] Matsuda K, Saiki T, Nomura S. et al. Near-field optical mapping of exciton wave functions in a GaAs quantum dot. Phys Rev Lett 2003, 91, 177401.

[29] Urban K, Kabius B, Heider M, Rose H. A way to higher resolution: Spherical-aberration correction in a 200 kV transmission electron microscope. J Electr Microsc 1999, 48, 821.

[30] Erni R, Rossell MD, Kisielowski C, Dahmen U. Atomic-resolution imaging with a sub-50-pm electron probe. Phys Rev Lett 2009, 102, 096101.

[31] Meyer G, Amer N. Novel optical approach to atomic force microscopy. Appl Phys Lett 1988, 53, 1045.

[32] Ruger D, Mamin HJ, Güthner P. Improved fiber-optic interferometer for atomic force microscopy. Appl Phys Lett 1989, 55, 2588.

[33] Blanc N, Brugger J, de Rooj NF, Dürig U. Scanning force microscopy in the dynamic mode using microfabricated capacitive sensors. J Vac Sci Technol B 1996, 14, 901.

[34] Barret RC, Tortonese M, Quate CF. Atomic resolution with an atomic force microscope using piezoresistive detection. Appl Phys Lett 1993, 62, 834.

[35] Mate CM, McClelland GM, Erlandsson R, Chiang S. Atomic-scale friction of a tungsten tip on a graphite surface. Phys Rev Lett 1987, 59, 1942.

[36] Giessibl FJ. Atomic resolution of the silicon (111)-(7 × 7) surface by atomic force microscopy. Science 1995, 267, 1451.

[37] Moreno C, Stetsovych O, Shimizu TK, Custance O. Imaging three-dimensional surface objects with submolecular resolution by atomic force microscopy. Nano Lett 2015, 15, 2257.

[38] Guo HX, Fujita D. Scanning Helium Ion Microscopy. In: Kaufmann EN ed, Characterization of Materials. 2nd. John Wiley and Sons, Hoboken, New Jersey, United States 2012, 2091.

[39] Fujita D, Sagisaka K. Active nanocharacterization of nanofunctional materials by scanning tunneling microscopy. Sci Technol Adv Mater 2008, 9, 013003.

[40] Martin Y, Wickramasinghe HK. Magnetic imaging by "force microscopy" with 1000 Å resolution. Appl Phys Lett 1987, 50, 1455.

[41] Schwarz A, Wiesendanger R. Magnetic sensitive force microscopy. Nanotoday 2008, 3, 28.

[42] Noda T, Ishida N, Mano T, Fujita D. Direct observation of charge accumulation in quantum well solar cells by cross-sectional Kelvin probe force microscopy. Appl Phys Lett 2020, 116, 163501.

[43] EW M, Panitz JA, McLane SB. The atom-probe field ion microscope. Rev Sci Instrum 1968, 39, 83.

[44] Fletcher JS, Lockyer NP, Vaidyanathan S, Vickerman JC. TOF-SIMS 3D biomolecular imaging of Xenopus laevis oocytes using buckminsterfullerene (C60) primary ions. Anal Chem 2007, 79, 2199.

[45] Seah MP, Dench WA. Quantitative electron spectroscopy of surfaces: A standard data base for electron inelastic mean free paths in solids. Surf Interface Anal 1979, 1, 2.

[46] Fujita D, Tanaka A, Goto K, Homma T. Ni-Co alloy as a new reference material for quantitative surface analysis. Surf Interface Anal 1990, 16, 183.

[47] Horiba K, Nakamura Y, Nagamura N. et al., Scanning photoelectron microscope for nanoscale three-dimensional spatial-resolved electron spectroscopy for chemical analysis. Rev Sci Instrum 2011, 82, 113701.

[48] Green IX, Tang W, Neurock M, Yates Jr. JT. Spectroscopic observation of dual catalytic sites during oxidation of CO on a Au/TiO$_2$ catalyst. Science 2011, 333, 736.

[49] Ohgi T, Sheng HY, Dong ZC, Nejoh H, Fujita D. Charging effects in gold nanoclusters grown on octanedithiol layers. Appl Phys Lett 2001, 79, 2453.

[50] Ohgi T, Fujita D. Consistent size dependency of core-level binding energy shifts and single-electron tunneling effects in supported gold nanoclusters. Phys Rev B 2002, 66, 115410.

[51] Xu MS, Fujita D, Sagisaka K, Watanabe E, Hanagata N. Production of extended single-layer graphene. ACS Nano 2011, 5, 1522.
[52] Anderson MS. Locally enhanced Raman spectroscopy with an atomic force microscope. Appl Phys Lett 2000, 76, 3130.
[53] Zhang R, Zhang Y, Dong ZC. et al., Chemical mapping of a single molecule by plasmon-enhanced Raman scattering. Nature 2013, 498, 82.

Part III: **Health, environment and industrial aspects of nanoethics**

Thomas H. Brock

7 Ethical aspects of nanomaterials in industry

Abstract: At the turn of the millennium, nanomaterials were on everyone's lips. A revolution was to cover all areas of technology; everything would be different and better. Indeed, this did not happen, but it is often overlooked that nanotechnologies have long been widely established, so the issues of health and safety by no means became obsolete. In this chapter, the current status of various strategies and measures for occupational health and safety with nanomaterials is discussed.

Keywords: nanomaterials, occupational health, occupational safety, measures, agglomerates, aggregates, nano-objects, exposure, exposure control, HARN, safety data sheets, risk assessment, measures, responsibilities, labeling, chemical Properties, dust, fume hood, air flows, cleaning, personal protective equipment, people with disabilities, respiratory masks, nanotubes, information

7.1 Introduction

Nanomaterials have had a firm place at the workplace for many years and are used in manufacturing, processing, and disposal or recycling. The volume and type of nanomaterials used has increased continuously during the course of the years and in many cases they are not even recognized as nanomaterials. They are deployed in workplaces encompassing laboratories, the production and processing of the nanomaterials, the user, and finally the disposal firm.

There is a general consensus that nanotechnologies are technologies that use structures within the range of approximately 1–100 nm. There are many attempts to establish a clear definition for this complex field, all of which more or less aim to narrow down the topic [1]. In order to determine and assess risks and derive protective measures, it has been proven useful to generously expand the upper limit toward the micrometer range (many measuring devices detect the range up to 1,000 nm) as the properties of the materials do not abruptly alter at 100 nm. In addition, larger secondary particles such as agglomerates and aggregates also play a significant role at the workplace. Although primary particles in the range of 1–100 nm will seldom be encountered at many workplaces, secondary particles made up of these primary particles

Thomas H. Brock, Professional Association for Raw Materials and Chemical Industry, Department of Chemistry and Biology, Competence Center for Hazardous Substances and Biological, Heidelberg, Germany

https://doi.org/10.1515/9783110669282-007

do occur, which may disintegrate into the primary particles again in the body. The application of strict limits would simply mask a significant part of this problem.

Nanomaterials occur more rarely than free, unbonded nano-objects (nano films and nanoplates, nanotubes, nanorods, or nanowires, as well as spherical nanoparticles). Alongside agglomerates and aggregates, composite materials are often used, for example, nanoparticles or nanotubes, which are integrated into a polymer matrix [2].

Furthermore, two types of material with nanoscale structures occur at the workplace (a third type would be ultrafine dusts from natural sources). Ultrafine aerosols are differentiated from the nanomaterials that are generally understood to be nanomaterials and are consciously and intentionally manufactured or used as such, as although they can share the same properties in principle, they can also be created unintentionally and cause exposure. Both types are to be considered equally for occupational health and safety measures. The ultrafine aerosols can be released by nanomaterials, although these are often created through process and processing steps on coarser materials. Although we will mainly discuss nanomaterials in the following, in most cases, the statements made can be applied to ultrafine aerosols, too, although the characterization of the dusts for risk assessment can be difficult due to the considerable time and effort involved. Assumptions about the nature and concentration of ultrafine dusts emitted during working procedures like milling, grinding, or laser ablation can be made but may be wrong. If no detailed guidelines for the safe processing are available, expert judgment is needed. In most cases, available technologies for an efficient exposure control will minimize the exposure to ultrafine dusts also, as long as filtered air is not blown back into the working area or this air is filtered properly (very efficient filters are available). Processing materials on the benchtop without further exposure control measures are possible according to expert judge mentor measurement at least of the particle number concentration in comparison with the background concentration.

The assessment and evaluation of the risks are hindered by the large number of nanomaterials and remaining gaps in our knowledge of the effects. The emergent properties of many nanomaterials also make this task difficult, as conclusions drawn on the characteristics of the nanomaterial on the basis of the properties of the coarser material can be misleading. For instance, nanoscale dusts from metals can be far more ignitable – in some cases even auto-ignitable – than their dusts in the micrometer range and of course the compact material. If this is not taken into account, explosions or even detonations can lead to serious consequences for people and the environment as well as severe material damage.

It will certainly be necessary to also consider other large molecules within this sub-1 nm range, for example, C_{60}, graphene, or carbon nanotubes. This also applies to ultrafine fractions of particles in coarse materials [3].

The primary particles have a tendency to congregate quickly to the less tightly bound aggregates or the polyvalent and very tightly bound aggregates. These may disintegrate under physiological conditions, releasing smaller or primary particles again, depending on the forces between the primary particles and the mechanisms of energy transfer (e.g., mechanical energy on gloves or the skin) or biological and chemical effects (e.g., separating and coating the smaller particles with proteins). So it is necessary to take these aggregates and agglomerates into account too. They may also function as carriers for molecules transporting them into biological structures.

Although many studies on effects of nanomaterials in humans have been published, it is absolutely necessary to keep in mind that there is no common behavior of all nanomaterials, that there is no phonotypical nanoparticle or nanomaterials. Hence, these results have to be interpreted with expert knowledge to yield a useful picture near to the truth.

7.2 Accident and illness prevention

The focus of occupational health and safety is to minimize risks As low as reasonably practicable (ALARP) but not necessarily to eliminate them entirely, as this is often not possible and remote from reality. In this context, the risk is often described as the product of the probability of occurrence of damage and its severity. The probability of occurrence is directly dependent on the exposure, which can be determined – albeit with some effort. In order to determine risks, however, the severity of the damage resulting from the respective scenario also needs to be quantified. This is considerably more problematic, as we are not fully aware of the mechanisms of possible nanospecific effects (insofar that these are relevant) and their effects on the organism. As such, if no sufficiently sound data are available on the effect, the risk level can only be controlled by limiting the extent of exposure. Several regulations or publications address these questions [4–17].

Performing a risk assessment is a common and proven strategy for handling hazardous substances. This requires the compilation of various kinds of information in order to at least make a qualitative assessment of the risks or, better still, perform liable estimation of the risks. Appropriate measures can then be derived from this and put in place.

The currently used nanomaterials are – with regard to the used quantities – predominantly substances that have been known for a long time, for example, carbon, silicon dioxide, or titanium dioxide. However, because the number of available types of nanomaterial is much larger, and the number of those that can be manufactured in theory is so enormous, there will be applications involving the handling of nanomaterials about which we know very little.

For this purpose, information will be required on various parameters. Some, but unfortunately not all, of this information can be found in the safety data sheets and technical data sheets (product specifications) if these are informative and complete.

With regard to nanomaterials or solid materials that contain a relevant proportion of nanomaterials, some gaps sadly remain. When handling these substances,

the raw and auxiliary materials, solvents, and reagents also have to be assessed. This includes in particular information and data on the following:
- Size distribution and shape
- Fire and explosion behavior, auto-ignitability
- Additions, coating, soiling, enclosures, adsorbates, "foreign atoms" incorporated through chemical bonds (e.g., metal atoms from the growth process)
- Chemical and physical–chemical properties, such as catalytic effects, reactivities, radical formation, surface activity, specific surface, ζ potential, absorptivity, kinetics of the agglomeration or aggregation, dustiness behavior, solubilities, aging, and stability
- Toxicology
- Exposure possibilities and levels (inhalation, dermal and oral)

At times, it can be difficult to recognize whether a product contains nanomaterials and whether such nanomaterials are released or created during (intended) use. Not all safety data sheets contain information on this; in case of doubt, the manufacturer or supplier should be consulted. The use of nanotechnology is occasionally advertised in products, in particular for the downstream user and consumer segment, even though it is of no significant relevance in the product itself. This is done simply for the reason that the label "nano" carries positive connotations, especially in some technically savvy markets. However, there are also cases where the contrary applies. This can make it much more difficult to gather information.

The toxicology of nanomaterials is not always analyzed sufficiently: in some cases, there are considerable data gaps that need to be circumvented as a precautionary measure. If a precautionary renunciation of certain materials or applications is not possible in a socioeconomic context, because this would deprive society of significant advantages, for example, in the therapy of illnesses, or the prevention of improvements in drinking water supply in water-scarce countries would be hard to justify in an ethical sense; prudent compromises must be made on the basis of a risk assessment with sensible and justifiable assumptions always keeping in mind that our knowledge is limited and there may be risks being not properly addressed, especially when it comes to new materials with new properties emerging from the (self) organization of nanomaterials. At the moment, such materials are used only in small quantities or are still in research and development stage.

If no usable and plausible data are available from the manufacturer or supplier or, if applicable, no dedicated examinations are available, the categorization as per the Technical Rule for Hazardous Substances 527 (TRGS 527) [4] represents a possible procedure for taking toxicology into account. This divides nanomaterials into four categories with regard to the toxic properties (Table 7.1).

If no conclusive data about a material are provided by a manufacturer or supplier, it can be allocated to one of the categories with some expert knowledge. The associated measures are categorized according to the time and effort involved. Category materials

Table 7.1: Categories of nanomaterials according to TRGS 527.

Cat. I	Soluble nanomaterials without specific toxic properties (solubility at least 100 mg/L water at room temperature): for example, many inorganic salts like sodium chloride and amorphous SiO_2
Cat. II	Soluble nanomaterials with specific toxic properties, for example, quantum dots, inorganic salts like heavy metal compounds, dendrimers, and lipid shells carrying toxic atoms or molecules inside
Cat. III	Granular biopersistent dusts without specific toxic properties, for example, carbon black and Al_2O_3
Cat. IV	Fibrous nanomaterials, for example, carbon nanotubes (nota bene not necessarily all carbon nanotubes)

require no special measures above and beyond the general protective and hygienic measures. Materials of categories II and III often have threshold values that must be adhered to. Category IV materials can be handled in the same manner as materials from one of the three previous categories if it has been proven – for instance, by the manufacturer – that these do not possess any asbestos-like properties. High aspect ratio nanomaterials (HARN, WHO fiber with at least one dimension between 1 and 100 nm) belongs to category IV.

The physical and chemical properties are often not well explored and taken into consideration as causes of possible risks, with the focus instead predominantly being on the toxic properties. These deficits in the assessment of risks can nevertheless cause immense damage to people, the environment, and assets. The large specific surface areas of nanomaterials often cause higher relativities and reaction speeds, which means that flammable materials in the nanoscale range can have significantly lower minimum ignition energies than coarser material or can even take on self-igniting property (Figure 7.1) [18].

7.3 Occupational health and safety with nanomaterials in practice

7.3.1 Risk assessment

One problem with the manufacturing and use of nanomaterials as well as the production of ultrafine aerosols is the sometimes insufficient data situation for assessing the risks. An important factor is the early acquisition or generation of data, at the latest when materials outside of the field of research are to enter the market. Extensive data often already exist for nanomaterials that are manufactured and used on a larger scale.

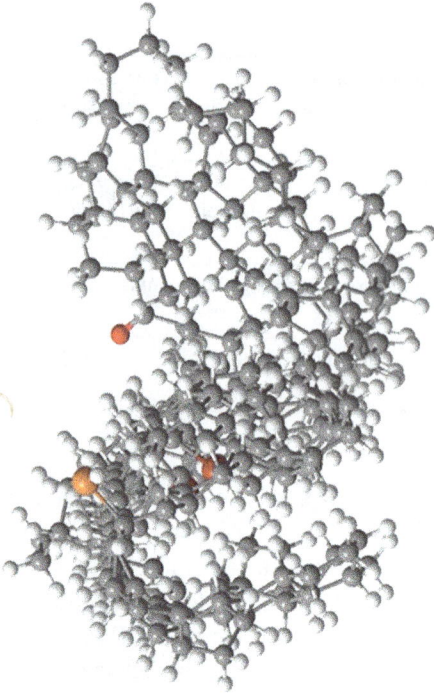

Figure 7.1: Molecular structure of a graphene flake.

Measurements can be performed easily for some properties, but can be difficult for others.

The source of knowledge for operational use is primarily the product information material provided by the manufacturer or supplier, that is, the safety data sheets and technical information sheets. Enquiries to the manufacturer or supplier can help close information gaps in the documentation. Sometimes it is necessary to show some perseverance when making enquiries to this end. All available information is not always provided immediately, but the remainder of a possible partial legal responsibility or at least voluntary obligations within the scope of responsible care can be very helpful here. Manufacturers have to gather information elsewhere, as of course they cannot simply refer to safety data sheets, but rather have to draw these up in full themselves. After all, it also has to be possible to determine the level of risk for internal use (Figure 7.2). Classification into one of the four categories must be possible; otherwise, the most dangerous properties must be assumed in each instance. This is often the case in research and development work, as there is frequently no data available. Here, however, is where the protective measures in laboratories come into play, which is also effective for other potentially very dangerous substances and have proven their worth over many years. These also include strategies for assessing exposure, for example, measuring options.

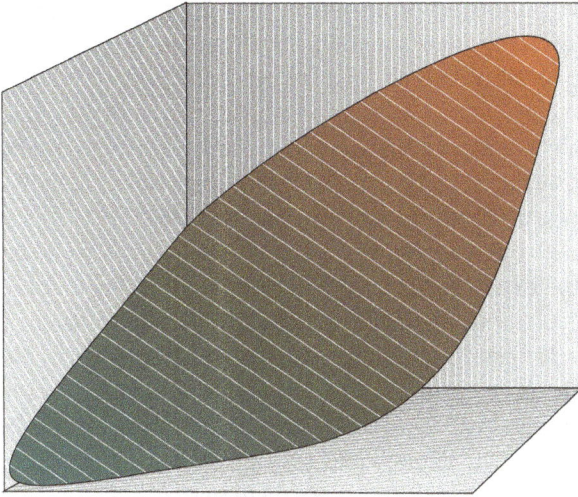

Figure 7.2: Risk correlating to quantity of the nanomaterial, the duration of handling and the dustiness of the material, reaching from green for a lower risk level to red for comparably higher risks.

The risk assessment must be performed prior to commencing the work, as otherwise risks may be recognized too late. Even if no damage occurs, costly retrofitting or delay to pending work is often necessary. When setting up production facilities and, if applicable, also laboratory and technical rooms, it is necessary to provide the technical and organizational prerequisites for safe working. When setting up a production facility, such planning errors can even lead to the cancellation of a project. In a laboratory that was set up for a specific purpose and where, for example, supply and exhaust air capacities were reduced for cost-related or other operational reasons (e.g., as the implementation of such reductions was deemed possible following an assessment of risks and specification of measures sufficient in this specific case), work on manufacturing nanomaterials will often not be possible without significant retrofitting of the ventilation technology. It may even be discerned that it is not possible to perform this retrofitting as, for example, no further supply or exhaust air capacity is available in the building and cannot be expanded, and procedures cannot be adapted to the extent that supply and exhaust air technology can be replaced by alternative exposure minimization methods. As a result, this would mean that the planned work – at least in this room or building – cannot be performed. If the work is still carried out, for example, due to economic considerations or in order to avoid endangering research results, serious – and even legal – consequences may be the result. The careful and thorough compilation of the risk assessment – or the prospective risk assessment, depending on the planning stage – and, where required, the immediate adaptation to change operational conditions such as the use of new chemicals with a risk potential that was previously not existent in this form and therefore not taken into account, therefore, sit at the heart of occupational health and safety. Of course, regulatory provisions also have to be observed.

All exposure routes are to be taken into account for determining the exposure situation and checking the effectiveness of the measures taken. The process of assessing the risks (Figure 7.3) may seem to be difficult and arduous, and this actually may be true in some cases. However, in many countries, this is not only a legal requirement but this is also a prudent strategy to avoid damage and woe.

All possible exposure routes – inhalation, dermal, and oral – must be taken into account for determining the exposure situation and checking the effectiveness of the measures taken. In particular, oral exposure to hazardous substances is often underestimated, as it is generally correctly assumed that hazardous substances are not intentionally or accidentally eaten or drunk. Due to hygienic deficiencies at the workplace, however, dermal and oral exposure can still occur, for instance, if substances are transported to the face by touching hand rails that have been contaminated by soiled gloves with the bare hand and from there get into the eyes or mouth, for example, by sweating.

The work procedure and the dustiness behavior of the material are significant for assessing the skin contact. Nanomaterials that are chemically very similar can differ greatly in terms of their dustiness behavior. For example, in the case of two iron oxides with similarly sized primary particles, one of them released large amounts of nanoscale particles when decanting and the other very little. If the documentation contains no details on the dustiness behavior and enquiries to the manufacturer or supplier do not yield the desired information, it is recommended to identify this data prior to their introduction into production at the latest. Standardized procedures are available for this.

With the data obtained in this manner and the exact knowledge of the work procedure and equipment to be used, it can be determined whether a relevant release of nano-objects will occur at all during the (intended) processes. Even if a nanomaterial does not release relevant amounts of nano-objects, this can still take place as a result of the work procedure, for example, through mechanical or thermal influences that destroy the matrix or split aggregates and agglomerates again. Even the drying of a solution or suspension can lead to nano-objects being released into the air from the dry residues. For some work procedures, further investigation is unnecessary: if a high-performance mill, which creates nano-objects out of larger material through intensive milling within the scope of a top-down process, is opened in a room after the milling process without any additional protective measures having been taken, one would generally determine that enormous quantities of nano-objects have been released.

As well as defining the technical (and other) measures, knowledge of the dust formation also allows the contamination risk to the skin presented by the deposit of released nano-objects to be assessed. Although in line with current knowledge it is believed that human skin provides a reliable barrier to nanomaterials when in good

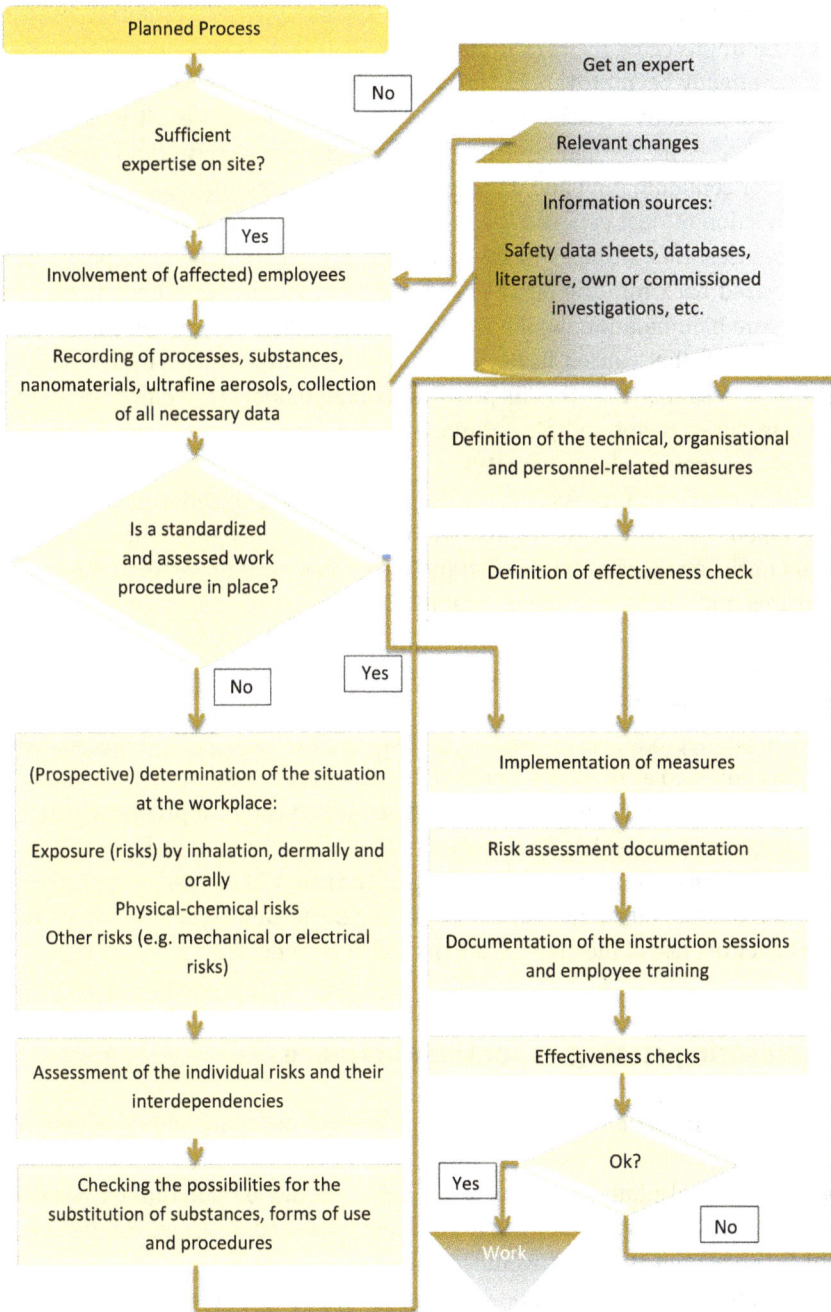

Figure 7.3: Process of risk assessment.

condition, this must not result in work being performed with unprotected hands, arms, or even unprotected facial skin due to the existing data gaps in this area, unless this has already been proven for the specific nanomaterial processed. It is necessary to wear protective gloves (in the case of disposable gloves, it is generally recommended to wear two pairs on top of each other as a precaution), regularly check them for contamination and damage, and change them regularly.

The collection of dust can also lead to deposits on the surfaces of work equipment and devices, and surfaces in the room and on clothing. These deposits can then be released back into the air in an uncontrolled manner at a later stage and cause exposure by inhalation, which can occur at a different place and time to the release of material that caused it. Because it is particularly difficult to detect such exposure, it is essential that this dispersal of contamination is avoided by taking technical measures. Swipe samples or direct reading measurements of manual resuspensions can provide clarity here [19].

The risk assessment process also involves documenting the effectiveness of the measures taken. This can be the ventilation test of the supply and exhaust air unit or regular checks by the supervisors that the employees have understood the specified measures and implement them reliably. The risk assessment also documents scopes of responsibility.

Quantifying the risks can be difficult for risks caused by toxicological effects due to a lack of data. Minimization of the exposure will minimize the risks too (for self-organizing or self-reproducing materials, this may be different). Risks of fires and explosions are addressed as for other combustible substances. The costs of the materials that can be very high or the amount that can be collected can be a problem for some standard procedures in need of large amounts of sample material, using a 20-L or a 1-m^3 explosion pressure test system, for example. The safety of the personnel carrying out these works has to be guaranteed also, cleaning the testing apparatus and its surrounding area from rest of the material and its reaction products.

7.3.2 Measuring strategies for the workplace

Measurements of nanomaterials can be very costly and time-consuming, which impedes broader application. A staggered approach has therefore proven useful for assessing exposure by inhalation. In a first step, it should be determined whether relevant exposure has occurred or can even be hypothesized. In an enclosed system operated in a vacuum, releases would not be expected as long as the system is sealed and monitored. The critical operating states in this case would be the extraction of the products, depending on how the filling process takes place, and in particular leaks and accidents as well as cleaning and maintenance work. For these, the exposure would have to be assessed separately across all exposure routes.

The measuring strategy contains a basic exposure assessment, which can be performed at all workplaces with limited effort, expenditure, and basic skills for handling the measuring equipment. The results of this assessment must then be compared with the background concentration that is to be determined in parallel. Of course, comparability must be ensured and it must be known whether the same nanomaterials in the air are being compared. Due to factory traffic, engine emissions can release large quantities of nanomaterials into the air in which comparative measurements are made, resulting in a background level that is very high but is not comparable to the nanomaterials to be measured at the workplace. Handheld devices are used as measuring equipment here. If the nano-objects deviate too significantly from the spherical shape, no useful measuring results are generally obtained. Further information can be found in [20, 21].

An example for measuring the emission from a fume hood is shown in Figure 7.4. While handling nanoscale TiO_2 in a fume hood (filling several grams from one bottle to another and back), the particle concentration in the air inside of the fume hood and outside at the researchers position, a high amount of TiO_2 nanoparticles was released, but no increase of the background concentration in the air of the laboratory is detected. The particle size was detected in a range between about 1 nm and about 1,000 nm covering agglomerates and aggregates too. To test the experimental setup, air from the inside of the fume hood was blown out waving a sheet of paper to give a short and small peak [22]. Precondition is the proper use of a well-maintained and tested fume hood (in this case, according to EN 14175 [23]) without air draughts in the laboratory atmosphere disturbing the air flows in the fume hood.

The technical expenditure is disproportionately higher within the scope of the expert exposure judgment and is therefore accompanied by high requirements on the people carrying out the measurements. This approach should be employed if it was not possible to achieve a sufficient level of certainty in the first stage. One scale for such measurements was provided by the Organization for Economic Cooperation and Development's tiered approach [24]. The mostly limited resources for carrying out such measurements can be better adapted to the level of difficulty of the enquiry and the scale of the problem in this way. These measurements allow reliable documentation of conditions at the workplace in accordance with the present knowledge, and it is possible to define measures appropriate to the scale of the problem.

There are few limit values for exposure to nanomaterials, and none based on health. However, several assessment values can be used as a basis for the assessment [25, 26]: although mass concentrations are traditionally used for assessment in the context of protection from hazardous substances, number concentrations and surface concentration are probably more meaningful for nanomaterials. For criteria for assessment of the effectiveness of protective measures, see [27].

As it often cannot be ruled out that in particular the less tightly bound aggregates of primary particles or fibrous nano-objects can be split open again in the body, the scale of dimensions to be considered should not be limited to 1–100 nm,

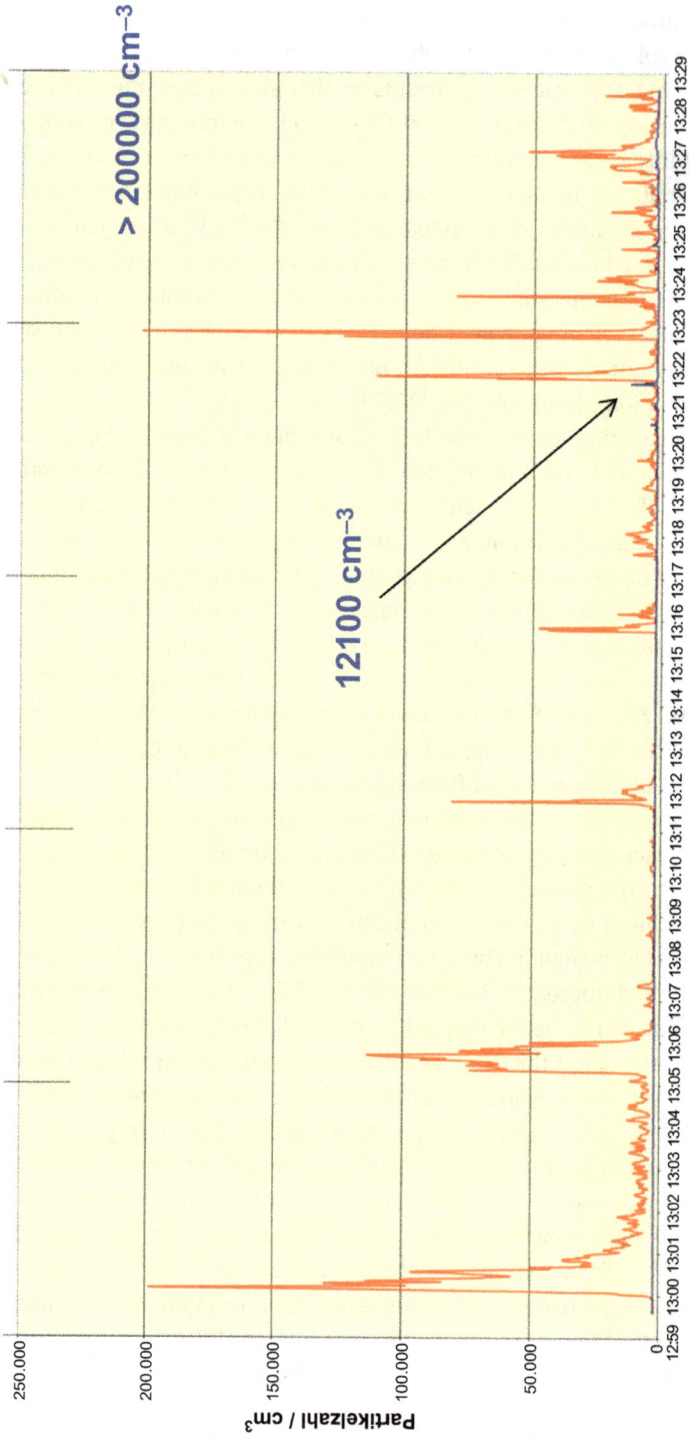

Figure 7.4: Particle concentrations measured inside and outside of a fume hood while handling nanoscaled TiO₂.

but rather also extend into the higher range, that is, up to the micrometer range. Many measuring devices automatically detect these sizes. In normal breathable air, nano-objects usually occur in close proximity to the place from which they were released and are almost exclusively gathered in aggregates or agglomerates having been attracted toward other coarser dust particles or droplets [28]. In a low-particle atmosphere, on the other hand, the nano-objects primarily agglomerate together due to a lack of other bodies to collide with.

7.3.3 Measures at the workplace

Typical work procedures that can lead to exposure to nanomaterials are:
- Manufacturing of primary particles and their agglomerates and aggregates
- Processing into products for the market
- Processing of coarser nanomaterials while releasing nano-objects (clogged nano-objects or nano-objects that have been removed from the matrix during processing, e.g., nanodroplets from polymers) or compound structures made up of nano-objects and matrix (e.g., carbon nanotubes that protrude from such polymer nanodroplets)
- Waste disposal
- Cleaning of systems, devices, and rooms (e.g., to repair a reactor)
- Accidents and leaks

It is advantageous to take the protective measures in the following order (Figure 7.5).

Figure 7.6 demonstrates the differences between the preparations of a nanomaterial, sun-blocking TiO_2 in this case. TiO_2 embedded in a matrix of water and oils as a sunscreen does not release any particles, and the TiO_2 isolated from the sunscreen by incineration in a crucible is agglomerated and does not release particles, as long it is not milled. Nanomaterials embedded in matrices are relatively safe to handle, since exposure via inhalation is not possible as long the material does not become dry or is not sprayed. Although exposure can occur through the skin or oral (hygiene), agglomerates and aggregates of primary nanoparticles may pose a danger when they are separated into smaller particles.

When handling other (hazardous) substances, field-tested and effective protective measures are available that also offer a high level of protection when working with nanomaterials. A possible risk can thus also be reduced by minimizing exposure. Therefore, it is usually not necessary to develop completely new protective measures, but rather to search the existing toolbox for the appropriate measures and apply these as intended. As such, technical ventilation measures not only help against toxic gases and dusts but also with nanomaterials, and their effectiveness always depends on whether they have been designed, built, and applied correctly. Sufficient protective measures can be specified using the available information from safety data sheets,

	Substitution of hazardous substances, use of applications or processes that lead to lower exposure, e.g.
	different nanomaterials that result in similar product properties
	chemical modifications of the nanomaterial that lead to lower toxicity or ignitability
Substitution	use of master batches or suspensions instead of pure nanomaterials
	use of modifications of the nanomaterial resulting in less formation of dust (e.g. through a coating)
Technical Measures	Technical Measures, e.g.
	using closed apparatuses and systems, working in the vacuum range, instead of (partially) opened systems
	cleaning systems before opening them by fitting cleaning devices
Organizational Measures	ventilation measures, measuring at the point of emission so that the breathing zone is not contaminated and no deposits occur at the workplace and in the neighbouring rooms
	Organizational Measures, e.g.
	staff training (see e. g.)
	spending as little time as possible in contaminated areas
Personal Protective Measures	minimize the number of people in these areas
	taking particular groups of people into account, e.g. young people
	hygiene policy for avoiding dispersal of contamination
	Personal Protective Measures, e.g.
	single-use protective suits, closed work clothes or laboratory coats that cover the body well
	protective gloves (with long cuffs if necessary)
	sleeve guards
	respiratory protection (not a permanent measure)

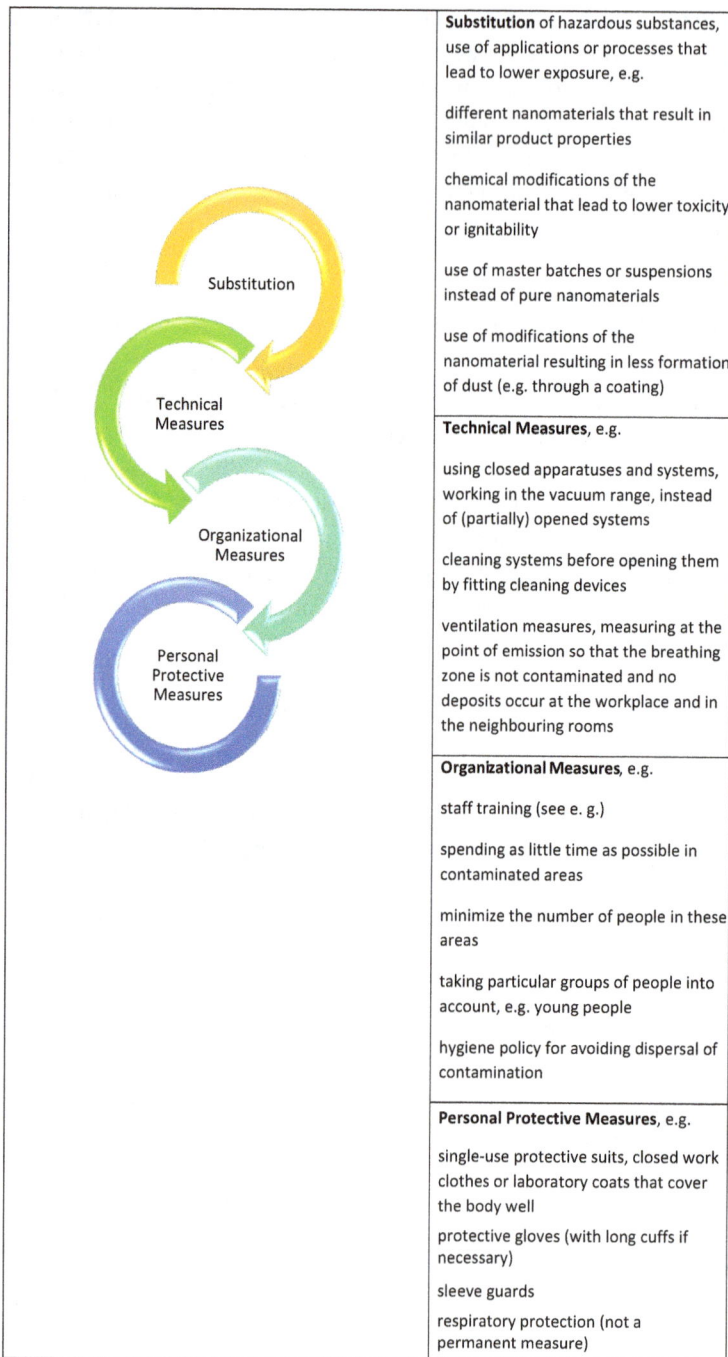

Figure 7.5: Strategy of minimizing the exposure; starting point of the process is the replacement of the material as far as it is reasonable.

Figure 7.6: (A) Typical sun screen (factor 50) with nano-TiO_2; (B) residue in the crucible after incineration; and (C) milled TiO_2.

other manufacturers' information, existing literature, or by contacting the manufacturer to make specific enquiries. The four categories provide assistance here in the event of data gaps but, if necessary, experts have to be consulted.

On the basis of several points it is possible to gain a first impression as to whether it is necessary to go into more detail in the risk assessment (see also [29, 30]).

Structural prerequisites: Appropriate construction, infrastructure, and fixtures and fittings of buildings and working areas with thorough planning documents, systems, and apparatuses built and operated in accordance with the state of the art.

Operational organization: Operation of buildings and systems with precise definition of responsibilities, in particular, in the case of facilities where different (legally separate) parties are responsible for ownership, technical operation, and utilization, supporting communication and control structures.

Occupational health and safety: Specification and awareness of responsibilities across all levels of the operational organization, expert safety and occupational medical advice available internally or externally, inclusion of employees in matters relating to occupational health and safety (e.g., in the risk assessment or in the selection and suitability test of personal protective equipment).

Agency workers and persons from outside the company or laboratory: Include cleaning staff, service technicians, and manual workers in the risk assessment, instruction, and monitoring of these persons by internal staff.

Gathering data for the nanomaterial: Plausible and complete information not only on toxicity but also on burning and explosion behavior, dangerous reactions (is it apparent that information is missing?), at least to a sufficient level, to categorize the material in accordance with four categories. This includes the up-to-date safety data sheet for the nanomaterial in question, not for similar products from other manufacturers or suppliers.

Informative labeling of the system parts and containers: In particular, exact designation of the nanomaterial, pictograms, H-phrases, and P-phrases as per the

Globally Harmonized System of Classification and Labelling of Chemicals (GHS) are necessary, and simplified labeling for own storage bottles in the laboratory is possible (Figure 7.7) [31].

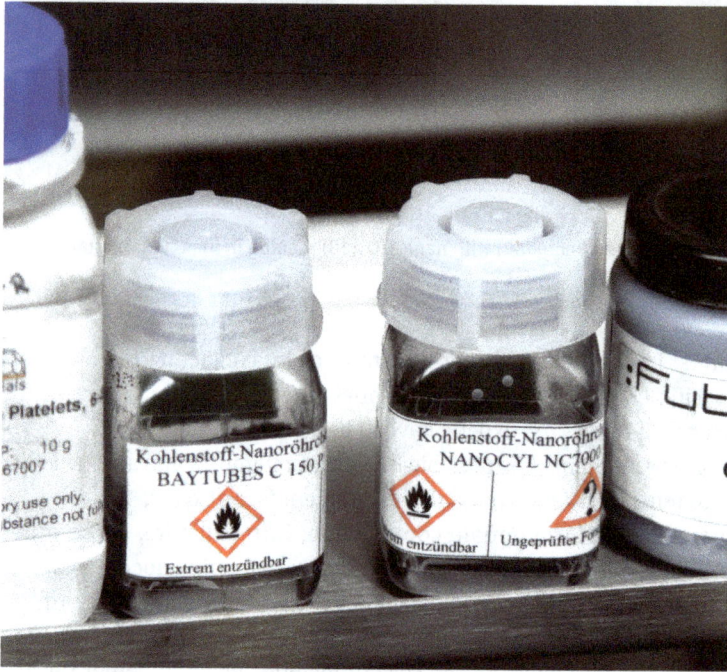

Figure 7.7: Example for a simplified labelling of a storage bottle with pictogram and short phrase.

Measures against fires and explosions: Inertization, vacuum, constructional explosion protection or avoidance of ignition sources, fire alarm systems, and fire fighting (even small quantities can lead to incipient fires or cause severe explosions when mixed with air).

Chemical Properties: For example, catalytic activity or auto-ignitability of the nanomaterials (when mixed with other substances, there can be a risk of a severe reaction or explosion; large surfaces can lead to accelerated reaction or self-ignition).

Protection against dust formation: If possible, enquire about or investigate (or commission an investigation of) the dustiness behavior and specify protective measures in accordance with this (significant quantities of dust can be formed, particularly when filling and decanting the nanomaterial, but this is not necessarily the case). If nanomaterials are handled openly during the activity, substance release and contamination of the working area are possible and there may be a risk of ignition. If suspensions nebulize, there is a risk of inhalation of the aerosol as well as of deposits on skin and mucous membranes and on the surfaces in the working area. This may also present a risk of ignition. Powders can give off dust more or less easily and can be blown away easily. Filling processes can take place in a practically exposure-free manner in (semi-)automatic weighing chambers for sacks, barrels, and other containers while extracting and separating escaping air contaminated with nanomaterials. If necessary, manual weighing processes can be performed in specially vacuumed and low-vortex exhaust booths, in sealed automatic scales, or through reweighing of sealed containers that are only opened in the fume hood (glove box, etc.). A multitude of further, in some cases, very simple technical solutions or minor but effective modifications to the workflow can be applied here. Dust can escape from unstable containers such as bags during handling. When compressed, air escapes with dust (this may even escape in the vacuum lock). You should therefore use solid, sealable containers, which should not be too large. Machines that can release nanomaterials (e.g., mills) must have a closed design or they must be enclosed and vacuumed. The extracted air can be cleaned well using high-performance filters.

Apparatuses and systems: These have to be engineered or designed in such a way that they have to be opened rarely and as little as possible (e.g., this can be achieved using automatic metering, filling, or cleaning devices).

Avoiding dust during work: Dust formation, nebulization, or dried-up leaks can be the result when performing work (dried-up substances may result in dust-generating solids); depending on the risk level, additional technical measures are to be taken. Contaminated auxiliary material for removing leaks can in turn release nanomaterials again, for example, damp clothes when drying. Contaminated auxiliary material must therefore be packed in a sealed container prior to this.

Air flows in the working area: For example, from supply or exhaust air units, or unintentional flows caused by draughts in the room. These must not carry the nanomaterials away and distribute or deposit them in suction channels or in the room. Some nanomaterials give off dust so easily that they need to be protected from air flows (e.g., in a glove box or with a decanting pipe from inert gas laboratory technique [32]) during handling. Flow conditions can be determined and assessed using battery-operated fog generators or larger mains-operated equipment for larger rooms, but test setups in fume hoods or systems in rooms with targeted ventilation can also be employed. Apparatuses and systems can block air flows or cause vortexes, and as a result unexpected exposure can occur.

Protection from dermal exposure: Wear protective gloves if there is a risk of contact with the solid or suspended nanomaterial, or a danger of suspensions dripping down, splashing, and nebulizing.

Processing materials and nanomaterials: Nanoscale aerosols can be created out of coarser or composite materials through milling, grinding, cutting, laser ablation, and so on. These can release free nanomaterials or nanomaterials completely or partially enclosed in a matrix. Nanomaterials can be formed and released from materials without nanostructures too.

Cleaning: Preclean contaminated equipment so that it can be cleaned by auxiliary staff without any danger. Systems such as reactors must be cleaned sufficiently that staff can also access them from the inside without any risk. Adhering nanomaterials can contaminate the surroundings and put people at risk. A release of dusts from contaminated equipment must be avoided until it is cleaned. The cleaning should therefore take place without delay. Until then, the containers must be sealed or placed in a clean enclosure.

Hygienic measures: These are personal measures for avoiding dispersal of contamination. Contaminated equipment and work tools, for example, writing equipment or contaminated gloves, must not be removed from the possibly contaminated area (e.g., the fume hood) before being cleaned (or being packed ready for disposal). Hand rails should be used but not touched with gloves. Sleeve guards can prevent nanomaterials clinging to the arms when working in the fume hood. Contaminated protective, work, and everyday clothes are to be packed in dust-free containers and given to the specialist cleaning company or disposed of. Used protective gloves must be disposed of if they cannot be cleaned. The dispersal of contamination can also be prevented by the appropriate design of workplaces, processes, and experiments. Do not interrupt work in the contaminated area unnecessarily, do not remove contaminated objects (e.g., gloves) from the area, and ensure that special attention is paid to hygiene when leaving the area.

Other substances: The other reagents, products, solvents, and auxiliary materials are to be taken into account in the risk assessment. The risks associated with these materials must also be taken into account, for example, handling in particle-filtered circulation operation is critical if gases or vapors also occur, or when an extraction box with absorption or adsorption filters is used and an opening cannot be detected in time.

Disposal: Collect and seal waste in suitable and labeled containers and dispose of it at appropriate intervals. The larger the container, the more dust can be released from it. The potential risk increases with the amount of waste stored, and in addition dangerous mixtures or aging processes can lead to gases and vapors being released, which in turn can cause fires or explosions.

Personal protective equipment: Alongside work clothes or laboratory coats, protective goggles and suitable shoes are to be worn. Protective gloves with low porosity – which also have to be resistant to the other hazardous substances they may come into contact with – protect against contamination. If there is a risk that the face, head, or body can become contaminated with nanomaterials, a protective screen, hood, or disposable suit (preferably made of closed textiles without seams or hollows) must be worn (define hygiene measures and disposal procedures).

Briefing and instruction: The properties of and risks posed by nanomaterials are not universally known and must be explained; staff must be familiarized with and trained in safe work processes and protective measures.

Further properties are to be taken into account in order to assess the risk, in particular, the ignition and burning behavior, aging and decay, and the catalytic effect.

Please take into consideration that people with disabilities at workplaces may need special equipment and assistance. For example, contamination may be of greater concern for a person in a wheelchair sitting at a bench or fume hood when traces of nanomaterials may be collected on the sleeves of the lab coat touching the benchtop. The following guidelines assign protective measures to the nanomaterials for orientation purposes; the risk assessment can provide a different result in specific cases (Table 7.2).

Table 7.2: Four categories of nanomaterials and assignment of typical protective measures.

Cat.	Measures
I	Basic protective and hygiene measures, avoidance of all unnecessary releases, and exposure
II	See measures of category IV.
III	In addition to the measures for category I, in particular, protective measures against exposure by inhalation are to be taken (effectively vacuumed enclosures, dust-free handling in fume hood, safety cabinet*, glove box or closed apparatus or systems also possible)
IV	– In addition to the measures for category I, in particular, protective measures for avoiding inhalative, dermal, and oral exposure are to be taken (closed systems, effectively vacuumed housings, dust-free handling in the tested fume hood, safety cabinet*, glove box, closed apparatus or system, preferably *clean-in-place* or *wash-in-place* in the case of a closed apparatus or system), personal protective equipment must be used, but respiratory protection is only necessary in exceptional cases – If it is proven that there are no asbestos-like properties, measures of the other categories are possible depending on the material

*If no protection is required from gases and vapors or from fragments or splashes propelled around the room.

The measures required in the individual case must be specified precisely within the scope of the risk assessment. Measures for avoiding unwanted contamination of work areas may also be necessary without a direct risk being established. Measures are also related to other hazardous substances used in the process (e.g., reactants, solvents, and auxiliary materials), equipment, or systems.

Particle-filtering respiratory masks offer high filtration efficiency for nano-objects and are particularly effective in the range below approximately 200 nm. However, respiratory protection measures should be used with care. The wearing of additional equipment can be a burden for employees, and it should be considered that, in contrast to volatile gases and vapors, a long-lasting contamination in the area of the workplace is possible, even at greater distances depending on the air flow conditions. Deposits can occur along the entire length of exhaust channels, particularly in places with little flow or where vortexes can form. This can be prevented through special structural design and expert construction.

Special care should be taken when handling nanotubes. They too can have a carcinogenic effect on humans if they have certain geometries, length–diameter ratios, and a rigid structure [33]. They often form tangled structures with individual nanotube ends protruding from them and can also cause inflammatory reactions. When handled, these sometimes release no nanotubes at all, although it is unclear whether this is also the case under physiological conditions. In this case, safe work can be performed using closed apparatuses, tested enclosures and extractions, and tested laboratory fume cupboards or glove boxes. The exhaust air can only be released into the work area after special cleaning. Overviews of these can be found in refs. [34, 35], for example.

Until now, the measures and effectiveness checks are focused on nanomaterials of the first two generations, and also on the third generation to a limited extent [36].

Many nanotechnological applications in the pharmaceutical industry are to be classified as passive (first-generation) or active (second-generation) structures. These are characterized by the fact that they behave like normal substances with regard to quantities, concentration, and dose. The systems begin to resemble biological systems to the extent that a multiplication that is actively controlled from the outside to an ever smaller extent can occur. Up to a point, this is already the case with the self-organizing processes of the third generation. However, whether multiplication mechanisms of the fourth generation can establish themselves outside of real biological systems is the subject of much debate, but it is not unthinkable. As NY City College's and NY City University's famous physicist Michio Kaku states: "it is very dangerous to bet against the future" [37]. In these cases, one would expect an independent increase in particle numbers, quantities, concentrations and doses, as is usual in nature. However, concepts for protective measures against such substances already exist; indeed, the methods are also available for safe handling of organisms and DNA. And like every self-reproducing system, unlimited multiplication is inconceivable as all systems of this kind reach the boundaries regarding available space, nourishment (reagents), and room for waste. With technological adjustments, this should therefore also make it possible to safely control nanomaterials of future developments [19].

Following these prudent practices and proven measures, it can be assumed that risks from handling nanomaterials are minimized. However, there is no such thing as zero risk. So there is no getting away from the fact that a society must itself define what minimum should be accepted.

7.4 Further needs

Continued safety research is required in the future. A number of issues remain open for future investigation, not only in toxicology but also for combustible materials and testing methods for smaller quantities. The efficiency of ventilation systems should be improved to reduce the risk of wrong usage (very often the wrong positioning), the costs for installing such systems, and also the need of energy for running the systems (electric energy for the electric motors or heating and cooling for the treatment of the fresh air).

One of the key components is materials for the information of workers and managers in small and medium enterprises, researchers, and all people who have to start handling nanomaterials without sufficient experience and knowledge about the safety procedures. It is also necessary to create the needed level of awareness that there might be risks associated with the planned work caused by nanomaterials (and ultrafine dusts).

This kind of information has to be in practical and concrete terms detailed enough to enable the reader to choose safety measures that are known to be effective and appropriate for the concrete work to be carried out. At the moment, most publications deal with the theory of nanomaterials and general principles of prevention, but the gap between this theoretical level and the needs of most users is too wide for them to be bridged by themselves. Here, a lot of practical and actionable information is needed. This includes print material with concentrated information (otherwise, people will not read it due to the common lack of time), useable also for training purposes. It has been shown that a combination of printed background information and a separate short concentrate from it in key phrases that are easy to be memorized is quite effective. The concentrated paper can be a folded leaflet to be carried in a pocket of the working clothes or the lab coat. An example is shown in Figure 7.8.

Especially people in research and development and younger people are addressed much easier by modern information technologies, tablet and personal computers, apps and Internet portals. This is quite costly but catches the interest and transports the right kind and level of information. An example is shown in Figure 7.9 [38].

7.5 Conclusion

If one penetrates beyond superficial discussion, a more complex story emerges. Since there is no such thing as a one-for-all-type "nanomaterial," the typical nano-effect, or a specific nanotoxicology one has to assess the hazards of the specific material using data from safety data sheets, publications, guidelines, and analogies. This looks more complicated than it actually is, at least in the majority of cases, since the materials widely used very often have properties that can be estimated by interpolation from data of coarser fractions and of single molecules or crystal units. For example, the increase of the surface area per gram ratio is no surprise but well known for all kinds of solid matter. Also the penetration of particles being small enough through biological structures is not a new finding. Adverse effects in the human body may be caused by such particles, but not as a general principle. Problems correlated with the toxicology of the substance per se or the problems caused by overload effects with dusts are known. However, in some cases, one detects new

Figure 7.8: Pocketable booklet on safe handling nanomaterials in laboratories.

properties emerging from the nanoscale, quantum effects in small portions of matter, for example. Whether these may be a problem is not always known. So toxic quantum dots (many are based on cadmium compounds) will show toxic effects due to cadmium, and quantum effects could be responsible for the catalysis of chemical reactions. On the other hand, iron oxide nanoparticles ("nanomagnets") are useful tools in cancer therapy, even if there might be still also some unknown effect on the body. But it would be not reasonable, maybe even ethically unjustifiable to withhold a cancer treatment due to this (distant) possibility.

Were we to face nanomaterials with self-reproducing properties in the future, we would have to take into account these "biological properties" too. But exposure control for microorganisms is a well-established practice already and should not be too difficult to adapt to such nanomaterials.

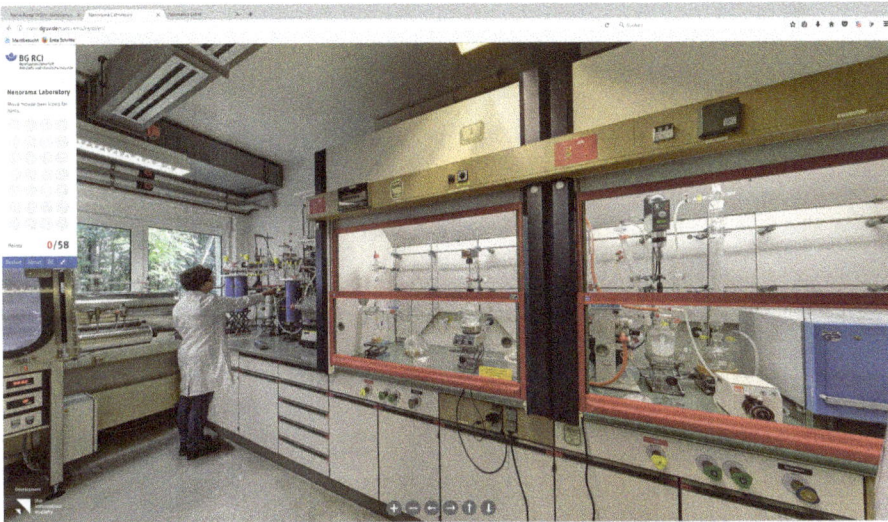

Figure 7.9: Web-based virtual laboratory for nanomaterials running in standard browsers which allows the user to walk around and investigate the different positions with information windows popping up and links to further reading material.

Minimizing the risks by exposure control apparently is the right strategy in most cases. The toolbox for exposure control is well equipped with strategies and technical solutions for controlling the risks with all kinds of substances, even the ones to be well known for their very dangerous properties. Measurement methods and devices have become available and help to clarify the situation at the workplace and to provide objective evidence to the discussion [39, 40]. Laboratories in research and development have taught us to cope with very toxic chemicals and new substances without data at all by using proper exposure control strategies – from the appropriate awareness of possible risks to the correct use of all protective measures. Decades of experience have shown that following these strategies the risks of accidents are comparably low to other workplaces. This leads to the conclusion that based on the actual state of knowledge, it is possible to work safely with nanomaterials and does not have to be no more complicated than working with other substances. In the light of the preceding discussion, it is clear that experiments with selected nanomaterials can be carried out even in school laboratories without posing risks neither to students nor to teachers. But the implementation of the appropriate measures in all relevant businesses and workplaces is *conditio sine qua non*. The established protective measures are generally well suited to reducing the possible risks to a responsibly low degree by minimizing exposure. Following these prudent practices and bestowing the appropriate care, one would not expect risks correlated with nanomaterials to be higher compared to other risks at workplaces.

References

[1]   Shatkin JA. Nanotechnology – Health and Environmental Risks. 2nd ed, Boca Raton, CRC Press, 2013.
[2]   ISO. Nanotechnologies – vocabulary – part 1: Core terms, ISO/TS 80004-1:2015, DIN CEN ISO/TS 80004-1: 2016-04.
[3]   COMMISSION RECOMMENDATION of 18 October 2011 on the definition of nanomaterial. Available at: http://eur-lex.europa.eu/legal-content/EN/TXT/?uri=CELEX:32011H0696. Accessed: 20 November 2015.
[4]   Committee on hazardous substances of the German Federal Ministry of Labour and Social Affairs: Announcement 527 "Manufactured nanomaterials". Available at: http://www.baua.de/en/Topics-from-A-to-Z/Hazardous-Substances/TRGS/Announcement-527.html;jsessio nid=C4F5BF7EE5E373563E80E35FA8461BC0.1_cid323. Accessed: 20 November 2015.
[5]   German Social Accident Insurance (DGUV). DGUV Information 213–851 Working safely in laboratories – basic principles and guidelines. Available at: http://www.guidelinesforlabora tories.de. Accessed: 23September 2020.
[6]   German Social Accident Insurance (DGUV). DGUV information 213–854 nanomaterials in the laboratory. Available at: http://www.dguv.de/medien/fb-rci/dokumente/info_zu_213_854.pdf. Accessed: 23September 2020.
[7]   Centers for disease control and prevention – national institute for occupational safety and health: safe nanotechnology in the workplace, 2008. Available at: http://www.cdc.gov/niosh/docs/2008-112/pdfs/2006-112.pdf. Accessed: 20 November 2015.
[8]   Centers for disease control and prevention – national institute for occupational safety and health: general safe practices for working with engineered nanomaterials in research laboratories, 2012. Available at: http://www.cdc.gov/niosh/docs/2012-147/pdfs/2012-147.pdf.Accessed: 20 November 2015.
[9]   US Food and Drug Administration. Considering whether an FDA-regulated product involves the application of nanotechnology, 2014. Available at: https://www.fda.gov/regulatory-infor mation/search-fda-guidance-documents/considering-whether-fda-regulated-product-in volves-application-nanotechnology. Accessed: 23September 2020.
[10] Italian Workers' Compensation Authority (INAIL). White Book, Exposure to Engineered Nanomaterials and Occupational Health and Safety Effects. Rome, INAIL, 2011.
[11] Australian Government Department of Health: Nanotechnology and therapeutic products, https://www.tga.gov.au/nanotechnology-and-therapeutic-products. Accessed: 23 September 2020.
[12] REACH nano consortium. Guidance on available methods for risk assessment of nanomaterials. Valencia 2015. Available at: http://www.invassat.gva.es/documents/161660384/162311778/01+Guidance+on+available+methods+for+risk+assesment+of+nano materials/8cae41ad-d38a-42f7-90f3-9549a9c13fa0. Accessed: 23 September 2020.
[13] Austrian Workers' Compensation Board (AUVA). Nanotechnologien. Vienna, 2012. Available at:https://www.auva.at/cdscontent/?contentid=10007.672853&portal=auvaportal. Accessed: 23 September 2020. [German language only]
[14] Ministry of Health, Labour and Welfare (Japan). Notification on precautionary measures for prevention of exposure etc. to nanomaterials, 2008. Available at: https://www.jniosh.johas.go.jp/publication/doc/houkoku/nano/files/mhlw/Notification_0207004_en.pdf. Accessed: 23 September 2020
[15] Institut national de recherche et de sécurité (INSR): Les nanomatériaux. Définitions, risques toxicologiques, caractérisation de l'exposition professionnelle et mesures de prevention,

Paris, 2012. Available at: http://www.inrs.fr/dms/inrs/CataloguePapier/ED/TI-ED-6050/
ed6050.pdf. Accessed: 23 September 2020

[16] Health and Safety Executive (HSE). Using nanomaterials at work, 2013. Available at: http://
www.hse.gov.uk/pubns/books/hsg272.pdf. Accessed: 23 September 2020.

[17] European Medicines Agency: Scientific Guidelines: Nanomedicines. Available at: http://www.
ema.europa.eu/ema/index.jsp?curl=pages/regulation/general/general_content_000564.
jsp&mid=WC0b01ac05806403e0. Accessed: 23 September 2020.

[18] Bouillard JX. Fire and Explosion of Nanopowders. In: Dolez PI editor, Nanoengineering –
Global Approaches to Health and Safety Issues. Amsterdam, Elsevier, 2015, 111.

[19] Brock TH. Occupational Safety and Health. In: Cornier J, Owen A, Arno K, Van de Voorde M
editors, Pharmaceutical Nanotechnology: Innovation and Production. Weinheim, Wiley-VCH,
2017, 331.

[20] DGUV: Ultrafeine Aerosole und Nanopartikel am Arbeitsplatz. Available at: http://www.dguv.
de/ifa/fachinfos/nanopartikel-am-arbeitsplatz/index.jsp. Accessed: 23 September 2020.

[21] European Chemicals Agency: REACH Guidance for nanomaterials published. Available at:
https://echa.europa.eu/de/-/reach-guidance-for-nanomaterials-published (in particular
https://echa.europa.eu/documents/10162/13643/appendix_r14_05-2012_en.pdf/7b2ee1ff-
3dc7-4eab-bdc8-6afd8ddf5c8d). Accessed:23 September 2020.

[22] Brock TH, Schulze B, Timm K unpublished.

[23] EN 14175: Fume cupboards – Part 1 & 2:2003, Part 3 & 4: 2004.

[24] Organisation for Economic Co-operation and Development: Harmonized tiered approach
to measure and assess the potential exposure to airborne emissions of engineered
nano-objects and their agglomerates and aggregates at workplaces. ENV/JM/MONO (2015) 19.

[25] Institute for occupational safety and health of the German social accident insurance
organisation (DGUV). Criteria for assessment of the effectiveness of protective measures.
Available at: http://www.dguv.de/ifa/Fachinfos/Nanopartikel-am-Arbeitsplatz/Beurteilung-
von-Schutzma%C3%9Fnahmen/index.jsp. Accessed: 23 September 2020.

[26] Van Broekhuizen P, Van Veelen W, Streekstra W-H, Schulte P, Reunderss L. Exposure limits
for nanoparticles: Report of an international workshop on nano reference values. Ann Occup
Hyg 2012, 56, 515.

[27] Schumacher C, Pallapies D Criteria for assessment of the effectiveness of protective
measures. Available at: https://www.dguv.de/ifa/fachinfos/nanopartikel-am-arbeitsplatz/
beurteilung-von-schutzmassnahmen/index.jsp. Accessed: 23 September 2020.

[28] Seipenbusch M, Binder A, Kasper G. Temporal evolution of nanoparticle aerosols in
workplace exposure. Ann Occup Hyg 2008, 52, 707.

[29] Brock TH. Safe Handling of Nanomaterials in the Laboratory. In: Vogel U, Savolainen K, Wu Q,
Van Tongeren M, Brouwer D, Berges M editors, Handbook of Nanosafety. Amsterdam,
Elsevier, 2014, 296.

[30] German Social Accident Insurance (DGUV). Sicheres Arbeiten in der pharmazeutischen
Industrie [German language only], Heidelberg, 2012.

[31] DGUV Laboratory Safety Expert Committee: Simplified labelling of laboratory containers.
Available at: https://www.bgrci.de/fachwissen-portal/topic-list/laboratories/guidelines-
forlaboratories/simplified-labelling-of-laboratory-containers/. Accessed: 23 September
2020.

[32] Shriver DF, Drezdzon MA. The Manipulation of Air-sensitive Compounds. 2nd ed, Hoboken,
Wiley-Interscience, 2017.

[33] Donaldson K, Poland CA, Murphy FA, MacFarlane M, Chernova T, Schinwald A. Pulmonary
toxicity of carbon nanotubes and asbestos – similarities and differences. Adv Drug Deliv Rev
2013, 65, 2078.

[34] Centers for disease control and prevention – national institute for occupational safety and health: current intelligence bulletin 65: occupational exposure to carbon nanotubes and nanofibers. Available at: http://www.cdc.gov/niosh/docs/2013-145/. Accessed: 23 September 2020.

[35] Lotz G. Arbeitsmedizinisch-toxikologische Beratung bei Tätigkeiten mit Kohlenstoffnanoröhren (CNT). Dortmund, Bundesanstalt für Arbeitsschutz und Arbeitsmedizin, 2015 [German language only].

[36] Roco MC. Nanoscale science and engineering: Unifying and transforming tools. AIChE J 2004, 50, 890.

[37] Kaku M. Physics of the Future. New York, Knopf Doubleday Publishing, 2011.

[38] DGUV Expert committee for dangerous substances. Available at: http://nano.dguv.de/en/nanoramas/, Accessed: 23 September 2020.

[39] Plitzko S, Meyer-Plath A, Dziurowitz N, Simonow B, Steinle P, Mattenklott M. Messung nano- und mikroskaliger faserförmiger Materialien an Arbeitsplätzen – Teil 1. Gefahrstoffe – Reinhaltung der Luft 2018, 78, 187 [German language only]. Available at https://www.baua.de/DE/Angebote/Publikationen/Aufsaetze/artikel1672.pdf. Accessed: 23 September 2020.

[40] Plitzko S, Meyer-Plath A, Dziurowitz N, Simonow B, Steinle P, Mattenklott M. Messung nano- und mikroskaliger faserförmiger Materialien an Arbeitsplätzen – Teil 2. Gefahrstoffe – Reinhaltung der Luft 2018, 78, 251 [German language only]. Available at https://www.baua.de/DE/Angebote/Publikationen/Aufsaetze/artikel1688.pdf. Accessed: 23 September 2020.

Ajey Lele and Kritika Roy
8 Militarization of nanotechnology

Abstract: The evolution of nanotechnology and nanoscience has been associated with bringing "the next industrial revolution" as it is capable of introducing better products and improvising manufacturing processes at several levels. Nanotechnology and nanoscience are interdisciplinary subjects and therefore, identifying even a minute detail about this technology continues to be a challenge. In all arenas, simultaneous work is in progress and accomplishments are at varying levels of success. In the military domain, nanotechnology could find several applications ranging from self-healing materials, medical aid (infection control) and nanoelectronics to sensing weapons of mass destruction and combatant protection kits (smart armor, active camouflage). With the dual-use nature of the technology, its impact on the offense–defense balance remains a matter of debate. This chapter explores the impact of nanotechnology on defense and concentrates on its current and futuristic applicability for military purposes. It also looks into the impact of nanotechnology on strategic stability and the way forward.

Keywords: Nanotechnology, dip-pen nanolithography, defense, depleted-uranium projectiles, risk

8.1 Introduction

The kind of advances science and technology is witnessing in the twenty-first century is nothing less than a sci-fi movie, especially with the cross-over of different technologies and the dual-use nature of most of them. Nanoscience and nanotechnology, in this context, are undergoing the emergence of an increasing number of new ideas and applications. These advances offer a unique opportunity for military applications which can indeed enable a class of lethal weapon that could alter the geostrategic landscape. This paper explores the impact of nanotechnology on defense and looks at its current and futuristic applicability for military purposes. It also looks into the impact of nanotechnology on strategic stability and the way forward.

Ajey Lele, Manohar Parrikar Institute for Defense Studies and Analyses, New Delhi, India
Kritika Roy, Hertie School Centre for International Security, Berlin, Germany

https://doi.org/10.1515/9783110669282-008

8.2 Origin of nanotechnology

Nanotechnology has been broadly defined as any technology that deals with size that is less than 100 nm. More specifically, nanotechnology refers to "designing and building machines in which every atom and chemical bond is specified precisely" [1]. In 1974, a professor at Tokyo Science University, Nario Taniguchi, brought forth the term "nanotechnology" to encapsulate machining in the 0.1–100 nm range.[1] He asserted that nanotechnology primarily involves the "processing, separation, consolidation, and deformation of materials by one atom or one molecule" [2].

In the early 1980s, nanotechnology was engendered by two researchers in IBM, Switzerland, who devised the scanning tunneling microscope, an equipment that produces extraordinarily thorough representations of electrically conducting atomic surfaces. The first evident example of nanotechnology was the spelling of the IBM logo in 35 xenon atoms which immediately became a popular symbol of nanoscale precision [3].

Nanotechnology is being developed at two stages:

8.2.1 Structural nanotechnology

It deals with minute structures, such as nanocrystals and complicated molecules, which is the focus of majority of nanotechnology researchers today. This field has recently become accepted as a field of research. However, the field is largely driven by applications in the commercial space. Additionally, structural nanotechnology can aid in optimizing existing procedures like speedy processors, more effective drugs, durable materials, and more efficient instruments and machines.

8.2.2 Molecular nanotechnology

This type has been named by Eric Drexler and is concerned with very small machines: robots, engines, and computers built atom by atom, smaller than a cell. Though this field is in its formative stages, the fears related to self-replication has already taken the center stage. However, with the current pace of development, it is difficult to contextualize the impact of obscure possibilities for their military utility. This paper does not look into such futuristic and scientifically uncertain aspects of nanotechnology.

Nanotechnology has often been associated with bringing "the next industrial revolution," as it is capable of introducing better products and improvising manufacturing

---

1 Under the most strict definition, nanotechnology is technology that operates anywhere within the nanometer length of scale. 1 nm is one-billionth (10^{-9}) of a meter. This is the realm of the atom, the smallest unit of an element.

processes at several levels [4]. Moreover, the core concept of nanotechnology can be integrated with any raw material and the potential applications are merely limited by imagination and research.

8.3 Current status of nanotechnology development

A two-pronged approach is being taken in advancing nanotechnology. First is the top-down approach, where researchers are improving existing techniques by shrinking the size of the already existing devices and machinery and incorporating recent technological developments like X-ray lithography and electron-beam writer. Additionally, nanotechnology can be used to produce nanoscale structures such as dip-pen nanolithography and incorporate nanoscale objects to larger objects to enable special functionality such as carbon nanotubes (CNTs) into electronic devices [5]. Second is the bottom-up approach, where the focus has been to "emulate nature by stimulating atoms and molecules to self-organize or self-assemble into complex systems that will function as devices" like cells in a human body. In this approach, the components arrange themselves by physical/chemical forces [6]. Currently, the research is trying to organize atoms and molecules in a particular manner to have nanomaterials of the required size and shape through controlled deposition on reaction parameters. This bottom-up means of curating "atom by atom" is still far from reality as for building electronic devices they cannot produce designed interconnected patterns.

Although each approach has its own challenges, the future practices might see a combination of both approaches that incorporates self-assembly into existing top-down technologies [7].

Nowadays, microelectronics can fabricate structures of about 100 nm. The age of nanofabrication, in a sense, is already here. Thus, it is often said that "the age of nanoscience has dawned, but the age of nanotechnology is finding practical uses of nanostructures" and is still in its formative stages [8]. Also, the development of nano-sized hybrid therapeutics and nano-sized drug delivery systems over the past decade has been remarkable. Implementing the use of knowledge and techniques of nanoscience in medical biology and disease prevention and remediation, that is, the utilization of nano-dimensional materials in nano-robots, in nano-sensors for diagnosis, delivery, and sensory purposes, and to actuate materials in live cells. For example, a nanoparticle-based method has been developed which combines both the treatment and imaging modalities of a cancer diagnosis [9].

Nanotechnology is an interdisciplinary subject and therefore, identifying even a minute detail about this technology continues to be a challenge. There are various physical, chemical, biological, and hybrid mechanisms available to incorporate nanomaterials. In all these arenas, simultaneous work is in progress and accomplishments are at varying levels of success.

8.4 Military applications of nanotechnology

Advances in nanotechnology would be enhancing various applications in exhilarating ways. Especially, in the military domain, nanotechnology could find several applications ranging from self-healing materials, medical aid (infection control), and nanoelectronics to sensing weapons of mass destruction and combatant protection kits (smart armor, active camouflage). With the dual-use nature of the technology, its impact on the offense–defense balance remains a matter of debate. Nanotechnology remains an evolving technology and many applications in the military domain are merely theoretical possibilities and require deeper research and development (R&D).

The following section explores the existing and futuristic nanotechnology developments and its impact on the military.

8.4.1 Electronics and computers

Computer and electronics is one area in which nanotechnology has gained considerable momentum. Nanotechnology applications will significantly enhance the performance of memory, solar-powered components, processors, and embedded intelligence systems while making them cost-effective [10]. Miniaturization would aid microprocessors to run faster, thereby facilitating speedy computations. Present techniques like photolithography, that is used for manufacturing chips to make structures smaller than 100 nm, is very expensive. There is an ongoing work to bring out more cost-effective options [11]. Additionally, the limitations of other technologies which are put together along with the nanotechnology-enabled equipment often undermine the progress made in the field of nanotechnology. It could be possible to fit nanotechnology-enabled electronic systems into a very small device, but power supply could pose a problem because battery size would not shrink in parallel. Thus, a micrometer-sized system would still need a centimeter-sized power supply. In cases such as large-distance communication by radio, huge antennas would be required [12]. In the near future, nanoelectronic systems may have to be produced within these limitations.

Nanotechnology could be the solution for the design and development of new forms of power sources. In general, nanotechnology has been anticipated to bring a revolution in the information technology (IT) sector and has been proposed to be the next wave of IT development in "pervasive computing."[2]

---

2 Pervasive computing implies an environment in which the dominant communications device is a descendant of today's smart phone, capable of serving as phone, broadband Internet device, video entertainment product, and accessing diverse sensor networks and databases. Like the current generation of broadband-connected desktops, the pervasive computing device will always be turned on and always hooked into cyberspace. It will bring the power of the broadband communications from the soldier to the shopper.

8.4.2 Sensors

Nanotechnology has larger utility in sensor manufacturing as it facilitates the design and production of miniature sensors of the size of micrometers instead of centimeters [13]. Nanomaterials-enabled smart technologies and advanced fabrication technology in electronics intersect toward creating a new set of sensors which can be used to enhance military intelligence gathering, damage detection, or border management.

Sensors made from nanocrystalline materials are highly susceptible to environmental changes. Applications for such sensors are detectors for smoke and ice on wings of aircraft, automobile engine performance sensors, and so on [14]. Nanosensors are also an innovative way to measure the air and water quality due to their size, quickness, and accuracy of measurements especially in areas where there is threat of chemical or biological attacks [15]. An example of this is detecting mercury in any medium (such as air and water) through the use of dandelion-like Au/polyaniline nanoparticles in conjunction with surface-enhanced Raman spectroscopy nanosensors [15]. A device for the detection of nerve gas agents in the atmosphere has also been developed based on nanotechnology applications [16]. This technology has been found useful in the production of mass spectrometers that are used to track biological and chemical warfare agents [17].

Toxicity sensor is another technology capable of having a single-cell microchip platform. This technology allows the insertion of molecular targets into the cell or the cell can simply be exposed to the environment while monitoring continuously for cell death – the readout is direct and virtually instantaneous. This platform could be used for medical and bio-warfare applications [18].

Cells, the powerhouse of life activities and the interactions of protein surfaces, are measured in nanometers. Many nations are researching on miniature machines and tools that can enter the human body. This is the "millionth-of-a-millimetre world" of biotechnology today. By using a person's saliva, body fluids, or blood, nanobiosensors can be created to reliably work with pathogens such as viruses and could be used for disease diagnosis [19].

In tissue engineering, a scaffold, measuring only 50 nms in diameter, can be built using nanofibers. The cost of drug and virus development be reduced by using nanochips to test various medications or a combination of chemicals and vaccines. Nanotechnology is demonstrating enormous ability in the inception of different direct and indirect applications useful for bio-defense purposes. Additionally, progress has been reported in various design methodologies for CNT-based biosensors and their employment for the detection of a number of biomolecules are being researched [20].

8.4.3 Stealth and camouflage

Stealth and camouflage play an important role in military affairs as it gives a strategic advantage in the battlefield. The employment of stealth technology in warship and aircraft construction to reduce radar cross-section, magnetic signature, and the infrared signature has been the ongoing endeavor of the designers. The advent of nanotechnology has given new concepts in the field of stealth technology [21].

Another feature is the "adaptive camouflage" where the material surface changes external appearance in response to a preprogrammed stimulus in the environment in which it works. Moreover, to attain adaptive camouflage, the material surface is enshrouded with thin plastic sheets, which have numerous embedded light-emitting diodes (LEDs). The colors and patterns displayed on the sheet are controlled by the LEDs with inputs from a camera which scans the surroundings. This adaptive change of color and pattern can be used as an effective deception tool in tactical situations. Several nanotechnology-enabled techniques are in the R&D stage and can be applied for adaptive camouflage.

8.4.4 Maritime applications

Nanotechnology has many potential applications in the maritime domain. Currently, there has been development in the next-generation all-electric warship that could revolutionize the navy's use of weaponry and manpower. Here, designers and manufacturers depend majorly on nano- and microelectronics technology. This technology is likely to be a critical component of a ship's system architecture. "These micro and nanoscale electronic packages are likely to maintain reliability under extremely harsh conditions resulting from concurrently acting vibrations, high-current density, high-power and high-temperature loads" [22].

Research is also underway to incorporate nano approaches into energetic materials on both the material-development- and material-production fronts. Products such as functionally graded nanocomposites[3] are examples of the potential for nanotechnology to bring innovations from the bench to the fleet [23]. Scientists believe that nanoparticles can be used to mark ships, fishing boats, navigable channels, and delimiting safe havens [3]. The crystals are soluble in paints, fuel, lubricants, specialty chemicals, glues, and so on but cannot be easily counterfeited, removed, or altered by anyone except the authorized agency which designed them.

---

3 Functionally graded nanocomposites offer higher wear resistance and higher toughness.

8.4.5 Space and other defense applications

Today, the new means of warfare necessitates accurate delivery of forces, while reducing the collateral damage. This can be facilitated by effective utilization of sensors and IT. This is further facilitated by the availability of durable, lightweight structural materials, and dependable explosives and propellants that release greater energy.

In satellite manufacturing, less vulnerable corrosive materials seem helpful. Nanostructural materials show tremendous promise for structural applications. Nanocomposites have already made their way into automobiles and are achieving 10% to 15% weight and strength improvements, with a promise of 20% to 25% of further improvements [24]. Also, such structural materials and the miniaturization are likely to play a vital role in the development of next generation of unmanned aerial vehicles/unmanned combat aerial vehicles (UAV/UCAVs) [25].

The current breed of satellites uses thruster rockets either to remain in orbit or to change orbit. This becomes necessary because of a variety of factors, including Newtonian compulsions. The amount of fuel they can carry onboard determines the life of these satellites. However, more than one-third of the fuel that is carried by the satellites is wasted by these repositioning thrusters due to incomplete and inefficient combustion of the fuel, such as hydrazine. This is because the onboard igniters wear out quickly and cease to perform effectively. Nanomaterials, such as nanocrystalline tungsten–titanium diboride–copper composite, are potential candidates for enhancing these ignitors' life and performance characteristics [25]. Apart from onboard fuel, satellites use solar power as a power source for various activities. Nanotechnology offers efficient way to the satellite designers to reduce the weight of such solar cells.

Nanomaterials are seen as alternative materials to their conventional counterparts by defense and space scientist. Lighter nanoporous materials like aerogels[4] [26] have wider applicability in spacecraft manufacturing and in the defense industry. Aerogels can be used to make special lightweight suits, jackets, and so on. With the advances in technology, military dependence on space assets is likely to increase as space-based systems would majorly control the communication and navigation aspects. Understanding of space weather in the near-earth and solar space environment would be crucial at any point of time. Nanostructured sensors are expected to play a fundamental role in obtaining information on the ionosphere and other regions of space [27].

With respect to the threats of chemical weapons terrorism, nanotechnology offers solutions against the usage of highly dangerous chemical agents like nerve

---

4 Aerogels are a low-density solid-state material derived from gel in which the liquid component of the gel has been replaced with gas. The result is an extremely low-density solid with several remarkable properties, most notably its effectiveness as an insulator. They are porous and extremely lightweight, yet they can withstand 100 times their weight. Various other techniques on similar lines are also being worked on.

agents or VX, GB, and so on. Nanoparticle oxides like CaO, Al2O3, and MgO interact with these chemicals much faster than microparticles and are suited for the rapid disintegration of these chemicals [28].

Depleted-uranium (DU) projectiles (penetrators) are known for their lethality against hardened targets and enemy armored vehicles. This is because DU has residual radioactivity, and hence is toxic (carcinogenic), explosive, and lethal to mankind and hence there has often been some discomfort towards their usage. However, there is no alternative for the use of DU penetrators as they possess a unique self-sharpening mechanism on impact with a target. Nanocrystalline tungsten heavy alloys, because of their unique deformation characteristics, such as grain-boundary sliding, lend themselves to such a self-sharpening mechanism. Hence, nanocrystalline tungsten heavy alloys and composites are seen to be potential alternatives to DU penetrators [14].

8.4.6 Conventional weapons/ammunition

Nanotechnology-based durable and lightweight materials would facilitate the development of conventional barrel-type weapons with reduced mass. It is envisaged that for miniature arms and lightweight weapons, barrels, locks and so on could be made of nanofiber composites. Even in ballistic and air-breathing missiles, the reduced mass would proportionally increase the speed, precision, or payload. It has also been estimated that nanotechnology-improved explosives and propellants would enter military domain within a decade's time [29].

8.5 Military investments in nanotechnology: marking the global progress

It is challenging to extract and compare the R&D investments made by each country as each nation employs a different basic structure and method of data aggregation in its policies on science and technology and industrialization. The following section provides a brief overview of global progress.

The US military has been involved in nanotechnology research since the early 1980s [30]. In the mid-1990s, the Department of Defence (DoD) identified nanotechnology as one of six "Strategic Research Areas" of interest.[5] Since the last decade, research and monitory investments by the DoD in the nanotechnology arena have become

---

5 The other five are bioengineering sciences, human performance sciences, information dominance, multifunction materials, propulsion, and energetic sciences.

more significant. The DoD nanotechnology program is grouped into seven programme component areas (PCAs),[6] which mirror the PCAs of the US National Nanotechnology Initiative (NNI) [31]. Currently, the President's 2020 budget requests over 1.4 billion USD for continued research, that is, cumulatively totaling nearly 29 billion USD since the inception of the NNI [30]. The "Institute for Soldier Nanotechnologies" at the Massachusetts Institute of Technology is also in place since 1998, which is doing R&D in several army-related nanotechnology issues [32]. DoD is investing in nanotechnology to advance both offensive and defensive military objectives. It is also looking at nanotechnology as a base technology for the production of soldier protection kits [33].

Additionally, nanotechnology being a dual-use technology, therefore, the advances made in the scientific and commercial arenas under civil funding are likely to find their way into defense applications. Even in the US, huge investments are being made in nanotechnology sectors dealing with issues other than defense. President Bush signed the twenty-first century nanotechnology Research and Development Act into law in the year 2003 [34].

European Union's (EU) technological framework, Horizon 2020, positions nanotechnology and advanced material technology as one of the key-enabling technologies . EU's total public spending on nanotechnology, when combining all its member states, rose to approximately 2.5 billion USD in 2014 and continued to increase by 9.8% in 2016. Other countries like Germany, France, the United Kingdom, and Russia [35] are investing in R&D of nanotechnology-based materials and systems for military utility. However, most Asian and European countries, with the exception of Sweden (a Swedish defense nanotechnology programme exists), do not run dedicated programs for defense nanotechnology research. Rather, they integrate several nanotechnology-related projects within their traditional defense-research structures, for example, as materials research, electronic devices research, or bio-chemical protection research [36].

China has invested heavily in nanotechnology in the past decades. It is one of the key areas of focus in the medium and long-term scientific programs between 2006 and 2020 [37]. China's National Centre for Nanoscience and Technology has more than 3,000 scientists working on various aspects of nanotechnology [38]. Another noteworthy facet has been the integration of elemental technologies of nanotechnology and materials with advances in artificial intelligence and machine learning. In Asian countries such as China, Taiwan, Korea, and Singapore, R&D facilities for nanotechnology were established to attract the world's R&D. In particular, China's huge R&D investment in this field is reflected in the recent rapid increase in academic papers [39].

---

6 The seven programme component areas are (1) fundamental nanoscale phenomena and processes, (2) nanomaterials, (3) nanoscale devices and systems, (4) instrumentation research, metrology, and standards for nanotechnology, (5) nanomanufacturing, (6) major research facilities and instrumentation acquisition, and (7) societal dimensions.

The United States, China, and Germany are the top owners of health-related nanotechnology patents filed internationally post 1975, holding 33%, 20%, and 13% of the total, respectively. However, despite being a strong leader in the field, China is not participating in international debates on the role of nanotechnology in sustainable development [40]. It is not possible to accurately comprehend China's military investments in the arena of nanotechnology, but going by China's past record there is a case to believe that it would have an interest in a military nanotechnology program as well.

India is carefully investing in the developments of nanoscience and nanotechnology. The Government of India initiated a Nanomaterials Science and Technology Mission in the tenth five-year plan. In accordance with this, the Department of Science and Technology started working toward evolving a framework for a national initiative in this field. India formally launched the Nanomaterials Science and Technology Initiative (NSTI) in October 2001.

For India, the field of nanotechnology is an arena of multidisciplinary research. The Department of Biotechnology also funds various projects on nanobiotechnology [41]. India has allocated INR 1,000 crores in the eleventh five-year plan for R&D in this field and every year, INR 200 crores would be allocated for this purpose. Various sectors of the Indian industry are investing in it in tune with their interests and requirements. However, Indian investments in nanotechnology are not very encouraging. In a report, Assocham noted that, the country is way behind many developing as well as developed economies in terms of R&D spending on nanotechnology, with annual investments touching a minuscule of 7 million USD, which is even below than that of Taiwan, which spends 104 million USD annually. In contrast, Germany spends 395.5 million USD, France 301.1 million USD, and the United Kingdom 180 million USD [42].

In the defense arena, India's Defence Research and Development Organization (DRDO) is working on areas like sensors, high-energy applications, stealth and camouflage, nuclear, biological, and chemical (NBC) attack protection devices, structural applications, nanoelectronics, and characterization. Presently, the focus has been on developing various types of sensors, NBC protection/detection devices and developing paint with camouflage characteristics and also integrating the technology to strengthen the Make in India initiative [43].

8.6 Nanotechnology and the future of strategic stability

Presently, the various aspects of the relevance and future of nanotechnology are being debated. Strategic stability has proven to be a tool for mitigating the risks of nuclear weapons. However, with technologies advancing at a rapid pace and their crossover which each other has brought this dynamic under the scanner. For instance, devices and component materials, which will play a key role in the coming Internet of Things/AI era, would be made of a lump of nanotechnology.

Imagine a nano-drone endowed with cameras and sensors to facilitate facial recognition and target detection along with a small quantity of shaped explosives. This miniaturized drone weapon would not only have the ability to react a hundred times faster than a human, but also could facilitate strikes of surgical precision. A swarm of such drones can unleash havoc penetrating critical infrastructures, or by just profile targeting the bad guys. The blend of miniturization and integration would obviously give nations strategic advantage and would pave the way for rapid development and further advances. All this essentially indicates that these weapons could be destabilizing in nature. This may lead to an arms race or just send the current strategic stability dynamics for a toss.

This has opened debates and deliberations among the experts. One opinion is that this technology is still nascent and may take a few more decades to mature. Hence, it is the correct time to ask for banning such technologies that may change the notion of deterrence in the foreseeable future. Additionally, there could be a possibility that the military doctrines of different countries could identify specific tasks and targets for such technology – like precision target attacking critical infrastructure. Also, there is a possibility that any use of, or threat to use such technology could increase the chances of war. At times, the opponent could even be forced to take irrational decisions. With this technology maturing, there exists a further possibility that such technology could facilitate the creation of weapons that may be capable of carrying tactical nuclear warheads.

With the USA and China pursuing the nanotechnology program with great vigor, it is but obvious that the major European and Asian states would accelerate its efforts to develop such technology. It is unlikely that this technology will be stifled before becoming fully operational in its weapons avatar. It has been estimated that this technology promises to revolutionize military affairs in a significant way. Although the cost has been a major factor in the development of nanotechnologies. However, the present trends, indicates that major states are keen to weaponize this technology and are investing in that effect.

8.6.1 The way forward

Applications of nanotechnology in the military domain are here to stay. States are likely to spend more on advancing this technology for reaping the benefits of the dual-use nature of this technology. Nanotechnology offers vital defense applications like the development of various sensors, soldier protection kits, improvement in command, control, communications, computers, intelligence, surveillance, and reconnaissance structures, and so on. Some areas of military motivated research also offer wider benefits in the civilian sphere. Investments in design and development of more powerful but lightweight batteries, smart fabrics, and so on could be a case in point.

Apart from materials and sensors/electronics, nanotechnology also has direct military applicability towards making imperishable armor, miniaturized surveillance devices, improved performance of UAVs/UCAVs and enhanced interfacing and targeting for soldiers and fighter/bomber pilots. Particularly for states like India, nanosensors would have great potential for real-time border management and surveillance through wireless networking.

Considering the dual-use nature of nanotechnology that can facilitate the wellbeing of human while at the same time become crucial war weapons, International cooperation and understanding is the only way forward. Although, the World Health Organization and the United Nation Industrial Development Organization's International Centre for Science and High Technology have initiated dialogues for regional networking between nations, scientific community and industry to begin work on nanotechnology related safety and governance issues [44]. However, a dedicated initiation is required by all the major powers as well as defense scientists and engineers working in the field of nanotechnology to come together and bring about an effective and globally acceptable policy for nanotechnology. Otherwise, an arms race could become inevitable in the foreseeable future.

References

[1] Storrs H. Nanofuture. New Delhi, Manas Publications, 2006, 21.
[2] Taniguchi N. On the basic concept of nano-technology, Proceedings of the International Conference on Production Engineering, Tokyo, Part II, Japan Society of Precision Engineering, 1974.
[3] Sean Randolph by the Bay Area Science and Innovation Consortium (BASIC). San Francisco, Nanotechnology in the San Francisco Bay Area: Dawn of a New Age, 2005, 9.
[4] Center for Responsible Nanotechnology on Meaning of Nanotechnology. (Accessed January 22, 2020 at http://crnano.org/whatis.htm)
[5] Sarangan A, in Fundamentals and Applications of Nanophotonics, 2016. (Accessed January 22, 2020 at https://www.sciencedirect.com/topics/chemistry/nanofabrication-process)
[6] Altmann J. Military Nanotechnology. London, Routledge, 2006, 20.
[7] Kulkarni S. Nanotechnology: Principles and Practices. New Delhi, Capital Publishing Company, 2007, Chari S., Info-Nano-Bio Technologies, Their Coming Convergence, and the Implications for Security, NIAS Report, 2003, p. 79.
[8] Whitesides G, Love C. The art of building small. Sci Am September 2001, 17, 40.
[9] Haba Y, Kojima C, Harada A, Ura T, Horinaka H, Kono K. Preparation of poly (ethylene glycol)-modified poly (amido amine) dendrimers encapsulating gold nanoparticles and their heat-generating ability. Langmuir, 2007.
[10] Foresight Institute, Nanotechnology Challenges and Problem. American Chemical Society (Accessed January 25, 2020 at http://www.foresight.org/challenges/infotech.html)

[11]   Whitesides G, Love C. The art of building small. Sci Am September 2001, 17, 12.
[12]   Altmann J. Military Nanotechnology. London, Routledge, 2006, 73.
[13]   Altmann J. Military uses of nanotechnology: Perspectives and concerns. Secur Dialogue 2004, 35, 1, 67.
[14]   Manasi K. Nanotechnology: Fundamentals and Applications. New Delhi, I.K, International Publishing House, 2008, 56.
[15]   Dahman Y. Nanosensors. In: Nanotechnology and Functional Materials for Engineers. Canada, Elsevier, 2017, 67–91.
[16]   Engineers Develop Biowarfare Sensing Device, Nanotechnology Now. (Accessed January 26, 2020 at http://www.nanotech-now.com/news.cgi?story_id=07929)
[17]   Larguinho M, Baptista P, Capelo J. Nanoparticles for Mass Spectrometry Applications. 2016. (Accessed January 26 2020 at https://www.researchgate.net/publication/292612596_Nano particles_for_Mass_Spectrometry_Applications)
[18]   Lele A. Strategic Technologies for the Military: Breaking New Frontiers. New Delhi, Sage Publications, 2009.
[19]   Prasad S. Nanobiosensors: The future for diagnosis of disease?, 26 March 2014. (Accessed January 26, 2020 at https://www.dovepress.com/nanobiosensors-the-future-for-diagnosis-of-disease-peer-reviewed-fulltext-article-NDD)
[20]   Cereijo M. Cuba's Killer Virus and New Nanotechnology. (Accessed January 27, 2020 at http://www.amigospais-guaracabuya.org/oagmc087.php)
[21]   Potential Applications of Nanotechnology in Maritime Environment (Accessed January 28, 2020 at https://www.ukessays.com/essays/engineering/potential-applications-of-nanotechnology-in-maritime-environment-engineering-essay.php)
[22]   Contrada J. New Technology Aids Navy. (Accessed January 28, 2020 at http://www.voyle.net/Nano%20Defence/Defence%202004-0022.htm)
[23]   Kavetsky R. Energetic Systems and Nanotechnology – A Look Ahead. Arlington, VA, Office of Naval Research, Accessed January 28, 2020 at, http://www.cecd.umd.edu/pdf/energsys.pdf.
[24]   Nanotechnology in the Aerospace and Defence Industry – Factors Driving Nanomaterial Developments. (Accessed January 28, 2020 at http://www.azonano.com/details.asp?Arti cleID=592)
[25]   Roco M, Bainbridge W eds., Nanotechnology: Societal Implications. The Netherlands, Springer, Dordrecht, 2007, 82–83.
[26]   Gash AE, Simpson RL, Satcher JH. Direct Preparation of Nanostructured Energetic Materials Using Sol-Gel Methods. In: Miziolek AW, Karna SP, Mauro JM, Vaia RA eds., Defense Applications of Nanomaterials. Washington, DC, ACS Books Department, 2004, 198–210.
[27]   Dressler RA, Ginet GP, Williams S, Hunt B, Nikzad S, Stephen TM, Karna SP. Nanotechnology Challenges for Future Space Weather Forecasting Networks. In: Miziolek AW, Karna SP, Mauro JM, Vaia RA eds., Defense Applications of Nanomaterials. Washington, DC, ACS Books Department, 2004, 46–62.
[28]   Kulkarni SK. Nanotechnology: Principles and Practices. New Delhi, Springer, 2015, 11, 258–59.
[29]   Altmann J. Military Nanotechnology. London, Routledge, 2006, 81–82.
[30]   United States Defense Nanotechnology Research and Development Program, 2009. (Accessed January 28, 2020 at https://www.nano.gov/sites/default/files/pub_resource/dod-report_to_congress_final_1mar10.pdf).
[31]   Roco MC. National Nanotechnology Initiative – Past, Present Future. Handbook on Nanoscience, Engineering and Technology. 2nd London, Taylor and Francis, 2007.
[32]   Altmann J. Military Nanotechnology. London, Routledge, 2006, 56.

[33] Kulinowski K. Nanotechnology: From "Wow" to "Yuck. In: Hunt G, Mehta M eds., Nanotechnology Risk, Ethics and Law, Earthscan. London, Taylor & Francis, 2006, 18.

[34] United States twenty-first Century Nanotechnology Research and Development Act, Public Law [108–153]. (Accessed January 30, 2020 at https://nifa.usda.gov/resource/21st-century-nanotechnologyresearch-and-development-act)

[35] Towards a European Strategy for Nanotechnology. European Commission Committee report, 2004. (Accessed January 31, 2020 at https://ec.europa.eu/research/industrial_technologies/pdf/policy/nano_com_en_new.pdf)

[36] Berger M. Military Nanotechnology – How Worried Should We Be? (Accessed January 31, 2020 at http://www.nanowerk.com/spotlight/spotid=1015.php)

[37] Qiu J. Nanotechnology development in China: Challenges and opportunities, National Science Review, March 2016, 148–52. (Accessed February 01, 2020 at https://doi.org/10.1093/nsr/nww007)

[38] China invests more than 800 mln yuan into nanoscience. People's Daily. (Accessed February 02, 2020 at http://english.people.com.cn/200506/10/eng20050610_189657.html).

[39] Towards a European strategy for nanotechnology. European Commission Committee report, 2004. (Accessed January 31, 2020 at https://ec.europa.eu/research/industrial_technologies/pdf/policy/nano_com_en_new.pdf).

[40] Nanotechnology and China, Responsible Nanotechnology. (Accessed February 02, 2020 at http://crnano.typepad.com/crnblog/2005/11/nanotech_and_ch.html)

[41] DST programmes boosting Make in India initiative (Accessed January 28, 2020 at http://dst.gov.in/about_us/ar01-02-sr-serc.htm) and based on a background note prepared by the Confederation of Indian Industry (CII) for its EmergeTech Conclave,September 28, 2006, New Delhi.

[42] India's nanotech spending below global levels, Economic Times. (Accessed January 30, 2020 at https://economictimes.indiatimes.com/news/economy/indicators/indias-nanotech-spending-below-global-levels/articleshow/2359978.cms)

[43] Mantha S, Vathsal S. The emission frequency of NANO DOTS. Int J Mater Res Sci 2007, 2, 1, 41–44.

[44] Sparls S. Nanotechnology: Business Applications and Commercialization. USA, 2012CRC Press.

Thomas Reuter

9 Nanotechnology in food: Ethics, industry practices, and regulatory frameworks

Abstract: What are the ethical implications of nanomaterials in food systems, given the potential of such a material to cause harm to human health and the environment? Following an outline of relevant ethical principles, this chapter charts the current use of nanomaterials in food and what we do and do not know about the risks associated therewith. Regulatory frameworks are then examined for their ability to mitigate risks.

Three recommendations are put forward. First, it is best to avoid all unnecessary food processing categorically; second, nano-processed food products should only enter the market when harmful impacts can be categorically ruled out on the basis of independent and in-depth research and where benefits are very significant; and finally, complete transparency on the use of nanomaterials and other additives is needed so that consumers can exercise individual discretion regarding their own exposure to nano-food products, even if they are safe, and the more so while any doubts remain about their safety.

Overall, the trend of the largely profit-driven global food industry has been and is still toward hyper-processing – despite consistent warnings of health professions about hyper-processed food. Nanotech takes this trend to a new level. Current voluntary producer ethics do not even guarantee transparency, let alone safety, except in jurisdictions where legislation demands it. While some nanomaterials may be beneficial and safe for some applications, industry self-regulation is not viable under these circumstances.

While regulations have been strengthened in some jurisdictions such as the European Union, regulators still struggle to catch up with the rapid development and application of ever-new nanotech products by the food industry. A restructure of our innovation systems is recommended so that all stakeholders are included in shaping its future direction from the start.

Keywords: engineered nanoparticles, ecologies, consumer, guidelines, ethics

A contribution to the book *Nanoethics: Giving Orientation to Societal Reflection* by Frans Brom, Rinie van Est, and Bart Walhout.

Thomas Reuter, World Academy of Arts and Science, University of Melbourne, Germany; University of Bonn, Germany

https://doi.org/10.1515/9783110669282-009

9.1 Introduction

The increasing presence of engineered nanoparticles (ENPs) in food is a problem that should first be approached from an ethical perspective. Ethics comes first and must not be reduced to an afterthought. In other words, we must look before we leap. This chapter therefore begins by outlining some of the relevant ethical principles that ought to be fully implemented in an ideal case scenario. The actual practice of contemporary nanotechnology use in food is then examined, together with the limited knowledge we have so far of the health consequences for consumers and environmental impacts. The aim is to inform the public not just of proven risks but also of "known unknowns" that render risk incalculable. We will then look at some of the gaps in monitoring regulations and labeling that limit consumer protection and/or the exercise of conscious consumer choice. A second aim is to provide the food industry with a straightforward and easy-to -understand ethical framework for evaluating their current use and future development of nanomaterials or, for that matter, any other innovative food processing technologies.

First, however, there is some need for transparency regarding the author's position. My background is in social science and my relevant expertise more specifically is derived from intensive ethnographic study of the changing moral economies of food systems in developing countries, and the role of moral economy in regulating the social impact and ecological sustainability of different regional food systems [1, 2]. This research has examined food systems in a highly holistic manner, showing them as systems with a specific history, politics, and culture of food production, distribution, and consumption, all with important implications for food security, food justice, diet, nutrition, and public health. Food systems are also a very central component of all "human ecologies." Human ecologies are locally diverse systems of human engagement with the environment that may or may not be sustainable, and thus can be in various states of degradation, stability, or regeneration. Other research has focused on the political economy of the global food system and the strategies and agendas of dominant actors therein, largely from the perspective of the impact of this system on consumers and the environment [3, 4]. I am not an expert or personally invested in nanotechnology research as such, but have a good knowledge of chemistry from an earlier career. While I am not the ideal person to discuss the technical intricacies of this fast-moving field, this is not an impediment, I trust, insofar as the aim of this book is precisely to explore the space of convergence between nanotechnology, a highly specialized and correspondingly esoteric field in its own right, and the wider industries in which it may be applied, often in tandem with a host of other disruptive technologies. An even wider convergence space is where new technologies such as nanoscience impact on human society and the world as a whole, a space that must be critically examined, not only as a matter of ethics but also in order to gage the genuine commercial viability of any innovation. The food industry is a pertinent example of a space in which such layered convergence prevails, and where more reflection, risk assessment, and accountability are required urgently if unintended harm is to be avoided.

Like other industries, the food industry – though participants may well experience it as a separate world of its own, with its own distinct internal logic of efficiency, quality control, productivity, and, above all, profitability – exists within a wider society that has little time for such insider concerns and is instead more interested to know whether the industry is acting in conformity with the law and in the best interest of consumers and the environment. With regard to the "spaces of convergence" in which nanotech exists, my research experience as a social scientist is very relevant. My personal ethical commitments, meanwhile, are to scientific neutrality and to advocacy for the victims of unethical actions by any party.

9.2 Ethics first! A framework for the ethical use of (nano)technology in the food industry

The twofold purpose of industrial ethics is to encourage reflection on issues surrounding the impact of technology currently utilized, and to give direction to the development or non-development of future technologies. The second purpose is to look toward the future, and this underlines how vital it is to think of ethics first, rather than after the creation of facts on the ground. For example, while the tobacco industry could profit from retrospectively engaging in ethical reflection, unimaginable and irreversible harm has already been done. In other words, while it is sometimes necessary to reflect on past wrongs, it is fundamentally unethical to engage in ethics after the fact. Genuine ethical action is necessarily forward looking, encouraging us to act cautiously and responsibly, rather than being retrospective, reactive, and focused on remedial action when things have gone wrong. This is why, in this chapter, the facts of nanotechnology use in food and associated risks will be examined second. The first task is to provide a framework that may assist in turning around established industry approaches to ethics: from being largely retrospective and reactive to become firmly visionary and proactive. Such a shift will not only enhance reputation but also secure long-term commercial success.

9.2.1 The precautionary principle

Legal systems and regulatory regimes differ around the world, and not all embrace a precautionary principle which, in the case of food, obliges a food producer to ensure that all their products have been proven to be harmless to consumers' health, or "safe to consume" before they enter the market. Even where such legislation exists, as in the case of the European Union (EU), the way this legal principle is defined is not entirely stringent. For example, a principle of *long-term harmlessness* is rarely considered. Long-term harmless means, in this case, that a food is not just safe to eat from a toxicological perspective, but that the natural nutritional value of a food has not been destroyed by processing or compromised by excessive storage or with inappropriate additives such as sugar, salt, or chemical additives. Food compromised with processing and additives, while it may qualify as passively harmless (not toxic in an acute sense), can be harmful to long-term health or even fatal if routinely and liberally consumed. Ideally, farmers and the food industry should strive to ensure the food they place on the market is harmless in the long term, and better still, actively health promoting.

In general, a precautionary principle is vital because producers are thereby compelled to take moral responsibility for any products they advertise and sell. Some jurisdictions have legislated this principle, which means that criminal liability is added to moral responsibility. Others lack such legislation and rely solely on the deterrence effect of post-harm compensation litigation, which is limited because large corporations have more resources to fight costly legal battles than affected consumers. Consumer health should not just be a legal concern, in any case. It should be a priority to any enlightened company leadership. Such leaders recognize the growing trend toward a multi-stakeholder-based mode of doing business, as opposed to a single-minded focus on profit, that is, on executive and shareholder benefits only. Businesses can be expected to do well and be future proof only if they realize that in recent years the financial industry (with support from the progressive vanguard of the corporate world), whether under the influence of public pressure or out of a growing sense of social and ecological crisis and felt responsibility for that crisis, is increasingly embracing a multi-stakeholder model for evaluating company performance. Today a poor social and ethical responsibility rating can make a company unattractive to investors and partners alike [5].

9.2.2 The transparency principle

The exercise of transparency is ideally a supplement to precaution only, but is necessary nonetheless due to the fact that most scientific assessments of food safety are by nature probabilistic. If a product is statistically very unlikely to cause harm it may be deemed harmless for all practical purposes by producers, researchers, and regulators alike. Consumers, nevertheless, have the right to apply a stricter personal standard of risk avoidance, if they choose to do so. Consumers thus must be given access to all information they require to avoid food additives that, in their opinion, pose too great a residual risk. In the absence of any fully conclusive assessments (below), this transparency principle would seem mandatory for all ENPs already present on the food market today.

The food industry produces and sells many processed food products today that, as medical science has proven beyond reasonable doubt, are harmful or outright deadly for consumers who are overly reliant on such foods in their daily diet. Such over-reliance is often not a choice but a consequence of the demands of work and other modern lifestyle pressures on people's time or their inability to access or afford more unadulterated, healthy foods. The justification of this industry practice, following an ethos of liberal individualism, is that consumers have every right to ingest substances with their food that may destroy their own health, if they find it pleasurable or convenient to do so. This argument certainly cannot provide ethical cover for acts of deliberate deception by producers. Moreover, the failure to convey pertinent information to consumers as to the content and hence the health risks of a food product needs to be regarded as deception by omission.

The sad reality is that failure to inform and even active deception are widespread practices in the food industry today. Apart from the numerous successful consumer complaints recognized in courts of law, consumer watchdog organizations continue to expose shocking behavior. The annual award of *Food Watch* for the most recklessly deceptive labeling and advertising of a food product is one example, providing ample testimony to this kind of malpractice by reckless producers.[1] The entire industry shares in the reputational damage caused by these worst cases. Whereas a high standard ideally should be assumed to be the industry norm, honest producers of healthy food are instead forced to spend a lot of money to prove their adherence to ethical standards, for example, by qualifying for organic certification. Along with consumers, they too are victims of unfair competition in jurisdictions that fail to monitor and penalize malpractice.

Food nanotech must be situated in this wider context, because the large manufacturers of hyper-processed food, along with some mass producers and distributors of fresh food, are most likely to adopt nanotechnology, and may do so with insufficient precaution and transparency. Note that while transparency empowers consumers and blatant deception disempowers them radically, there is a large gray area in between where consumers are partially informed with the express purpose to displace responsibility onto them. This is a subtle and dangerous game that leaves the consumer semi-informed but (presumably) fully self-responsible.

9.2.3 The principle of minimalism

Innovation for the sake of innovation, without foresight and in the absence of a clear vision of the future we wish to create, is exactly what has led humanity to a current multidimensional ecological crisis, wherewith the extinction of our species

---

1 See https://www.foodwatch.org/en/foodwatch-international/; and specifically on deceptive labelling (in German): https://www.foodwatch.org/de/aktuelle-nachrichten/2020/gruenlaender-kaese-von-hochland-erhaelt-den-goldenen-windbeutel-2020/

has now become a distinct possibility [6]. Like the magician's apprentice, we have proven ourselves to be too inventive and smart for our own good or, to put it another way, not ethically smart enough to tame our inventive minds and the technologies they create so they will actually serve our objective, long-term collective needs and best interests. Blind confidence in the promise of a technological utopia has been a central pillar of the modernist project from the beginning [7]. An element of hubris within the spirit of industrialism was noted early on, a deeply inhumane and even demonic streak. From the time of the Romantic Movement onward, there have been periods of intense public criticism of industrialization. The work of writers like Charles Dickens and many others uncovers this darker side of progress and its injustices. Charlie Chaplin's *Modern Times* (1936), a silent movie produced long before today's acute crisis, is an excellent characterization of technological overreach and the risk of dehumanization.

In this modality, safety is reduced to a miniscule concern and innovation becomes an uncontrollable juggernaut – comparable to a high-powered Formula 1 racing car equipped with no more than a bicycle brake (Figure 9.1). The allocation of funding to safety in the US nanotech sector aptly illustrates this distortion and the unacceptable marginality of ethical concerns.

Environmental, Health and Safety Funding as a
Percentage of Total NNI Funding (US) between 2005–2010

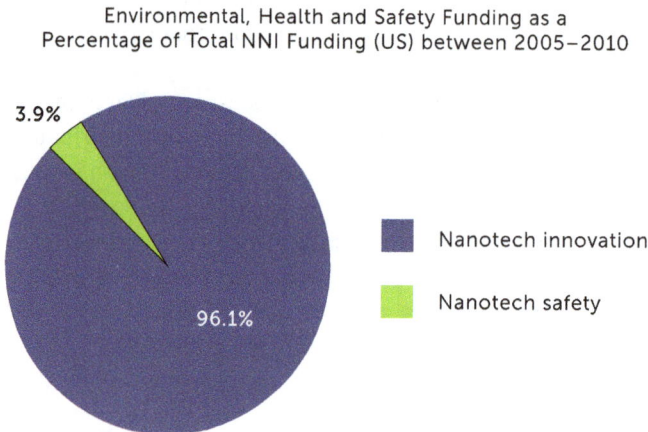

Figure 9.1: The (nano)-innovation juggernaut.
Adapted from *Nano Exposed: A Citizen's Guide to Nanotechnology*. Report, International Centre for Technology Assessment, December 2010.

While advocates of industrialization will hasten to point to the accompanying rise in gross domestic product (GDP) and individual "living standards," which is indeed a well-documented fact – at least for some populations – we must ask, at what cost? More fundamentally, we find today a growing international debate about the relevance of conventional measures of economic performance to actual life quality, measures on which the "success story" of industrialism heavily relies on. Instead we are witnessing a gradual shift in focus away from GDP to more holistic measurements of economic success wherein human health, wellness, social inclusion, and ecological sustainability are core criteria [8].

The lesson we ought to have learned from the history of three centuries of industrialization by now is quite clear: Technological innovation can cause collateral damage, and yet innovations often have been embraced without due exercise of foresight and responsibility. Known and avoidable negative consequences are allowed to unfold. In short, the precautionary principle is not applied. The problem is deeper, however. There is an inherent unpredictability in technology that cannot be remedied by foresight, due to the inherent limitations of the latter. This is the realm of unknown unknowns.

Even a responsible innovator may fall victim to the unintended and all but unforeseeable consequences of their innovation, which may unfold in close interaction with other simultaneous innovations. An illustrative example is provided by the cluster of innovations that led to industrial mass usage of chlorofluorocarbons (CFCs) in a variety of new applications (as refrigerants, propellants, foaming agents, etc.). This inadvertently led to the depletion of the ozone layer from the late 1970s onward, until 1987, when the Montreal Protocol was agreed upon to curb CFC emissions. This came too late for countless victims of skin cancer (above the long-term case average). The example shows that our capacity for foresight can be insufficient. Perhaps the well-known reactivity of CFCs should have been considered more carefully as a risk factor, but perhaps it is not overly surprising that the ozone layer, 20 km above their heads, was not foremost on the minds of personal deodorant producers. How can this deeper problem be avoided?

One aspect of the problem, which is rampant in food nanotech, is that known unknowns are allowed to persist even though the capacity to shed light on them through research does exist. The reason may be that once a producer is aware, or can be reasonably expected to be aware of a risk, it becomes a known risk with obvious consequences for liability. In the end, however, such a strategy is both unethical and a commercial risk. Insurers are beginning to issue warnings about this risk. One of the world's largest insurers, Allianz, published a report in 2017 that warns of the implications of safety risks for the $1 trillion forecast contribution of nanotech to industrial agriculture: If "nanoparticles leach from products to accumulate in human bodies or in the environment there may be costs associated with medical treatment of affected persons and/or remediation of environmental conditions," while "product recall due to research findings indicating that a product is a hazard" may mean that "insurance policies may [have to] cover the costs of recalling products, the interruption to the business and the detrimental impact to the companies brand, among other things" (p. 7) [9]. The insurance industry thus advocates for more research to better quantify or rule out unknown risks and, in the meantime, suggests that some risks are already known for some products presently on the market.

Another aspect is the question of genuine utility. Minimalism, our third ethical principle, may at first seem like an impediment to progress. Real progress is more than just change, however – it is change for the better. To justify the pursuit of a particular innovation, the need it addresses must therefore be great, and there must be no substantive risks. Where an innovation is not strictly necessary, it is better to be completely risk averse and "first do no harm," an ethical rule well known from the medical profession's oath of Hippocrates.

None of the nanotech innovations in food are strictly necessary, however, or *Homo sapiens* would have been unable to live without them for the last 300,000 years. Is the technology then, at least, of great benefit? Has it become necessary now because there is indeed a looming global food shortage? Or has it simply become technically possible and potentially profitable to a single stakeholder, namely, the producer?

While basic food processing by way of cooking, drying, fermentation, and other means has been practiced from prehistoric times, and has been of vast benefit in the sense that it has greatly facilitated human dispersion and population growth, industrial processing is a different matter. The medical profession worldwide has consistently issued dire warnings against the consumption of hyper-processed foods [10]. Many conventional industrially manufactured food articles are a source of serious, and indeed deadly, long-term health risks, and the liberal addition of nanotech could exacerbate these risks. Broadly speaking, the only advocate for even more food processing is the industry itself. In short, in the absence of a definitive and substantive need, to be considered case by case, technological innovations should be seen as a source of unnecessary risk.

The recommendation of public health agencies to consumers is to consume fresh foods, or foods that have been processed in a gentle way. In addition, there are ever louder calls to reduce the length of supply chains as a necessary measure to avoid the environmental impact of extremely carbon emission-intensive practices – such as airfreighting food from one continent to another or excessive packaging with non-reusable materials. These are practices for which nanotech may serve as an unwitting enabler, for example, by slowing down or detecting food spoilage. This is not to deny that there are problems in the global as well as national and local food systems that need fixing, such as food wastage, or that achieving complete food sovereignty in localities at various scales may be impractical, or the need for farmers to remain economically viable and able to produce at prices consumers can afford. These issues involve a range of different value considerations and are thus complex, with much scope for disagreement about best practice.

For the purpose of this chapter, it may suffice to say that the safest solution for future-proof food systems, from the field to the plate, may be to deindustrialize and regenerate them. That is, if public health and the environment are our paramount concern. According to recent research, it is a realistic and ecologically sound option to supply the entire world with organic food [11]. And even if food systems are not deindustrialized, or not immediately and completely, the way technology is used within them must become much more cautious, smart, and low impact. Nanotechnology could be of value in this context, at least temporarily, for example, by increasing the effectiveness of chemical fertilizer and thus reducing its overall use and impact, until it can be replaced with circular, organic options.

Unfortunately, in the food sector and in many other branches of industry, innovation is not subjected to a minimalist principle. Careful evaluation of our genuine needs is not the driving concern, nor are practices based on the Hippocratic call for harm avoidance. Innovation is instead fetishized as an icon of modern life and driven ever on, blindly and at great speed, by a very different, reckless logic of "stay ahead of your enemies or perish." Like many other new technologies, the possibility of military applications for nanotechnology and an associated arms race has been a driving force in the sector's development [12]. This military logic of innovation can and must be questioned, however, even on its own ground, because there are other and better options for building human security, such as multilateral arms restriction treaties. Instead of reforming the security industry and its mode of operation, alas, its bloody-minded military logic of innovation has been extended haplessly to other fields, most notably to market economics, where fierce competition is often cited as a legitimate cause for companies engaging in similarly reckless, technologic alarm races among themselves. This view of economics as a war zone does not hold water because the pursuit of exorbitant profits is hardly a battle of life and death and does not deserve the same ethical concessions. And, of course, even in a military confrontation, the avoidable killing of civilians is considered a war crime under the Geneva Convention.

A ruthlessly competitive attitude is certainly not at all appropriate for the food sector. Nutrition and public health should not be a battle zone in which private profit is maximized at the externalized cost of massive "collateral damage" (measured in human life years lost), which health systems must try to ameliorate at great expense to the public. Food is similar to oral medicine in that it is ingested and has properties that are in essence "medicinal" (health promoting). In many cases, such as herbs and spices, the boundary between what constitutes a food and a medicine can be difficult to find. In the specific case of nanomaterials, medicinal food supplements such as vitamins constitute a major application. While these products are marketed as non-prescription medicines, the Hippocratic principle certainly should be applied.

Despite these ethical caveats, we find that nanotechnology is being introduced liberally into agricultural production processes, food ingredients, food processing, food packaging, and the delivery of nutritional supplements, with insufficient knowledge available to rule out harm, and with sufficient knowledge available to suggest that harm could indeed occur. Those in the industry who cannot bring themselves to think about ethics for its own sake should consider the reputational and liability risks. Until then, it is a matter of consumer beware, or in more progressively governed countries, regulators beware.

It is time to look at these risks and unknowns in more technical detail.

9.3 Nano materials in food today: Known risks and blind spots

The range of nano materials used in agriculture, animal feed, human food, and food packaging is enormous and growing [13]. The proliferation of Encapsulated Nano Particle (ENP) applications shows significant variation across jurisdictions, however, depending on the strictness of regulations (below). Among other uses, ENPs may be food additives, vehicles for the so-called smart delivery of vitamins and other nutrients via nano-encapsulation, anti-caking or antimicrobial agents, or fillers to increase the tensile strength and durability of packaging material [14]. Food nano-sensing is another application, whereby spoilage or pathogen presence in food can be detected to enhance food safety. Note that ENPs used in primary production, food packaging, and animal feed may not be subject to the same degree of regulatory control, but may nonetheless find their way into the human body in some cases, though likely in smaller quantities.

The task of assessing the risk profile of diverse ENPs across multiple applications, sometimes in combination with other additives and ENPs, is daunting. Each has its own unique risk profile, and potential interactions remain largely unstudied. Some interactions are of serious concern. For example, "a food product that contains lipid nanoparticles (such as a beverage, sauce, dressing, or cream) may increase the bioavailability of hydrophobic pesticides on fruits or vegetables consumed with them" (p. 9) [15]. The pathways of potential harmful effects are indeed very complex and numerous, including interference with normal gastro-intestinal tract (GIT) function, accumulation in organs and other tissue, cytotoxic effects on cellular function, altered location of bioactive release, enhanced oral bioavailability, interference with microbiota in the gut, and allergic immune responses.

Assessment is not made any easier by the fact that food nanotech is situated at the intersection between several fields separated by substantial disciplinary boundaries, including medicine, human biology, dietary studies, and materials engineering. There are few experts available who have the in-depth training in all fields required to envisage and conduct research on potential risk pathways in a comprehensive manner. Toxicity and pathways of ENPs differ from those of their macroequivalents. Nanomaterials "have unique properties unlike their macroscale counterparts due to the high surface-to-volume ratio and other novel physiochemical properties like colour, solubility, strength, diffusivity, toxicity, magnetic, optical, thermodynamic, etc." [16].

The difficulty of assessing ENP impacts on health leads to contradictory or inconclusive results. Causes of discrepancies include the testing of different forms of the same ENPs with different physicochemical properties, different test methodologies, and failure to take into account the interactive effects of other additives, dietary patterns, the food matrix, and the fate of ENPs' passage through the complex system of the GIT. Furthermore, while the majority of toxicity studies are based on administration of large single doses (gavage studies, largely conducted on rats), there is a lack of long-term dietary impact studies in rodents [17], and rodent-model studies in turn may not be sufficient to extrapolate accurately to ENPs' effect on humans [18].

The general risk spectrum of ENPs has been known for some time. A 2007 report already observes that the ENPs found in and around food can enter the body through the skin, by ingestion and or by inhalation (the latter is an issue mainly for workers in ENP manufacture); they can be acutely toxic or cause chronic disease, including inflammation, disrupting organ chemistry or causing cancer [19]. According to the same report, ENPs can enter cells and organelles within cells; ingested they can contribute to Crohn's disease or colon cancer; via skin exposure or implants there may be autoimmune responses and allergies, urticaria, and vasculitis; in the lungs, they may cause asthma, bronchitis, emphysema, or cancer; in the circulatory system arteriosclerosis, hypertension, or thrombosis and heart disease, and in the brain Alzheimer's and Parkinson's disease. Heavy metal release is also a major concern [20], and there may be other risks as yet unknown.

In general, the state and conclusiveness of research remains inadequate, and much investment is needed into targeted, independent research projects. Some of the very significant cost of such independent research should be borne by those who intend to commercialize the technology, while safeguarding the independence of the researchers involved in safety assessment. Laboratory methods for assessing nanotoxicity are also still in need of development, reflecting an incomplete understanding of possible pathways by which ENPs may impact on various parts of the body. Oxidative stress is one important pathway that has been documented [21]. The principle of precaution applies in the meantime: a lack of research means that we are dealing with a "known unknown," an evident but unquantifiable risk.

Some early concerns about ENPs in food have since been all but ruled out, while in other cases, the verdict remains unclear or assessments are not yet available at all. Some examples will illustrate the complexity of the science. One type of ENP commonly used as an anti-caking agent is nano-amorphous silica ($nSiO_2$). There have been studies suggesting the product can be considered safe, with a reasonable degree of certainty, and regulators in the EU have assumed this to be the case. But not all consumers will be satisfied with these reassurances because residual doubts remain. For example, a study commissioned by Food Standards Australia and New Zealand concedes that "Some types of nano-SiO_2 can cause chromosomal damage to mammalian cells in in vitro test systems" (p. 5) [17]. Another study claims that $nSiO_2$ can be cytotoxic to human lung cells by causing oxidative stress upon exposure [22]. Comparable or greater uncertainties remain with regard to nano-titanium dioxide ($nTiO_2$) and nanosilver (nAg), according to the same authors. nAg is finding a growing use as an antimicrobial, anti-deodorant, and supplement, and yet animal studies show that nAg particles can accumulate in organs after ingestion, including the liver, kidneys, spleen, stomach, and small intestine [23]. Meanwhile, tumor-like changes have been observed in human cells exposed to $nTiO_2$ under in vitro conditions [24]. There are countless other products under research, fully developed and some already on the market, for which there is a lack of adequate safety data. For now, only partial lists of these products and their manufacturers and applications are available [25].

The presence of known, low-level, or "tolerable" risks demands for the application of the principle of transparency. Consumers have a right to know that they are ingesting a substance that is legally assumed safe but not safe beyond any doubt. Appropriate labeling should be a voluntary industry ethics standard for all ENPs, and preferably, labeling ought to be made mandatory as a global standard. Such disclosure must be followed by further research that helps regulators and/or consumers to better estimate the residual risk. Life itself is risky, of course, and taking small risks is a part of everything we do. And sometimes one risk needs to be weighed up against others. Nevertheless, it must also be considered that we are not all the same. Some consumers may be more vulnerable to ENPs than others, for example, if they have a pre-existing health condition. I have not found any data on the potential implications of such vulnerabilities. Perhaps scientists and regulatory bodies should work more closely with consumer protection NGOs (non-government organizations) and journalists and food-related consumer advice publications, such as *Food Safety Magazine*[2] and *Mother Earth News.*,[3] This would help communicate the level of risk accurately to the public, rather than manufacturers or regulators intervening in public debate in a post hoc fashion.[4] Until now, the reception of food nanotech has been marked by public suspicion, in part because companies are hesitant even to speak about their use of these technologies to journalists.[5]

Innovative methods for risk assessment and management and new laboratory testing standards to measure nano-toxicity are required, and still lacking. Collaboration between innovators and regulators from the early stages of technology development would be helpful in this regard and in general. And increasingly, the need for a sustainable, circular economy needs to be considered. What happens to nano-enhanced food packaging after use? Can it be safely disposed, recycled, or reused, and what is the cost? If ENPs are excreted by the human body, what is their fate in the environment? Are they isolated in sewerage treatment or are they released? Are aquatic life forms more sensitive to ENPs than us? In short, nanotech food manufacturers and regulators need to consider safety and ecological impact across the entire product life cycle as a design principle, and not just test finished products prior to release. If such value considerations are an afterthought only, moral hazard is very likely to follow: It is difficult for a company to ditch a finished product that has already received a large

---

2 https://www.foodsafetymagazine.com/magazine-archive1/februarymarch-2017/nanotechnology-in-the-food-industry-a-short-review/
3 https://www.motherearthnews.com/real-food/food-policy/nanotechnology-zmgz12amzrog
4 For a case of post hoc intervention, see https://www.foodstandards.gov.au/consumer/foodtech/nanotech/Pages/Sydney-Morning-Herald-nanotechnology-response.aspx
5 See www.theguardian.com/what-is-nano/what-you-need-know-about-nano-food

investment of funds in the course of its development. Ethics thus should guide the design process from the start.

9.4 Regulatory frameworks: The European example

Governments of some developed countries, like Canada, Japan, and Australia, began to consider regulation of nanotechnology in food in earnest in the early 2000s. A few, including the EU, have now adopted legally binding provisions (below). Others lack even a regulation to enshrine precaution as a general principle and duty for food producers. International collaboration and guidelines are only recently beginning to emerge. In 2009 and 2012, the Food and Agriculture Organization and the UN held expert meetings followed by a report on nanotech and food safety in 2012,[6] and a technical paper on safety measures used by a wide range of jurisdictions in 2013.[7] Efforts are also under way under the auspices of the Organization for Economic Cooperation and Development (OECD) and its "Working Party on Manufactured Nanomaterials." The International Organization for Standardization and others share information and harmonize assessment guidelines and regulations [26]. Nevertheless, national and international NGOs are often ahead of government institutions, and among them, the efforts of the Institute of Food Technologists and the International Life Sciences Institute (ILSI) deserve special commendation.[8]

Stark differences in regulations across jurisdictions remain. This raises serious issues for international food trade, and specifically for imports into more strictly regulating jurisdictions, such as the EU, from countries where regulations are still underdeveloped, such as Indonesia. The United States, to name another example, does not apply precautionary principles to food safety the way the EU does, and also has favored voluntary, non-binding guidelines for nanotechnology disclosure in food. In 2008, Friends of the Earth found 100 food or food-related products that contained nanoparticles, ranging from beer to baby drinks, but had not undergone safety testing by the US Food and Drug Administration (FDA) or other agencies, and did not feature a warning label. Moreover, while the US government in the preceding year had spent 1,4 billion USD on nanotech research, only 40 million USD were spent on safety research [27]. While some progress has been made, US nano-food regulations are still weaker than those of the EU. To the best of my knowledge, the FDA does not publish a list of nanomaterials that it has assessed (status 2011).

A brief history of the EU case may serve as an example of a relatively advanced and still developing regulatory framework. An early watershed was a 2008 report

---

6 http://www.fao.org/fileadmin/templates/agns/pdf/topics/Nano_Report_Final_20120625.pdf
7 http://www.fao.org/docrep/018/i3281e/i3281e.pdf
8 https://www.ift.org and https://ilsi.org

assessing the capability of existing, general food safety regulations to deal with the specific challenges of nanotechnology in food [28]. The report concluded that the regulations were relevant in many ways but insufficient to answer these challenges, triggering a call for new research and a cascade of subsequent regulatory responses. Since 2010, nano-engineered versions of previously licensed food additives have to be re-evaluated before they can be used.[9] In 2016, EFSA scientists provided a state of the science overview on microplastics and nanoplastics as contaminants in food.[10] General guidelines for safety assessment of ENPs in feed and food were released by the European Food Safety Authority (EFSA) in 2011 and a revised draft version in 2018. The public consultation on this draft closed recently, in September 2020. These guidelines concern human health risk assessment of new products such as novel foods, food contact materials, food/feed additives, and pesticides. In addition to novel foods, the mandate covers other food products, because the definition of engineered nanomaterials under Regulation (EU) 2015/2283 is also directly or indirectly applicable to food flavorings, food additives, feed additives, and food supplements such as vitamins, minerals, or other nutrients. The comprehensiveness of the recommended product safety tests nevertheless remains debatable, given the dynamic state of scientific knowledge on this issue. EFSA itself concedes there are uncertainties related to the identification, characterization, and detection of ENPs because of a lack of suitable and validated test methods, and recommends additional research to address the many current uncertainties and limitations.[11]

The 2018 guidelines are referred to as "Part 1," while the preparation of "Part 2" "is ongoing and will cover guidance on environmental risks assessment of nanomaterials and nanotechnologies under EFSA remit" (p. 4) [29], thus responding to a major concern, voiced earlier, regarding the need to monitor risks along the full life cycle of ENPs – including recycling, disposal, and entry into the environment. Note that guidelines are not at all the same as practices. The actual use of EFSA guidance by EU Member States is currently "still quite low" according to the report just cited (p. 8).

To address knowledge gaps and harmonize the approach to ENP safety issues across the EU, EFSA established a scientific research network among institutes across EU Member States in 2010, known as the "Scientific Network of Risk Assessment of Nanotechnologies in Food and Feed" (the Nano Network).[12] The network has global influence, given that its annual meetings are also attended by observers, such as OECD and USFDA representatives.[13] Furthermore, the 2019 meeting of the Nano Network was embedded in the Global Summit "Regulatory Science 2019: Nanotechnology

---

9 http://eur-lex.europa.eu/LexUriServ/LexUriServ.do?uri=CELEX:32008R1333:EN:NOT
10 https://www.efsa.europa.eu/en/news/microplastics-and-nanoplastics-food-emerging-issue
11 https://www.efsa.europa.eu/en/topics/topic/nanotechnology
12 http://www.efsa.europa.eu/en/supporting/pub/362e.htm
13 https://efsa.onlinelibrary.wiley.com/doi/abs/10.2903/sp.efsa.2018.EN-1393

and Nanoplastics."[14] The minutes of the network's meeting are published, ensuring transparency. These meeting records show the struggle to catch up with a flood of "innovations" entering the market. Not all are necessarily intentional nano-food additives. For example, nano-sized fertilizer use is expanding rapidly, and new assessments are needed to see how it may impact on the environment and food chain.[15] EFSA has also scheduled a scientific colloquium to assess the human health risks of micro- and nano-plastics in food for May 2021.

Apart from issuing risk assessment guidelines, the explicit listing of nano-ingredients on the product label (e.g., "contains: silicon dioxide [nano]") has been mandatory in the EU since 13 December 2014.[16] Overall, the EU is thus leading in responsible regulatory processes. A complete analysis of this history well deserves a separate paper, but this condensed account will suffice for present purposes. It shows that even the most responsible regulators remain on the back-foot, rather than directing innovation from the design stage. EFSA does conduct public consultations and also held a stakeholder workshop on nanoscience and nanotechnology in Parma, Italy, in April 2019, but these consultations are still after the fact of innovation. Major structural changes would be needed to change this scenario in the direction of greater stakeholder participation in the steering of processes of industrial innovation. Meanwhile, industry is creating a future for us all that primarily reflects its own interests. While this pattern persists, risk avoidance is hardly possible and risk mitigation will remain an uphill battle to catch up with the facts created by industry.

9.5 Recommendations: Promoting ethical conduct in the food industry

In the short term, ethics does not seem to pay, while unethical business behavior can produce enormous profits. The trafficking of deadly drugs of addiction is a pertinent example of the scope for profit open to organized criminals who are willing to disregard the welfare of others, and sometimes the damage done by corporate crime can be of comparable magnitude. In most industries, however, the reality is that most CEOs respect the law but are hard pressed to find compromises or loopholes that strike a perceived "balance" between profitability and ethical responsibility. Where this balance is located matters greatly. Were a major soft drinks producer to reduce the amount of sugar in their products by 5%, how many healthy life years

---

14 https://ec.europa.eu/jrc/en/event/conference/gsrs19-global-summit-regulatory-science-2019-nanotechnology-and-nanoplastics
15 https://efsa.onlinelibrary.wiley.com/doi/epdf/10.2903/sp.efsa.2020.EN-1784(see p.9).
16 http://eur-lex.europa.eu/LexUriServ/LexUriServ.do?uri=OJ:L:2011:304:0018:0063:EN:PDF

would be saved? Would the company's profits really plummet? Or would they increase, as a consequence of long-term reputational gain, avoided liability, and general changes in nutritional awareness within the population?

The new stakeholder capitalist model of doing business, now being promoted by the World Economic Forum (as mentioned earlier), suggests that CEOs need to think more long term and come to understand themselves as part of a system, the social and environmental sustainability of which they are morally co-responsible to uphold. Ethical concerns are system concerns, and ethical behavior is rational because the elements in a system are interdependent. In this information age, moreover, a violation of stakeholder concerns is unlikely to go unnoticed for long, and likely to be punished by ever more conscious and well-organized consumers. Importantly, investors too are increasingly aware of systemic concerns, and are moving their funds in the direction of sustainable, responsible, well-governed and farsighted companies. This in turn may reduce the pressure on CEOs to pursue short-term profits ruthlessly and by any means. Innovative legislation concerning the regulation of financial markets as well as changes in the laws concerning corporate and investor liability, both with the aim of discouraging short-term profiteering, are also potential tools for a systemic transformation toward a more ethical and inclusive innovation process.

Failing this, regulations can be strengthened, especially pre-market regulations that are able to stop harmful products from reaching consumers or the environment in the first place. Rigorously applied, stringent pre-market regulations may also encourage food producers to open up to the idea of a more inclusive design process that would place consumer NGOs, government regulators, and other societal actors in front of the wheel of industrial change, rather than behind it.

References

[1]   Reuter TA. Understanding food system resilience in Bali, Indonesia: A moral economy approach. Culture, Agriculture, Food and Environment 41, June 2019, 4–14. Published online on 3 June 2018 at http://dx.doi.org/10.1111/cuag.12135. Last Accessed date: 1/3/2021.
[2]   Reuter TA, MacRae G. Regaining Lost Ground: A Social Movement for Sustainable Food Systems in Java, Indonesia. Anthropology of Food. Published online on 18 July 2019 at https://journals.openedition.org/aof/10292. Last Accessed date: 1/3/2021.
[3]   Reuter TA. Seeds of life, seeds of hunger: Corporate agendas, seed sovereignty and agricultural development (Indonesia, East Timor). Anthropology of Food, December 2017, 1–18. Published online 2017. https://aof.revues.org/8135. Last Accessed date: 1/3/2021.
[4]   Reuter TA. The Struggle for Food Sovereignty: A Global Perspective. In: Reuter TA ed, Averting a Global Environmental Collapse: The Role of Anthropology and Local Knowledge. London, Cambridge Scholars, 2015, 127–47.

[5]   World Economic Forum. Measuring Stakeholder Capitalism: Towards Common Metrics and Consistent Reporting of Sustainable Value Creation. White Paper, September 2020. Published online at http://www3.weforum.org/docs/WEF_IBC_Measuring_Stakeholder_Capitalism_ Report_2020.pdf. Last Accessed date: 1/3/2021.

[6]   Reuter TA. Principles of Sustainable Economy: An anthropologist's perspective. CADMUS – J World Acad Art Sci 3, 2 May 2017, 131–49.

[7]   Reuter TA. Rebuilding Sustainable Ways of Life: Economic anthropology's contribution to a better understanding of the climate change crisis. Econ Anthropol 8, 1 January 2021, 175–79.

[8]   Fox J. The economics of well-Being. Harvard Business Review, January–February 2012. Published online at https://hbr.org/2012/01/the-economics-of-well-being. Last Access date: 1/3/2021.

[9]   Allianz Global Corporate & Specialty and Predicate. Emerging liability risks: Nanotechnology in food. Risk Bull 1, 2017, 1–8.

[10]   Elizabeth L, Machado P, Zinöcker M, Baker P, Lawrence M. Ultra-processed foods and health outcomes: A narrative review. Nutrients 2020, 12, 7, 1955. Published online 30 June 2020.

[11]   Muller A, Schader C, Scialabba NE-H. et al., Strategies for feeding the world more sustainably with organic agriculture. Nat Commun 2017, 8, 1290.

[12]   Altmann J. Military Applications of Nanotechnology: Preventive Limitation Is Needed. Controv Sci Tech 3, 276–91. Published Online: 25 Aug 2010.

[13]   Cerqueira MÂ, Pastrana LM. Does the Future of Food Pass by Using Nanotechnologies? Frontiers in Sustainable Food Systems. Published online, 26 March 2019. Last Access date: 1/3/2021.

[14]   Sekhon BS. Food nanotechnology – an overview. Nanotechnol Sci Appl 2010, 3, 1–15.

[15]   McClements DJ, Xiao H. Is nano safe in foods? Establishing the factors impacting the gastrointestinal fate and toxicity of organic and inorganic food-grade nanoparticles. Sci Food 2017, 1, 6, 1–13.

[16]   Singh T, Shukla S, Kumar P, Wahla V, Bajpai VK, Rather IA. Application of Nanotechnology in Food: Perception and Overview. Frontiers in Microbiology. Mini Review, published online 7.8.2017, 1.

[17]   Drew R, Hagen T, Tox Consult Pty Ltd. Potential Health Risks Associated with Nanotechnologies in Existing Food Additives. Report commissioned by Food Standards Australia New Zealand, May 2016. https://www.foodstandards.gov.au/publications/Docu ments/Safety%20of%20nanotechnology%20in%20food.pdf. Last Access date: 1/3/2021.

[18]   Seyed MA, Marzieh G, Mohsen K. Safety of nanotechnology in food industries. Electron Phy 6, 4 Oct-Dec 2014, 962–68.

[19]   Buzea C, Pacheco I, Robbie K. Nanomaterials and nanoparticles: Sources and Toxicity. Biointerphases 2007, 2, MR17-71.

[20]   He X, Hwang H-M. Nanotechnology in food science: Functionality, applicability, and safety assessment. J Food Drug Anal 2016, 24, 671–81.

[21]   Dimitrijevica M, Karabasila N, Boskovica M, Teodorovica V, Vasileva D, Djordjevicb V, Kilibardac N, Cobanovica N. Safety aspects of nanotechnology applications in food packaging. Procedia Food Sci 2015, 5, 57–60.

[22]   Athinarayanan J, Periasamy VS, Alsaif MA, Al-Warthan AA, Alshatwi AA. Presence of nanosilica (E551) in commercial food products: TNF-mediated oxidative stress and altered cell cycle progression in human lung fibroblast cells. Cell Biol Toxicol 2014, 30, 89–100.

[23]   Gaillet S, Rouanet JM. Silver nanoparticles: Their potential toxic effects after oral exposure and underlying mechanisms – a review. Food Chem Toxicol 2015, 77, 58–63.

[24] Magdolenova Z, Collins AR, Kumar A, Dhawam A, Stone V, Dusinska M. Mechanisms of genotoxicity. A review of in vitro and in vivo studies with engineered nanoparticles. Nanotoxicology 2014, 8, 233–78.

[25] Nile SH, Baskar V, Selvaraj D, Nile A, Xiao J, Kai G. Nanotechnologies in Food Science: Applications, Recent Trends and Future Perspectives. Nano-Micro Lett 2020, 12, 45.

[26] Organisation for Economic Co-operation and Development. Regulatory frameworks for nanotechnology in foods and medical products: Summary results of a survey activity, 2013. https://www.oecd-ilibrary.org/docserver/5k47w4vsb4s4-en.pdf?expires=1620041587&id= id&accname=guest&checksum=C20A26B08E370CDB32C47AFB4ADC44EE. Last Access date: 1/3/2021.

[27] Biello D. Do Nanoparticles in Food Pose a Health Risk? Scientific American, online edition, 13 March 2008.

[28] EU Commission. Communication from the Commission to the European Parliament, the Council and the European Economic and Social Committee: Regulatory Aspects of Nanomaterials. Brussels, 17 June 2008. https://eur-lex.europa.eu/LexUriServ/LexUriServ.do? uri=COM:2008:0366:FIN:EN:PDF. Last Access date: 1/3/2021.

[29] European Food Safety Authority (EFSA). Annual report of the EFSA Scientific Network of Risk Assessment of Nanotechnologies in Food and Feed for 2019. Published online on 18 February 2020. https://doi.org/10.2903/sp.efsa.2020.EN-1784. Last Access date: 1/3/2021.

Jean Pierre Massué

10 Ethical challenge of nanomedicine

Abstract: "Man is only a reed, but he is a thinking reed. The whole universe must not arm itself to crush it; a vapour, a drop of water is enough to kill it. But when the universe would crush him, man would be even more noble than what kills him, since he knows that he is dying and the advantage that the universe has over him. The universe knows nothing of it." (Blaise Pascal, "Pensées" in Œuvres Complètes, p. 528)

Keywords: nanotechnology, dignity, bioethics

10.1 Introduction

Nanomedicine is linked with the application of nanotechnology to health. It exploits the improved and often novel physical, chemical, and biological properties and organic drafted molecules of materials at nanometric-scale.

The objectives of nanomedicine can be presented as having a potential impact on the prevention, early and reliable diagnosis, and treatment of diseases:
- The 3-F model:
  i. **FIND,**
  ii. **FIGHT,**
  iii. **FOLLOW.**

Find concept corresponding to the use of nanotechnology. The ultimate goal of diagnosis is to identify disease at the earliest stage possible, ideally at the level of a single cell.

Fight concept using nanoparticles as therapeutic agents such as drug carriers for the effective transport of therapeutics or for the benefit of regenerative medicine. For example for drug delivery to the appropriate site which could be especially useful to treat more effectively cancer.

Follow concept using nanoparticles to follow the delivery of drug and the effect of treatment.

Note: Deceased, November 2020
We would offer our special thanks to Prof. Massue who is although no longer with us, but will continue to inspire us through his dedication toward his work. The present work is among his last contributions. The text is not modified and we respect his contribution. No changes have been made.

Jean Pierre Massué, Member European Academy of Sciences and Arts, Saltsburg, Austria

https://doi.org/10.1515/9783110669282-010

Ethics: Nanomedicine as an enabling technology for many future medical applications touches on:
– issues such as genetic information
– the new dimension to the bio "human" and non-bio "machine"

The ethical analysis of nanomedicine has to be based on the fundamental principles of:
– **Human dignity** and derived ethical principle
– **Non-exploitation**: using individuals as means
– **Privacy:** not invading the right of a person for privacy,
– **Non-discrimination**: people deserve equal treatment
– **Informed consent:** people cannot be exposed to treatment or research without their free and informed consent
– **Equity**: everybody should have fair access to the benefit under consideration
– **The precautionary principle:** which entails the moral duty of continuous risk assessment with regards to not fully foreseeable impact in the human body

The most significant ethical issues at present relating to nanomedicine involve risk assessment and risk communication in clinical trials management.

10.2 The international approaches

At United Nations level, UNESCO [1] acts as a forum for multidisciplinary, multicultural, and pluralistic ideas on bioethics and on the ethics of sciences and technology, through four mechanisms:

The International Bioethics Committee (IBC), created in 1993, is a permanent committee comprising 36 independent experts appointed by the Director-General of UNESCO. Each member is elected on the basis of equitable geographical representation for a four-year term. IBC promotes refection on the ethical and legal issues raised by research in the life sciences and their applications to ensure respect for human dignity and freedom.

The Intergovernmental Bioethics Committee(IGBC), created in 1998 with 36 Member States elected by the General Conference of UNESCO, is convened at least once every two years to examine the advices and recommendations of IBC, informing the IBC of its opinions and submitting their proposals and points of view.

The World Commission on the Ethics of Scientific Knowledge and Technology (COMEST) is the third consultative organ, comprising 18 specialists from different world regions named by the Director-General of UNESCO.

UNESCO acts as the Secretariat of the United Nations Inter-Agency Committee on Bioethics established in 2003. This committee assures coordination between intergovernmental organizations dealing with issues related to ethics.

Concerning bioethics, UNESCO has focused on the social and political considerations convinced that the resolution of ethical issues raised by the use of science and technology determines the way we live together and the choices made by the society affecting our future.

In October 2005, the General Assembly of UNESCO adopted by acclamation the Universal Declaration on Bioethics and Human Rights. Together with the declaration, the General Conference of UNESCO adopted a resolution which calls upon Member States to make every effort to give effect to the principles set out by the declaration.

10.3 At European level

European Union and Bioethics: contribution of nanomedicine to Horizon 2020 [2].

Nanotechnology has already provided many different medical solutions both for therapeutics and diagnostics. Nano-delivery of drugs, for example, has and will provide new products to address unmet medical needs in cancer and other diseases. In addition, many nano-features, such as new imaging agents or smart materials will be prerequisites for implementation of personalized medicine.

The introduction of nanotechnologies into medical applications requires nanomedicine stakeholders to understand and apply the process of open innovation, which is essential for translation to the clinic. The best option to successfully implement the open innovation model in Europe consists in establishing a supply chain providing nanomedicine products compatible with industrial processes and strategies.

The creation of this new supply chain requires a major change in thinking of all stakeholders and the empowering of an organization to actively manage the translation effectiveness of its members (academics and SMEs) to help revitalize industry in Europe.

Key tasks of this organization will be the set-up of:

- A translation advisory board with experienced industrial experts, who will apply horizontal innovation filters on R&D proposals from academics and SMEs to select, guide, and push forward the best translatable concepts toward funding and clinical proof of concept.
- New infrastructures supporting the translation of nanomedical materials:
  - A European nano-characterization laboratory,
  - A European pilot line for GIMP manufacturing of batches for clinical trials,
  - A European network of preclinical centers of excellence,
  - A European coordinated effort on nanomedicine with clinical organizations.

10.4 Council of Europe

Parliamentary Assembly Recommendation 2017 [2013, 3].

Nanotechnology: balancing benefits and risks to public health and the environment Author(s): Parliamentary Assembly

Origin: Assembly debate on April 26, 2013 (18th Sitting)

In the report presented to the Parliamentary Assembly of the Council of Europe, nanotechnology is presented as the manipulation of matter on an atomic and molecular scale. Nanomaterials involve structures having dimensions of nanometers (nm), that is one billionth (or 10–9) of a meter, typically between 1 and 100 nm in size. At such dimensions, materials can show significantly different physical, biological, and/or chemical properties from materials at bigger dimensions, which opens up a range of new possibilities not only for technology but also for problems,

Nanotechnology and its myriad applications have the potential for enormous benefits (in particular in the field of "nanomedicine"), but also for serious harm. As with most emerging technologies, many risks, both to public health and to the environment, are as yet poorly understood. However, commercial applications of nanotechnology are already in widespread use. Regulations have struggled to keep up with the pace of scientific innovation.

For years, the Parliamentary Assembly and the Committee of Ministers of the Council of Europe have been advocating the need for a culture of precaution in incorporating the precautionary principle into scientific and technological processes, with due regard for freedom of research and innovation. In 2005, the Heads of State and Government of the Council of Europe gave undertakings in the final Declaration of the 3rd Summit of the Council of Europe "to ensure security for our citizens in the full respect of human rights and fundamental freedoms" and to meet, in this context, "the challenges attendant on scientific and technical progress."

The Assembly believes that, in keeping with these undertakings, the Council of Europe, as the only pan-European body with a human rights protection mandate, should set legal standards on nanotechnology based on scientific knowledge and the precautionary principle which will protect 800 million Europeans from risk of serious harm, while encouraging nanotechnology's potential beneficial use.

The Assembly thus recommends that the committee of ministers work out guidelines on balancing benefits and risks to public health and the environment in the field of nanotechnology which:

– Respect the precautionary principle while taking into account freedom of research and encouraging innovation;
– Allow for consistent application to all nanomaterials under regulation across borders and regardless of their origins (synthetic, natural, accidental, manufactured, and engineered), functional uses, or biological fate;
– Seek to harmonize regulatory frameworks, including in the areas of risk assessment and risk management methods, protection of researchers and workers in

the nanotech industry, consumer and patient protection and education (including labeling requirements taking into account informed consent imperatives), as well as reporting and registration requirements, in order to lay down a common standard but difficult to reach . . . are negotiated in an open and transparent process, involving multiple stakeholders (national governments, international organizations, the Parliamentary Assembly of the Council of Europe, civil society, experts, and scientists) in the framework of a dialogue which transcends the geographical area of the Council of Europe;

Can be used as a model for regulatory standards worldwide:

- First take the form of a Committee of Ministers' Recommendation, but it could also be transformed into a binding legal instrument if the majority of Member States so wish, for example in the form of an additional protocol to the 1997 Convention for the Protection of Human Rights and Dignity of the Human Being with regard to the Application of Biology and Medicine: Convention on Human Rights and Biomedicine (ETS No. 164, "Oviedo Convention").
- Allow for the creation of an international, interdisciplinary center to be the world's knowledge base in the field of safety in the near future, without prejudice to the continued support, even in financial terms, for ongoing research projects aimed at determining potential risks of nanomaterials.
- Promote the development of an assessment system of ethical rules, advertising materials, and consumer expectations regarding research projects and consumer products in the nanotechnology field impacting human beings and the environment.

The assembly recommends that the Council of Europe's Committee on Bioethics (DH-BIO) be entrusted with a feasibility study on the elaboration of possible standards in this area, based on paragraph 5 of the present recommendation, as a first step in the start of negotiations on the topic with a multi-stakeholder approach. This study should include, in any case, ongoing scientific research at international level to learn about the risks of nanotechnological material. Thus, the scientific community will be actively involved in the drafting of any proposal for standardization and/or legislation.

10.5 WHO: World Health Organization

The Research Ethics Review Committee (ERC) is a 27-member committee established and appointed by the Director-General. Its mandate is to ensure WHO only supports research of the highest ethical standards. The ERC reviews all research projects, involving human participants supported either financially or technically by WHO.

While the majority of the committee consists of WHO staff, international external individuals are also appointed as committee members.

Members bring with them valuable and extensive experience and knowledge in research in different fields, and receive appropriate training in research ethics before commencing their role within the ERC. The broad range of research expertise of committee members, together with the ethics training, ensures that all proposals are thoroughly and fairly reviewed for ethical research conduct.

The ERC reviews and advises on research: fully or partially funded by WHO and managed by WHO, in which WHO is either a partner or collaborator.

The ERC is guided in its work by the World Medical Association Declaration of Helsinki [1964, 4], last updated in 2008 as well as by the International Ethical Guidelines for Biomedical Research Involving Human Subjects (CIOMS 2002) and the International Guidelines for Ethical Review of Epidemiological Studies (CIOMS 2009). According to these guidelines, all research involving human subjects should be carried out in accordance with the fundamental ethical principles of respect, beneficence, nonmaleficence, and justice: World Medical Association Declaration of Helsinki (1964), International Ethical Guidelines on Epidemiological Studies (CIOMS 2009).

In the book on human rights, ethical and moral dimensions of health care [5], resulting from the research of the European Network of Scientific Cooperation on Medicine and Human Rights set up under the auspices of the Parliamentary Assembly of the Council of Europe [5], it was underlined that human rights affirm the principle of the safeguard of life and are thus directly concerned when we analyze the problems posed by therapeutics, namely the use of the methods made available to us by science and, in particular, medical science in the service of human health. Religions and philosophies in general have been playing a major role for a very long time in the establishment of basic rules for codes of medical ethics. We have to remember the Hippocratic Oath:

> I will prescribe regimen for the good of my patients according to my ability and my judgment and never do harm to anyone. . . . In every house where I come I will enter only for the good of my patients, keeping myself far from all intentional ill doing and all seduction

10.6 The human rights and ethical dimensions of health care

One hundred and twenty cases examined from the standpoint of legal norms, international and European ethics and the Catholic, Protestant, Jewish, Muslim, Buddhist, and agnostic moralities are presented as an aid to decision-making and teaching by the European Network of Scientific Cooperation on Medicine and Human Rights of the European federation of Scientific networks.

Council of Europe, March 1998. Printed in Germany.

Similarly, the prayer composed in the twelfth century by Moses Maimonides says:

"Support the strength of my heart so that it will always be ready to serve the poor and the rich, the friend and the enemy, the good and the bad. Let me see only the man in he who suffers."

The creation of national consultative ethics committees represents a big step forward. They deal with subjects which were long considered taboo, such as the problem of sterilizing the mentally handicapped.

Among the major areas of important contribution of nanomedicine in the field of health can be underlined:

- Methods for improving diagnostics
- The possibility of bringing therapeutic agents directly into contact with specific targets
- An important contribution to the improvement of cancer therapy

A. The diagnostics

The method used for the diagnostic function is not based on the marking of antibody molecules by radioactive substances, but, for example, on the marking of antibodies by magnetic nanoparticles. The marked antibodies are introduced into a sample of tissue, which is then exposed to a strong magnetic field. Antibodies that have come into contact with a specific antigen then generate a magnetic field that can be measured indicating the presence of a pathogenic element.

B. The therapy

A nanobot with nano-therapeutic agent has chemical sensors that can unambiguously recognize fluid-borne or in cyto arbovirus particles and, once recognition has occurred, destroy them, and also reverse the cellular damage. Population of this size should be able to destroy all viral particles, effect needed repairs in at most an hour, destroy them, and also reverse the cellular damage.

C. Validation and prophylaxis

A proper therapeutic protocol will include a procedure for follow-up to ensure that the prescribed treatment was correctly executed with good results.

Nanomedical treatments will require supervision.

Validation is to be considered as a post-treatment re-diagnosis to ensure that no disease remains present in the patient.

J. C. Bennett describes the implicit social contract between doctor and patient that will still apply in the nonmedical era, as it does today:

> To receive medical care, patients must trust their bodies and their very lives to physicians, and so to be in an honest position to give medical care, physicians must earn such radical trust. Mere technical treatment of disease does not suffice. Patients must be able reasonably to believe that their physicians care about them in an extraordinarily personal way. This exchange of care for trust, while not identical to friendship or love, is equally binding. From it develops an interdependence that is far from unwholesome; rather, it potentiates care and promotes healing.

10.7 Long-term prospective

10.7.1 Genetic manipulation

The human genome is our genetic identity and is undoubtedly one of the keys to tomorrow's medicine. Recent advances in this area will have a real impact on how to treat diseases and other pathologies, especially in terms of prevention and also of personalized treatments.

In this context, nanotechnology is an articulated arm of this new medicine, and again, the possibilities are multiple.

Tiny, intelligent nanoparticles will be able to "deliver" drugs at specific points, reducing the required doses as they pass, and crisscross our bodies in search of irregularities or risk areas. A kind of meticulous technical control that will make the work of doctors more efficient could prove to be a new weapon of choice in the fight against cancer.

10.7.2 The human rights: ethical and moral dimensions of nanomedicine

Twenty years ago, I used to coordinate the drafting of the book *Human rights, Ethical and Moral: Dimensions of Health Care* (Council of Europe publishing). The aim is to identify case studies describing representative situations encountered by the medical doctors in the exercise of their function and to make an evaluation of these case studies from the standpoint of legal, deontological, and ethical points of view and that of various moralities: Catholic, Protestant, Jewish, Muslim, Buddhist, and agnostic Morality.

Let's present a brief summary of the analysis of one case study concerning genetic manipulation:

> **The case study:** An ethics committee is asked to authorize the modification of a genome: manipulation of a chromosome fragment and creation of a new transmissible characteristic on an animal and on man.

From the standpoint of international law: Recommendation 1100 (1989) of the Parliamentary Assembly of the Council of Europe, such project shall be permitted and subjected to approval by national or regional multidisciplinary "bodies" to be set up as a matter of urgency by Member States.

10.8 The human rights and moral dimensions of health care

One hundred and twenty practical case studies are examined from the standpoint of legal norms, international and European ethics, and the Catholic, Protestant, Jewish, Muslim, Buddhist, and agnostic moralities.

More particularly, for the European Parliament, the absolute preconditions for the use of genetic manipulation are:

- Genetic analysis and genetic counseling must be designed exclusively for the well-being of those concerned and be based exclusively on voluntary agreement, and the results of an examination must be communicated to those concerned at their request.
- They must on no account be used for the scientifically dubious and politically unacceptable purpose of positively improving the population's gene pool, negatively selecting genetically undesirable characteristics, or laying down "genetic standards."
- The principle of a patient's right to self-determination must have absolute precedence over the economic pressures imposed by health-care systems since every individual has an inalienable right to know his genetic make-up or to remain in ignorance.
- The establishment of individual gene maps may only be carried out by a doctor; the forwarding collection, storage, and evaluation of genetic data by government or private organizations shall be prohibited.
- The development of genetic strategies for the solution of social problems must not be allowed since it would undermine our ability to understand human life as a complex entity which can never be compassed entirely by any scientific approach.
- The knowledge acquired by means of genetic analysis must be absolutely reliable and enable unequivocal statements to be made about precisely defined medical facts, knowledge of which may directly benefit the health of the patient affected.

From the ethical standpoint: The WMA adopted in Madrid in 1987 a declaration pointing out that genetic manipulation must take place in the conditions laid down in the Declaration of Helsinki and envisages such for the purposes of genetic therapy.

In its Declaration of Marbella (1992) on the human genome project, the WMA draws attention, in particular, to the danger of eugenics in genetic therapy and the use of genes for nonmedical purposes.

From the standpoint of religious moralities:

–   Catholic: All attempts to interfere with man's chromosomal or genetic heritage which are not therapeutic but tend toward the production of human being selected according to sex or other pre-established qualities are contrary to the personal dignity of the human being, his integrity, and his identity. Creation of new transmissible characteristics in animal can be ethically acceptable if its purpose is the real good of mankind and on condition that it should not disturb the biological balance of nature.
–   Protestant: All the texts are categorically opposed to this possibility. By conceiving itself as the object of possible manipulations and experimentations, mankind makes itself an object and loses the very meaning of what gives it human reality.
–   Jewish: Genetic research, like all medical research can be undertaken and pursued if it has a therapeutic purpose and above all if its rise is aimed at curing transmissible hereditary diseases like hemophilia or Tay-Scachs diseases. In this case, it may be undertaken only after the tacit and informed agreement of the patient or his legal representative and only in this case where no other therapy exists as yet.
–   Muslim: The modifications of the human genome which are transmissible to the person's descendants is utterly forbidden in Islam as it leads to the creation of an organism which differs from divine creation. Transplantation of embryos between humans and animals must be prohibited from the point of view of Muslim morality, as the result is the production of organisms different from divine creation.
–   Buddhist: Genetic engineering opens up new therapeutic possibilities and prospects for improving human life, but can also bring uncontrolled abuse which could endanger the future of the human race; the greatest caution is thus called for.

From the standpoint of agnostic morality: As things stand, this genetic manipulation would not be prohibited. Authorization to proceed to the modification of the human genome can be requested under certain conditions:

–   Modification of somatic cells.
–   Modification of the germ cells up to the fifteenth day only.
–   It can only be carried out for preventive and curative therapeutic purposes, to eradicate certain diseases, with the free and informed consent of the patient or his legal representative.

10.9 Innovative nanomedicine at the occasion of the COVID-19

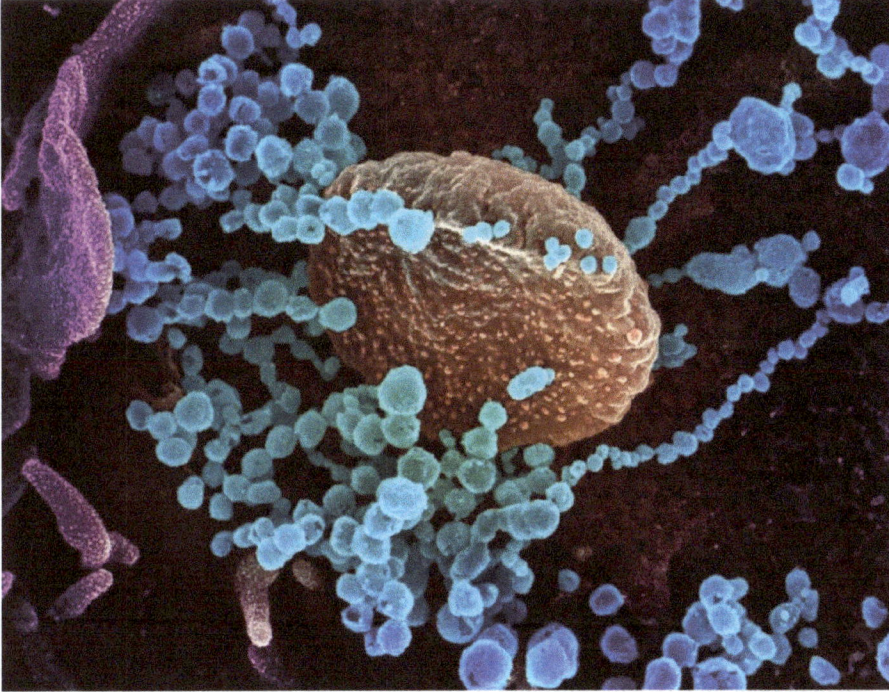

Figure 10.1: Novel Coronavirus SARS-CoV-2.

This scanning electron microscope image shows SARS-CoV-2 (round blue objects) emerging from the surface of cells cultured in the lab. SARS-CoV-2, also known as 2019-nCoV, is the virus that causes COVID-19. The virus shown was isolated from a patient in the United States. Image captured and colorized at NIAID's Rocky Mountain Laboratories (RML) in Hamilton, Montana. Credit: NIAID.

At the occasion of the Coronavirus outbreak has been declared by the World Health Organization: WHO underlining that: "it is crucial to rapidly gain a better understanding of the new identify virus especially in relation to potential clinical and public health measures that can be put to immediate use to improve patients health and or contain the spread of COVID-19."

10.10 Can nanomedicine through the manipulation of matter or at atomic, molecular, and supra-molecular scale help in the fight against COVID-19?

Nanomaterials scientific community can contribute in the fight against COVID-19. Indeed nanomaterials can be used for the development of diagnosis, carriers for therapeutics, and vaccine development.

Detection:

Nanotechnology brings interesting prospects in order to develop affordable detection methods. Nanosensors are a reality showing the possibility to detect viruses at low concentration and consequently warning clinician very soon.

Prevention:

SARS-CoV is affecting human through tiny droplets of viral particles coming from breathing, sneezing, coughing, and breathing and entering the body through the eyes, mouth, and nose. Those germs can survive for four days.

The nano-filter reusable masks help to ensure filtering efficiency.

Therapy:

Nanotechnology can be used to deliver drugs at the right place.

Nanoparticles have the ability to intercept the pathogens and viruses due to the potential surface modification. It is possible to modify nanoparticles in order to dissolve the lipid membrane of the viruses or even bind them to the spike protein or penetrate into the envelope and encapsulate the nucleus-capsid and RNA.

In conclusion I would like to underline that:

Regulations of nanomedicine, aside from the very important contribution nanomedicine is providing to the society, remain controversial, as we lack clear frameworks for the use, disposal, and recycling of nanomaterial being a considerable problem to be solved.

Certain nanommaterials are known to cause harm to humans and the environment. Predicting how nanoparticles would react under different environmental conditions has remained difficult. In addition, risks of toxicity and other hazards of nano-waste are not well understood.

But it is clear that the specific and unique contribution nanomedicine is providing to the world society affected by the COVID-19 is the best demonstration of its importance and consequently the necessity to support the development of research in the area of nanotechnology and in particular for nanomedicine.

References

[1] Nanotechnologies, éthiques et politique, édité par Henk A.M.J. ten Have UNESCO Collection Ethique, Secteur des Sciences Sociales et Humaines, Division de l'Ethique des Sciences et des Technologies.

[2] Horizon 2020, The biggest E.U. Research and Innovation Programme 7 years 2014–2020.

[3] Doc. 13117, Report of the Committee on Social Affairs, Health and Sustainable Development, rapporteur: Mr. Sudarenkov. *Text adopted by the Assembly* on 26 April 2013 (18th Sitting).

[4] Adopted by the 18th WMA General Assembly, Helsinki, Finland, June 1964 and amended by the: 29th WMA General Assembly, Tokyo, Japan, October 1975 35th WMA General Assembly, Venice, Italy, October 1983 41st WMA General Assembly, Hong Kong, September 1989 48th WMA General Assembly, Somerset West, South Africa, October 1996.

[5] The Human rights, ethical and moral dimensions of health care: One hundred and twenty cases examined from the standpoint of legal norms, international and European ethics and the catholic, Protestant, Jewish, Muslim, Buddhist, and agnostic moralities, presented as an aid to decision-making and teaching by the European Scientific Co-operation network on "Medicine and Human Rights" of the European federation of Scientific networks. Council of Europe Publishing, March 1998.

Oluwatosin Ademola Ijabadeniyi
11 Nanofood and ethical issues

Abstract: An ethical and sustainable food system with potential to enhance food security could be achieved if nanotechnology techniques or tools are properly applied
during the cultivation, production, development, and packaging of foods. For this
to be realized, nanoagriculture and nanofoods must be protected from nanomaterials that may cause harm to human health and the environment. Stakeholders in
food nanotechnology should ensure that ethical considerations are at the top of
their priority so as to promote public health. The chapter concludes that a satisfactory food nanotechnology ethical analysis approach should focus on autonomy,
non-maleficence, beneficence, and justice. Other ethical principles that are equally
important to achieve food security include respect for human dignity and sustainable food production. The onus is also on nanofood scientists to operate with this
principle "everything that scientists are capable of, shouldn't be done if it fails ethical standard."

Keywords: nanoagriculture, nanofoods, ethical

11.1 Introduction

11.1.1 Background information about nanotechnology and definition

Nanotechnology is the manufacture and application of materials and structures at
the nanometer scale [1]. It has also been defined as the study of manipulation of
matter in atomic and molecular structure [2]. Currently, nanotechnology has developed into a multidisciplinary research sector encompassing physical, chemical, biological, engineering, and electronic sciences with a lot of potential for industrial
applications [3, 4]. The main products of nanotechnology, that is, nanomaterial and
nanoparticles have been described by FAO/WHO [5] as any material that has one or
more dimensions in the nanoscale range (1–100 nm) and discrete entity that has all
three dimensions in the nanoscale respectively. There are numerous examples of
nanomaterials and nanoparticles; however, those that are intensively studied include gold nanoparticles, semi-conductor quantum dots, polymer nanoparticles,
carbon nanotubes, nanodiamonds, and graphene [6].

Oluwatosin Ademola Ijabadeniyi, Department of Biotechnology and Food Science, S9 L1, Durban
University of Technology, Steve Biko Campus, Durban, 4001, South Africa;
oluwatosini@dut.ac.za, tosynolu@yahoo.com

https://doi.org/10.1515/9783110669282-011

Nanotechnology is one of the key technologies of the twenty-first century with great potential to economically and scientifically transform lives [7]. The first commercial application of nanotechnology was in material science, microelectronics, aerospace, and pharmaceutical industries [8]. It was also initially employed to manufacture construction materials for floors and walls as well as constituents for cosmetics, sporting equipment, and wastewater treatment [9]. According to Dingman [10], nanotechnology was used to manufacture self-cleaning glass and army uniforms that monitor the health of the wearer to camouflage those changes to match its surroundings. Recent applications of different nanomaterials are in sample preparation techniques [11], highly active photocatalysts [12], biosensors [6], biofuel production [13], flexible and stretchable bio-electronic devices [14], and electrochemical detection of heavy metals [15].

Nanotechnology has also found application in the food industry recently. Food produced through direct or indirect application of nanotechnology or nanomaterials has been referred to as nanofood. A food is termed "nanofood" when nanoparticles, nanotechnology techniques, or tools are used during cultivation, production, processing, or packaging of the food. Also, nanofood can be derived after foods, seeds, chemical pesticides, and food packaging have been broken down and manipulated at the micro-scale level through nanotechnology [4, 16, 17]. Table 11.1 presents examples of nanofoods. The benefits of food nanotechnology are discussed in the next chapter.

Table 11.1: Examples of foods, food packaging, and agriculture products that now contain nanomaterials [4, 18].

Type of product	Product name and manufacturer	Nano-content	Purpose
Beverage	Oat chocolate and oat vanilla nutritional drink mixes; toddler health	300 nm particles of iron (SunActive Fe)	Nano-sized iron particles have increased reactivity and bioavailability
Food additive	Aquasol preservative; AquaNova	Nanoscale micelle (capsule) of lipophilic or water insoluble substances	Nano-encapsulation increases absorption of nutritional additives and increases effectiveness of preservatives and food processing aids. Used in wide range of foods and beverages
Food additive	Bioral™ Omega-3 nanocochleates; BioDelivery Sciences International	Nano-cochleates as small as 50 nm	Effective means for the addition of highly bioavailable Omega-3 fatty acids to cakes, muffins, pasta, soups, cookies, cereals, chips, and confectionery

Table 11.1 (continued)

Type of product	Product name and manufacturer	Nano-content	Purpose
Food additive	Synthetic lycopene; BASF	LycoVit 10% (<200 nm synthetic lycopene)	Bright red color and potent antioxidant. Sold for use in health supplements, soft drinks, juices, margarine, breakfast cereals, instant soups, salad dressings, yoghurt, crackers, etc.
Food contact material	Nanosilver cutting board; A-Do Global	Nanoparticles of silver	"99.9% antibacterial".
Food contact material	Antibacterial kitchenware; Nanocare Technology/NCT	Nanoparticles of silver	Ladles, egg flips, serving spoons, etc., have increased antibacterial properties
Food packaging	Food packaging Durethan® KU 2–2601 plastic wrapping; Bayer	Nanoparticles of silica in a polymer-based nanocomposite	Nanoparticles of silica in the plastic prevent the penetration of oxygen and gas of the wrapping, extending the product's shelf life. To wrap meat, cheese, long-life juice, etc.
Food packaging	Nano-ZnO plastic wrap; SongSing nanotechnology	Nanoparticles of zinc oxide	Antibacterial, UV-protected food wrap.
Plant growth treatment	PrimoMaxx, Syngenta	100 nm particle size emulsion	Very small particle size means mixes completely with water and does not settle out in a spray tank

11.2 Application of nanotechnology in food production, processing, and packaging

Nanotechnology has diverse application in the food sector at the moment and it may likely change the whole agrifood sector in the nearest future [4, 19]. According to Sekhon [20], food nanotechnology has the potential to open up a whole new set of possibilities for the food industry [20]. Detailed examples of nanofood applications are presented in Figure 11.1. FSAI [21] has also summarized their applications into the following six categories:
1. sensory improvements (flavor/color enhancement and texture modification)
2. increased absorption and targeted delivery of nutrients and bioactive compounds
3. stabilization of active ingredients such as nutraceuticals in food matrices

4. packaging and product innovation to extend shelf-life
5. sensors to improve food safety and
6. antimicrobials to kill pathogenic bacteria in food.

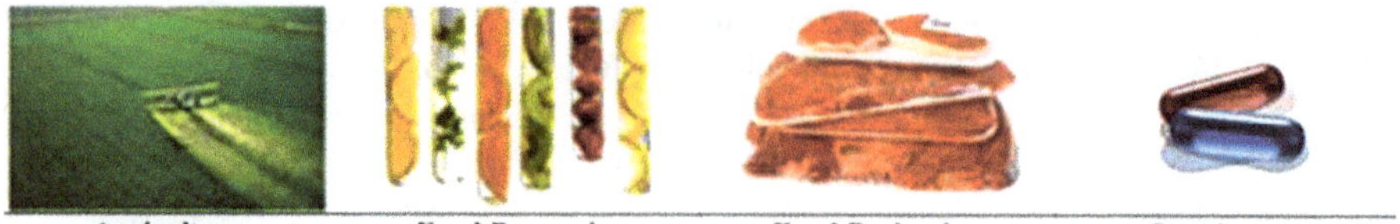

Agriculture	Food Processing	Food Packaging	Supplements
• Single molecule detection to determine enzyme/ substrate interactions • Nanocapsules for delivery of pesticides, fertilizers and other agrichemicals more efficiently • Delivery of growth hormones in a controlled fashion • Nanosensors for monitoring soil conditions and crop growth • Nanochips for identity preservation and tracking • Nanosensors for detection of animal and plant pathogens • Nanocapsules to deliver vaccines • Nanoparticles to deliver DNA to plants (targeted genetic engineering)	• Nanocapsules to improve bioavailability of neutraceuticals in standard ingredients such as cooking oils • Nanoencapsulated flavor enhancers • Nanotubes and nanoparticles as gelation and viscosifying agents • Nanocapsule infusion of plant based steroids to replace a meat's cholesterol • Nanoparticles to selectively bind and remove chemicals or pathogens from food • Nanoemulsions and –particles for better availability and dispersion of nutrients	• Antibodies attached to fluorescent nanoparticles to detect chemicals or foodborne pathogens • Biodegradable nanosensors for temperature, moisture and time monitoring • Nanoclays and nanofilms as barrier materials to prevent spoilage and prevent oxygen absorption • Electrochemical nano- sensors to detect ethylene • Antimicrobial and antifungal surface coatings with nanoparticles (silver, magnesium, zinc) • Lighter, stronger and more heat-resistant films with silicate nanoparticles • Modified permeation behavior of foils	• Nanosize powders to increase absorption of nutrients • Cellulose nanocrystal composites as drug carrier • Nanoencapsulation of neutraceuticals for better absorption, better stability or targeted delivery • Nanocochleates (coiled nanoparticles) to deliver nutrients more efficiently to cells without affecting color or taste of food • Vitamin sprays dispersing active molecules into nanodroplets for better absorption

Figure 11.1: Examples of nanofood applications [16].

From the categories listed above, it can be inferred that nanotechnology will improve food processing, packaging, and safety; it will enhance flavor and nutrition; it will lead to production of more functional foods from everyday foods with added medicines and supplements and; it will result in increased food production and cost effectiveness [4]. Furthermore, the unique properties of nanomaterials have opened up possibilities of addressing food safety and quality problems in a new way and from a totally new perspective [22].

However, nanotechnology also has an immense application on the farm. Nanotechnology has the potential to mitigate the negative effects of chemical agrochemicals along with ecosystem biomagnification caused by them [23]. According to Mukhopadhyay [24], "Nanotechnologic intervention in farming has bright prospects for improving the efficiency of nutrient use through nanoformulations of fertilizers, breaking yield barriers through bionanotechnology, surveillance and control of pests and diseases, understanding mechanisms of host-parasite interactions at the molecular level, development of new-generation pesticides and their carriers." Other applications on the farm according to Mukhopadhyay [24] include clay-based nanoresources for precision water management, reclamation of salt-affected soils, and stabilization of erosion-prone surfaces. Nanotechnology may therefore be one of the technologies that can successfully mitigate against the challenges of climate change and food insecurity because it can help to address the issue of sustainability [25].

The leading application of nanotechnology in the food industry is in packaging. Hybrid nanostructured materials with improved mechanical, thermal, and gas properties have been used to package the food. The nanomaterials were able to increase its shelf life and also provide a more environmentally friendly solution because of reduction on reliance on plastics as packaging materials [4, 26, 27]. Many nanomaterials are also environmentally friendly. Zein, a bionanocomposite, can give rise to biodegradable zein films with good tensile and water-barrier properties when dissolved in ethanol or acetone [27]. In addition, nanomaterials used in packaging have antimicrobial properties. Emamifar et al. [28] showed that packaging materials made from nanocomposite film containing nanosilver and nanozinc oxide were significantly able to reduce microorganisms that could cause spoilage in orange juice. Novel food packaging technology may in fact be the most promising benefit of nanotechnology in the food industry in the near future, and food companies are said to have started producing packaging materials based on nanotechnology that are delaying spoilage and improving microbial food safety [16].

Another area of interest and equally of great importance for the food industry is food safety and preservation [4, 29]. Nanosensors in packaging materials serving as "electric tongue" or noses can help to detect food pathogens in food [30]. According to [31], nanosensor is reported to have the capability to recover even just one *E. coli* bacterium located in ground beef [31]. Furthermore, nanosensor is able to measure safety at real time and the procedures are quick, sensitive, and less labor-intensive [32]. The future outlook of biosensors however is shifting to usage in detecting allergens, monitoring fermentations, and product quality [33].

11.3 Risks of nanofood

As discussed above, food nanotechnology provides opportunities and benefits such as lower pesticide use, improved traceability, and safety of food products, yet there are some risks associated with food nanotechnology [3]. Table 11.2 shows examples of food nanotechnology-related potential risks and hazards but also benefits.

The risks described in the table above are a cause of concern. Nanoparticles when used in food may cause harm to consumers, workers, and the environment because of their unique properties. According to Maria et al. [34], nanomaterials have very different properties such as greater reactivity, strength, fluorescence, and conduction compared to micrometer-sized materials.

Table 11.2: Food nanotechnology – benefits and risks.

Benefits	Risks
Lower pesticide use	Potential human health, e.g oxidative damage &
Improved traceability & safety of food products	inflammation of GI; acute toxic responses
Reductions in fat, sugar, salt, and preservatives	(cancers, lesions of liver & kidney)
Enhanced nutritional value of food/beverages	Concerns for workers health & safety
Novel flavors & textures	Potential harmful effects to the environment
Maintenance of food quality & freshness	
More hygienic food processing	
Extended product shelf life	

Adapted from Handford et al. [25]

Also, many other researchers have previously reported about concerns of food nanoparticles. According Das et al. [32], unexpected toxicity may occur as a result of nanoparticles interacting with living systems. Some nanoparticles are able to cross biological barriers, for example blood-brain barrier, eventually accessing cells and organs [1]. The possibility of nanomaterials presenting different hazards from those of the same material in a micro or macro form have also been reported [21]. Cheftel [35] concluded that the use of nanoparticles in foods or food contact materials are associated uncertainties and safety concerns.

Furthermore, in-depth potential risks of nanomaterials to human health and even to the environment are still unknown [4, 36, 37]. This is because there is little or no scientific information on the effects of nanotechnology applications on human or animal health and also the environment [4, 38]. It is therefore essential to apply appropriate precautionary principles for certain potential nanoparticles that can cause severe and irreversible harm to humans [7].

Concerns that may arise as a result of consumption of nanofoods have been grouped into three major areas, they are, least concern, some concern, and major concern [4]. Processed food with nanomaterials that are not biopersistent when digested or solubilized in the gastrointestinal tract are classified as area of least concern [39]. Food products that contain nonbiopersistent nanomaterials but carry across the gastrointestinal tract are areas of some concern. The areas of major concern, however, are where foods include insoluble, indigestible, and potentially biopersistent nano-additives (e.g. metals or metal oxides) or functionalized nanomaterials [39]. Such applications may pose a risk of consumer exposure to "hard" nanomaterials – the ADME profile (adsorption, distribution, metabolism, and elimination) and toxicological properties of which are not fully known at present [39]. Also since most nanomaterials used in foods are organic moieties, they may present some concern because they are able to carry other foreign substances into the blood through the nutrient delivery system (2001) [40].

Powell et al. [41] has also reported that it is possible under certain conditions for very small nanoparticles to gain access to the gastrointestinal tissue via paracellular transcytosis across tight junctions of the epithelial cell layer. However, whether there are realistic situations of nanoparticle exposure that lead to significantly abnormal reactive oxygen species (ROS) and inflammasome activation responses in vivo in the gut remains have not been established [4]. Lack of knowledge regarding the effect on pharmoacokinetics and bioavailability of changes in the physicochemical properties of normally inert and non-biodegradable materials such as inorganic particles, for example titanium dioxide, and biological polymers in moving to the nanoscale has been reported [21]. This is concerning because changes may occur with potential cascade effects on cellular homeostatis when they get into the body system [21]. Nanomaterials also pose a major risk to the environment especially through nanopackaging. According to Fabrega et al. [42], nanosilver in packaging could leach into the environment and in the long run contaminate the food supply.

The potential risks associated with nanotechnology applications in agrifood may therefore lead to similar pushback as observed with the GMO in Europe [43].

11.4 The role of ethics in responsible nanofood innovation

Food ethics, which is the interdisciplinary field that ensures decent analysis and guidance from farm to fork, is necessary for food innovation. Ethics however do not operate in isolation. They are influenced by numerous factors. For example, values such as moral, cultural, religious, and political influence ethical behaviors of actors in the food industry. According to Kaiser and Lien [44], the ethics and the politics of food cannot be separated so like other values. The increasing gap or distance between production processes and consumption has made many consumers prioritize ethics when choosing or purchasing food [45].

Not all the assumptions about the risk and dangers of nanomaterials in food may be accurate, however, it is ethically necessary to fill the present knowledge gaps through more research and thorough risk assessment of nanofoods and nanopackaging materials [4]. Ethics matter in responsible innovation, that is, innovation that is not only focused on profit but also people's well-being and environmental sustainability. A satisfactory ethical analysis approach for food nanotechnology should therefore be of autonomy, non-maleficence, beneficence, and justice [1, 46, 47]. Other ethical principles that are equally important to achieve food security include respect for human dignity and sustainable food production [38]. Ethical matrix analysis developed by Mepham [48] described autonomy as self-determination, non-maleficence as no harm, beneficence as do good, and justice as fairness.

Out of all these four categories, the key for food industries applying nanoparticles in their products or materials is non-maleficence. This is because of the degree of unknown risk associated with nanotechnology application in food and the possibility of harm for consumers, animals, plants, and even future generation [1, 49], where it is known that certain nanomaterials could be hazardous to humans, plants, or the environment; it will be unethical for such nanomaterials to be used directly or indirectly in food production. For example, cerium dioxide (nano-CeO_2) has been shown to cause leaf stress and damage to N_2 fixation in soybean. Zinc oxide nanoparticle (nano-ZnO) treatment also gave rise to gene toxicity [50]. The two nanoparticles could also change the nutritional quality of soybean invariably affecting plants', humans' and animals' health [51]. Other crops in which adverse effects caused by nanoparticles have been observed include kidney beans [52], lettuce [53], and watermelon [54].

The fact that food nanotechnology is able to do some good (beneficence) cannot be overemphasized however non-maleficence factor should be prioritized. Risk/benefit balance analysis should first and foremost be determined by manufacturers, processers, or producers. Equally important is complete transparency and risk/benefit communication about potential risks and benefits to consumers and other stakeholders [1]. Unfortunately, food manufacturers have not been transparent enough and also not communicating enough to consumers. This could be the reason why there is a lack of consumer trust in the food and beverage market [55]. Lack of consumer trust was evident in a survey conducted by the Center for Food Integrity in 2017 and 2018. Consumers were asked to respond to the question, "Am I confident in the safety of the food I eat." In 2017, 47% of consumers strongly agreed with the statement while only 33% agreed with it in 2018 (Grylls). An ethical manufacturer should therefore seek to instill trust to customers and consumers reassuring them that their products are safe, healthy, and sustainable.

Food manufacturers or producers should also ethically consider the justice or fairness factor. Those exposed to risks from nanomaterials should be made aware and should be compensated after being exposed to risks. Equally important is the ethical principle of autonomy, that is, the right of consumers to choose whether they wish to be exposed to unknown potential risks. It also involves the capacity of a rational individual to make an informed, uncoerced decision. Central to this is the provision of adequate and appropriate labeling to assist consumers or customers with identifying ingredients, products along with processes used in food production and processing [1, 56]. Furthermore, open discussion and public debate about the opportunities and threats should be allowed so that people can have independent opinions. For example, the European Commission has directed Member States to strengthen public debate on benefits, risks, and uncertainties related to nanotechnology [57].

Manufacturers and producers alike should also make good governance and commitment to food regulation part of their core values since they are essential attributes of an ethical behavior. It must be emphasized that there is little or not enough nanofood regulation. It was only the European Union that had adopted a mandatory labeling regulation requiring food ingredients to be listed as "nano" if they fit with their definition of engineered nanomaterials; however, the regulation was not detailed [58–60]. Government and other relevant bodies should develop national and global legislation for regulation of nanotechnology in food so as to enable proper development of nanofood. Regulatory development for nanofoods should be carried out through a three-phase process described by [61] which includes: (1) the use of research and development database to assess applications of nanotechnology to food and agriculture; (2) selecting particular products to assess and identify the risks and benefits; and (3) extrapolating to analyze appropriate regulatory or non-regulatory governance systems for the applications of nanotechnology in foods. Furthermore, criteria such as particular size range and measure, physical and chemical properties, and processing and safety concerns are needed to be considered for the development of the standard, definition, control measure, and regulation of nanofood [62].

Risk governance of nanomaterials in the agro and food industry is vital [29]. There should be proper risk-benefit assessment and risk management. It is also important that nanotechnology innovation does not proceed ahead of nanoagriculture and nanofood policy and regulatory system [63]. A situation where nanoparticles are deliberately included in food without approval by the regulatory system should be discouraged. According to [64], nanoparticles are intentionally added in different food products in Australia despite repeated claims by Australia's food regulator.

Different governmental and international organizations therefore have roles to play in nanotechnology governance. They must ensure that the nanofoods are produced, developed and marketed in an atmosphere of adequate and unrestrictive regulation. A restrictive regulation discourages innovation and development thereby negatively affecting the nanofood industry [29]. A direct form of regulation however can be implemented, such as premarket-market authorization, mandatory labelling and the establishment of a public register of nanofoods and producers [65].

International cooperation (among States within the international organizations such as OECD, EU, AU, WHO, FAO, ISO, USFDA, Codex, etc.) is required to develop a proper risk governance for nanofood and nanoagriculture. International Organization of Standardization (ISO) in particular has an important role to play in the governance of food nanotechnology. Examples of current initiatives in ISO relevant for governance of nanotechnology include: the technical committee on nanotechnologies (TC229) and the guidance on social responsibility (ISO 26000) [66].

Furthermore, food manufacturers should show a commitment to corporate social responsibility (CSR). Such sincere commitments will involve following high ethical standards. Not only would CSR enhance the company's overall reputation but it may lead to improved profitability as a result of consumers' preference for products from companies with perceived fulfillment of legal expectations of CSR [67]. The conceptual framework for CSR corporate identity, governance, and nano-technology is shown in Figure 11.2. The model showed that CSR corporate identity antecedents and components have roles to play in enhancing the legitimacy of nanotechnology and nanofoods.

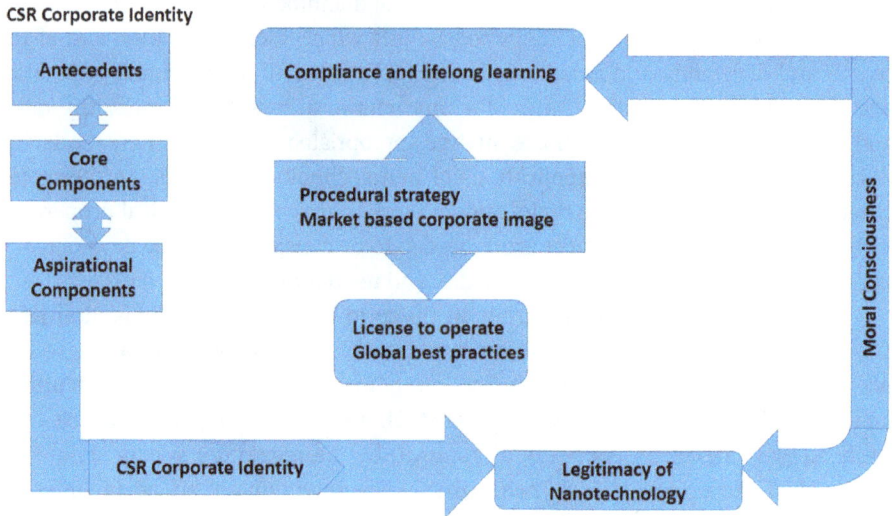

Figure 11.2: Conceptual framework for CSR corporate identity, governance, and nanotechnology. Adapted from: Ijabadeniyi and Ijabadeniyi [33]

11.5 Ethical standards for nanofood researchers and developers

Scientists and developers of nanofood have responsibility to be aware of and adhere to regulations related to nanotechnology. They should work with this principle: "everything that scientists are capable of, shouldn't be done if it fails ethical standard." All stakeholders should ensure an environment that encourages ethical practices through education, stewardship, and clear and fair policies and practices that promote ethics during research and development of nanofoods as well as integrity and compliance. Wellness of people and the planet should be prioritized above abnormal profit and irresponsible innovation, which do more harm than good.

Iavicoli et al. [19] have also recommended adequate risk management strategies for agriculture workers directly and indirectly applying nanotechnology. The authors also emphasized the importance of occupational safety practices and policies.

Nanofood researchers should adhere to these four ethical duties when carrying out research: (a) the duty to show respect for people; (b) the duty to alleviate suffering; (c) the duty to be sensitive to cultural differences and different cultural perspectives which individuals might bring to questions of health and health care; and (d) the duty to not exploit the vulnerable or weak for own advantage [68].

Furthermore, nanofood developers and researchers must accept that though food nanotechnology has a lot of potential, societal and ethical issues have major roles to play for successful adoption. According to [69], issues such as sustainability, naturalness, risk management, innovation trajectories, and economic justice may present objections to industrial biotechnology innovation. Similar issues may also affect food nanotechnology progress if commitment to responsible research and innovation is lacking in addition to the absence of principles that promote trust among stakeholders in the food supply chains.

11.6 Conclusion

Stakeholders in food nanotechnology must make sure that ethical considerations are at the top of their priority so as to allow for acceptance from the public. Not all the negative assumptions about the risk and dangers of nanomaterials in food may be accurate; however, it is ethically necessary to fill the present knowledge gaps through more research and thorough risk assessment of nanofoods and nanopackaging materials. A satisfactory food nanotechnology ethical analysis approach should focus on autonomy, non-maleficence, beneficence, and justice. Other ethical principles that are equally important to achieve food security include respect for human dignity and sustainable food production. Scientists and developers of nanofood should strive to have ethical values. Responsibility is on them to be aware of and adhere to regulations related to nanotechnology.

References

[1]   Coles D, Frewer LJ. Nanotechnology applied to European food production – A review of ethical and regulatory issues. Trends Food Sci Technol 2013, 34, 32–43.
[2]   Chellaram C, Murugaboopathi G, John AA, Sivakumar R, Ganesan S, Krithika S, Priya G. APCBEE Procedia 2014, 8, 109–13.
[3]   Handford CE, Dean M, Henchion M, Spence M, Elliott CT, Campbell K. Implications of nanotechnology for the agri-food industry: Opportunities, benefits and risks. Trends Food Sci Technol 2014, 40, 226–41.
[4]   Ijabadeniyi OA. Quality and Safety of Nanofood, in Nanotechnology in Agriculture and Food Science. Axelos MA, Van de Voorde MH eds. Weinheim, Germany, Wiley-VCH Verlag GmbH & Co. KGaA, 2017. https://doi.org/10.1002/9783527697724.ch17. Accessed March 8 2020.

[5] FAO/WHO expert meetings on the application of nanotechnologies in the food and agriculture sectors: Potential food safety implications. Meeting report 2010. Available from www.fao.org/docrep/012/i1434e/i143e00.pdf. Accessed February 24 2020.

[6] Holzinger M, Goff AL, Cosnier S. Nanomaterials for biosensing applications: a review. Front Chem 2014, 27. https://doi.org/10.3389/fchem.2014.00063. Accessed March 4 2020.

[7] Bachmann A, Diskurs EI. 2007. Synthetic nanoparticles and the precautionary principle, an ethical analysis. www.ekah.admin.ch/inhalte/_migrated/content_uploads/e-Gutachten-Synthetische-Nanopartikel-2007_01.pdf. Accessed February 14, 2020

[8] Weiss J, Takhistoy P, McClements DJ. Functional materials in food nanotechnology. J Food Sci 2006, 71, R107–R116.

[9] Doyle ME. 2006. Nanotechnology: A brief literature review. Food Research Institute briefings. University of Wisconsin. Available from http://fri.wisc.edu/docs/pdf/FRIBrief_Nanotech_Lit_Rev.pdf, Accessed March 7 2020.

[10] Dingman J. Nanotechnology: It's Impact on food safety. J Environ Health 2008, 70, 47–50.

[11] Tian J, Xu J, Zhu F, Lu, Su C, Ouyang G. Application of nanomaterials in sample preparation. J Chromatogr 2013, 1300, 2–16.

[12] Kandiel TA, Feldhoff A, Robben L, Dillert R, Bahnemann DW. Tailored titanium dioxide nanomaterials: Anatase Nanoparticles and brookite nanorods as highly active photocatalysts. Chem Mater 2010, 22, 2050–60.

[13] Verma ML, Puri M, Barrow CJ. Recent trends in nanomaterials immobilised enzymes for biofuel production. Crit Rev Biotechnol 2014, 36, 108–19.

[14] Choi S, Lee H, Ghaffari R, Hyeon T, Kim D. Recent advances in flexible and stretchable bio-electronic devices integrated with nanomaterials. Adv Mater 2016, 28, 4203–18.

[15] Palisoc S, Vitto RIM, Natividad M, (2019). Determination of Heavy Metals in Herbal Food Supplements using Bismuth/Multi-walled Carbon Nanotubes/Nafion modified Graphite Electrodes sourced from Waste Batteries. Scientific Reports 9. https://www.nature.com/articles/s41598-019-54589-x. Accessed March 2, 2020.

[16] Garber C, (2007). Nanotechnology food coming to a fridge near you. http://www.nanowerk.com/spotlight/spotid=1360.php. Accessed June 11, 2011.

[17] Scrinis G, (2010). Nanotechnology: Transforming Food and the Environment. http://www.foodfirst.org/en/node/2862. Accessed June 15, 2011.

[18] FOE (Friends of the Earth) 200). Out of the laboratory and onto our plates: Nanotechnology in food and agriculture. Available at http://midgetechnology.com/Documents/Nano%20Out%20of%20the%20Lab%20On%20To%20Our%20Plstes.pdf (Accessed June 16, 2011).

[19] Iavicoli I, Leso V, Bee DH, Shvedova AA. Nanotechnology in agriculture: Opportunities, toxicological implications, and occupational risks. Toxicol Appl Pharmacol 2017, 329, 96–111.

[20] Sekhon BS. Food nanotechnology – an overview. Nanotechnol Sci Appl 2010, 3, 1–15.

[21] FSAI. The Relevance for Food Safety of Applications of Nanotechnology in the Food and Feed Industries. Food Safety Authority of Ireland Abbey Court eds. Dublin, Food Safety Authority of Ireland, 2008, 82.

[22] Kalita D, Baruah S. The Impact of nanotechnology on food. Adv Nanomater 2019. https://doi.org/10.1016/B978-0-12-814829-7.00011-2. Accessed March 6 2020.

[23] Baker S, Volova T, Prudnikova PS, Satish S, Nagenda PMN. Nanoagroparticles emerging trends and future prospect in modern agriculture system. Environ Toxicol Pharmacol 2017, 53, 10–17.

[24] Mukhopadhyay SS. Nanotechnology in agriculture: prospects and constraints. Nanotechnol Sci Appl 2014, 7, 63–71.

[25] Fraceto LF, Grillo R, De Medeiros GA, Scognamiglio GR, Bartolucci C. Nanotechnology in agriculture: Which innovation potential does it have? Front Environ Sci 2016, 4. https://doi.org/10.3389/fenvs.2016.0020. Accessed March 2 2020.

[26] Perch H, (2007). How is Nanotechnology being used in Food Science? http://www.understandingnano.com/food.html. Accessed June 11, 2011.

[27] Sozer N, Kokini JL. Nanotechnology and its applications in the food sector. Trends Biotechnol 2009, 27, 82–89.

[28] Emamifar A, Kadivar M, Shahedi M, Zad-Soleimanian S. Effect of nanocomposite packaging containing Ag and ZnO on inactivation of Lactobacillus plantarum in orange juice. Food Control 2011, 22, 408–13.

[29] Ijabadeniyi OA, Ijabadeniyi A. Governance of Nanoagriculture and Nanofoods in Nanoscience and Nanotechnology, Advances and Developments in Nano-sized Materials. Van de Voorde M eds. Berlin, De Gruyter, 2018, 88–100.

[30] Bhattacharyya A, Datta PS, Chandhmi P, Barik BR (2011). Nanotechnology- A new frontier for food security in socio economic development. http://disasterresearch.net/drvc2011/paper/fullpaper_22.pdf. Accessed June 15, 2011.

[31] Lilie M, Cantini AA, (2001). Nanotechnology in agriculture and food processing. Conference proceedings, University of Pittsburgh, Eleventh Annual Freshman Conference, April 9. 1–9.

[32] Das M, Saxena N, Dwivedi PD. Emerging trends of nanoparticles application in food technology: Safety paradigms. Nanotoxicology 2009, 3, 10–18.

[33] Warriner K, Reddy SM, Namvar A, Neethirajan S. Developments in nanoparticles for use in biosensors to assess food safety and quality. Trends Food Sci Technol 2014, 20, 183–99.

[34] Maria L, Moses JA, Anandharamakrishnan C. Ethical and Regulatory Issues in Applications of Nanotechnology in Food in Food Nanotechnology, Principles and Applications. Anandharamakrishnan C, Parthasarathi S eds. London, Taylor and Francis, 2019.

[35] Cheftel CJ, 2011. Emerging risks related to food technology. Advances in Food Protection. NATO Science for peace. http://www.springerlink.com/content/p585630701412061/. Accessed June 17, 2011.

[36] Chaudhry Q, Scotter M, Blackburn J, Ross B, Boxall A, Castle L, Aitken R, Watkins R. Applications and implications of nanotechnologies for the food sector. Food Addit Contam 2008, 25, 3, 241–58.

[37] Dowling AP. Development of nanotechnologies. Mater Today 2004, 7, 30–35.

[38] Casabona CMR, Epifanio LES, Cirion AE 2010. Global food security: ethical and legal challenges. http://doi.org/10.3920/978-90-8686-710-3. Accessed February 14, 2020.

[39] Chaudhry Q, Castle L. Food applications of nanotechnologies: An overview of opportunities and challenges for developing countries. Trends Food Sci Technol 2011, xx, 1–9.

[40] Mukul D, Ansari KM, Anurag T, Dwivedi PD. Need for safety of nanoparticles used in food industry. J Biomed Nanotechnol 2001, 7, 13–14.

[41] Powell JJ, Faria N, Thomas-mckay E, Pele CL. Origin and fate of dietary nanoparticles and microparticles in the gastrointestinal tract. J Autoimmun 2010, 34, J226–J233.

[42] Fabrega J, Luoma SN, Tyler CR, Galloway TS, Lead JR. Silver nanoparticles: behavior and effects in the aquatic environment. Environ Int 2011, 37, 517–31.

[43] Malsch I, Hvidtfelt-Nielsen. 2010. Nanobioethics: Observatory nano 2nd annual report on ethical and societal aspects of nanotechnology. https://www.researchgate.net/publication/266088759. Accessed February 14, 2020.

[44] Kaiser M, Lien ME 2006. Ethics and the politics of food. https://doi.org/10.3920/978-90-8686-575-8

[45] Korthals M (2006). Ethics of Food Production and Consumption. https://doi.org/10.1533/9781845692506.5.624. Accessed March 8 2020.

[46] Ebbesan M, Andersen S, Besenbacher F. Ethics in nanotechnology: starting from scratch? Bull Sci Technol Soc 2006, 26, 451–62.

[47] Kuzma J, Besley JC. Ethics of risk analysis and regulatory review: from bio-to nanotechnology. Nanoethics 2008, 2, 149–62.

[48] Mephem B. A framework for the ethical analysis of novel foods: The ethical maxtrix. J Agric Environ Ethics 2000, 12, 165–67.

[49] Reddy VLP, Viezcas HJA, Peralta-Videa JR, Gardea-Torresdey JL. Lessons learned: Are engineered nanomaterials toxic to terrestrial plants? Sci Total Environ 2016, 568, 470–79.

[50] Priester JH, Moritz SC, Espinosa K, Ge Y, Wang Y, Nisbet RM, Schimel JP, Goggi AS, Torresdey-Gardea JL, Holden PA. Damage assessmemt for soybean cultivated in soil with either CeO2 or ZnO manufactured nanomaterials. Sci Total Environ 2017, 579, 1756–68.

[51] Peralta-Videa JR, Hernandez-Viezcas JA, Zhao L, Diaz BC, Ge Y, Priester JH, Holden PA, Gardea-Torresdey JL. Cerium dioxide and zinc oxide nanoparticles alter the nutritional value of soil cultivated soybean plants. Plant Physiol Biochem 2014, 80, 128–35.

[52] Majumdar S, Peralta-Videa JR, Bandyopadhyay S, Castillo-Michel H, Hernandez-VIEZCAS J, Sahi S, Gardea-Torresdey JL. Exposure of cerium oxide nanoparticles to kidney bean shows disturbance in the plant defense mechanisms. J Hazard Mater 2014, 278, 279–87.

[53] Trujilo-Reyes J, Majumdar S, Botez CE, Peralta-Videa JR, Garden-Torresdet JL. Exposure studies of core-shell Fe/Fe3O4 and Cu/CuO NPs to lettuce (Lactua sativa) plants: Are they a potential physiological and nutritional hazard? J Hazard Mater 2014, 267, 255–63.

[54 Wang Y, Hu J, Dai Z, Junli L, Huang J. In vitro assessment of physiological changes of watermelon (Citrullus lanatus) upon iron oxide nanoparticles exposure. Plant Physiol Biochem 2016, 108, 353–60.

[55] Grylls B. 2020. Are you confident in the safety of our food? https://www.newfoodmagazine.com/article/103050/are-you-confident-in-the-safety-of-our-food/. Accessed January 16, 2020.

[56] Mwaanga P, 2018. Risks, uncertainties and ethics of nanotechnology in Agriculture. https://www.intechopen.com/books/new-visions-in-plant-science/risks-uncertainties-and-ethics-of-nanotechnology-in-agriculture. Accessed March 2, 2020.

[57] Baran A. Nanotechnology: Legal and ethical issues. Econ Manage 2016, 8, 47–54.

[58] European Parliament and the Council, 2011. Regulation (EU) No 1169/2011 of 25 October 2011. Official Journal of the European Union. L304/18. http://faolex.fao.org/docs/pdf/den108120.pdf. Accessed May 27, 2016.

[59] Gruere GP. Implications of nanotechnology growth in food and agriculture in OECD countries. Food Policy 2012, 37, 191–198.

[60] Marrani, D. Nanotechnologies and novel foods in European Law. Nanoethics 2013, 7, 177–188.

[61] Jones PBC. A nanotech revolution in agriculture and the food industry. Available at http://www.isb.vt.edu/news/2006/artspdf/jun0605.pdf 2006, Accessed April 13, 2020.

[62] Chau C, Wu S, Yen G. The development of regulations for food nanotechnology. Trends Food Sci Technol 18, 2007, 269–280.

[63] Renn O, Roco MC. Nanotechnology and the need for risk governance. J Nanopart Res 2006, 8, 153–191.

[64] Lyons K, Smith N. Governing with ignorance: Understanding the Australian food regulator's response to nano food. Nanoethics 2018, 12, 27–38.

[65] Sodano V. 2015. Regulating food nanotechnologies: ethical and political challenges in Food ethics and innovation. Edited by Dumitras DE, Jitea LM and Aerts S. https://doi.org/10.3920/978-90-8686-813-1, February 21, 2020.

[66] Forsberg, E 2010. The role of ISO in the governance of nanotechnology. http://dx.doi.org/10.13140/RG.2.2.29759.10409. Accessed March 31, 2020.

[67] Ijabadeniyi A, Govender JP. Coerced CSR: Lessons from consumer values and purchasing behaviour. Corp Commun: Int J 2019, 24, 515–31.
[68] HPCSA. Health Professional Council of South Africa. 2008. General ethical guidelines for biotechnology research in South Africa. https://jutapharmapedia.co.za/files/media/10470/9cc22047-b081-a50f-7cc1-a4e7f466bbe6.pdf. Accessed February 25, 2020.
[69] Asveld L, Osseweijer P, Posada JA. Societal and Ethical Issues in Industrial Biotechnology. In: Fröhling M., Hiete M. (eds) Sustainability and Life Cycle Assessment in Industrial Biotechnology. Advances in Biochemical Engineering/Biotechnology, Springer, Cham, 173, 2019, https://doi.org/10.1007/10_2019_100. Accessed March 20, 2020.

Part IV: **Nanotechnology governance: Societal and legal aspects**

Ilise L. Feitshans

12 Nanoethics for safe work: philosophical foundations of safer nanodesign protecting workplace health

Abstract: Sound occupational health risk management programs are the grease for the machinery of powerful economic engines. Information provided through occupational health programs helps employers survive because accidents and disease are not simply expensive but wasteful. No one can afford waste in our economy. The fat to be trimmed, however, is not the same as the grease for the wheels and machinery that makes smooth commerce.

Keywords: occupational health law, human rights, health law, global health, universal rights, human right to health

12.1 International efforts to harness nanomaterials under law

The notion that no one will regulate nanotechnology is a concept from the past.[1] Popular demand for some type of nanotechnology regulation has caused a rethinking of the role in society played by regulatory governance of risk and risk management programs. New issues range from the questions of standardization using robust science to the policy opportunity for consciously removing embedded sexism and racism from regulatory frameworks. Despite a wide variety of opinions about the type of risk and methods for prevention, experts agree that precautionary approaches are necessary. The new paradigm for labor law in the twenty-first century must reflect that reality. Bringing together public health principles and existing workplace standards will require education, awareness, and outreach to nontraditional groups, such as professional associations, nongovernmental organizations (NGOs), and research institutes. Nanotechnology's revolution for commerce has resulted in an international

---

1 "Nanotechnology Law for Commercialization of Nano-Enabled Products" the Journ Applied Material Science & Engineering Research Vol 3 Issue 2. https://www.opastonline.com/current-issue-amse.

Ilise L. Feitshans, Director, ESI Safernano, Fellow in international law of Nanotechnology, European Scientific Institute Archamps France (ESI) and Guest Researcher Center for Biomedical Law, Faculty of Law University of Coimbra Portugal and Invited Professor ISTerre University of Grenoble France, e-mail: Ilise.feitshans@gmail.com

https://doi.org/10.1515/9783110669282-012

call for partnerships between governments and the producers of nanotechnology. In response, there have been rapid developments toward the codification of nanotechnology regulations on several levels throughout the world: governments within their own nations or working with several governments, under the auspices of international organizations, and trade associations and NGOs also preparing text for use as guidelines, model acts, and laws preventing not only liability but also a variety of additional legal concerns. Every nation and several regional organizations offer a legislated nanotechnology program whose underlying enabling legislation authorizes appropriations following a strategic agenda. These developments will call into question concepts of statehood and governance, asking whether these concepts are antiquated or, instead, are still so vibrant that they can apply to the rules that will govern nanotechnology. It remains unclear whether national governments or the superstructure of international governmental organizations in the United Nations (UN) will be the correct place to address emerging laws that will govern nanotechnology.

12.2 Principles of exposure to risk: employer responsibility for recognized hazards

Across the world, hundreds of federal, state, and international bodies of law are bottomed upon the philosophy that work-related illnesses are an avoidable aspect of industrialization and that consumer protection against risky products is a cornerstone of public health. Employer's acceptance of responsibility to provide safe and healthful working conditions is part of the paradigm in which an employer makes the choice to run an enterprise and employ people. But it is also clear that not every event that follows from the decision to grant employment is also a product of clear and discernible choice. It makes sense, however, that employers should be accountable and be included within society's repository for liability. Therefore, tools are needed to assist in the achievement of goals made by choice and reducing the risks that follow from those choices.

Occupational health and safety management laws are implemented through several channels for intervention to prevent workplace harms and thereby reduce corporate liability, such as risk management programs. Risk managers, industrial hygienists, and many additional health professionals use well-honed prevention strategies to stop problems from becoming catastrophically large. Such in-house programs, and the supporting infrastructure of regulatory agencies, work together to do much more for the economy than merely reduce the costs of accidents and the overall burden of disease in society. Nanotechnology laws, whether licensing new products or setting forth criteria for risk governance, can provide an opportunity to redirect the resources within these systems and thereby provide a more flexible but comprehensive system for protection of health than civil society has experienced in prior generations.

12.2.1 Emerging global regulatory principles governing risk at work

Looking back with twenty-first century hindsight, we may quibble whether the nineteenth and twentieth-century laws went far enough to protect enough people; whether slaves were the beneficiaries of these laws; whether poor people or other vulnerable populations were unduly expendable; or whether the high economic costs of compliance actually retarded the progress of society. The twenty-first century's emerging style of governance has already demonstrated that a unique admixture of public power, combined with a strong infusion of procedures, data, and policies from corporate and private resources, has profound implications for future models of governance. Throughout the world, people are questioning whether democracy remains a vital form of governance or merely an antiquated path for those whose societies are rooted in the heritage of ancient Greece antecedent to the Roman Empire. Meanwhile, new forms of governance, including royal-based governance with democratic components, are taking hold across the world. The question which, if any, of these styles will be the mode of fashion or the enduring model?, creates jobs for political theorists.

These questions are surprisingly relevant for the creation of a regulatory regime that will foster the development of nanotechnology while protecting public health because the selection of the forum to create that regime may be as important as the text of the law itself. In the twentieth century, dynamic forces accomplished sweeping social changes that brought governmental funding of programs involving colossal unquantified risks, such as the risk of civil war that followed the creation of nationalized civil rights under law; mass social protection programs offering the middle class pensions, health care, and aid to people living in poverty; and the risks associated with funding "Big Science," including extravagant weapons. In the realm of science, dynamic social forces precipitated the development of programs designed to create, detonate, manufacture, store, decommission, and dismantle nuclear weapons – an expensive endeavor that consumed three lifetimes of research, from the 1930s to the present, including the Cold War. The thriving status of both the military–industrial complex and the world's civilian population testifies to the ability to balance profits and risks while promoting commerce and protecting public health. In the twenty-first century, several innovative approaches have been taken to introduce the information and concerns of end users, who are corporations, workers, multinational enterprises, multinational governments, national governments and federations, and consumers, into the discourse about nanotechnology's potential risks and how to address those risks under law. NanoImpactNet of the European Union (EU), for example, aspired to create "responsible development of manufactured nanoparticles" and to support the definition of regulatory measures and implementation of appropriate legislation in Europe by developing "strong two-way communication" to stakeholders and the EU. These developments and the growing trend toward multinational treaties and agreements governing applied nanotechnology represent an end to traditional control by governments in policymaking and implementation. This new era in governance started small, with attention to developments for terminology and regulations but foretells the possibility of a transnational approach to regulation of business and global commerce.

12.3 Nanotechnology impacting health and work as part of the human condition

A unique feature of emerging nanotechnology regulatory frameworks is their transnational collaboration, both within the EU and inside privately funded NGOs such as the International Organization for Standardization (ISO). Achieving the goal of including stakeholder views in nanotechnology risk governance, various legislative initiatives target key problems and then evaluate the inadequacies and gaps in compliance with a goal of demonstrating how performance can be improved through positive incentives for compliance. Model legislation features awards for excellence, preferential treatment favoring program participants, reduced sentencing, special risk pools in insurance, and protection of information generated by internal audits through the use of the self-evaluative privilege. Legislation enabling corporations and end users of nanomaterials to enjoy self-evaluative privilege gives discretion to protect special information concerning intellectual property and trade secrets, while making good faith efforts to obey the law. As such, self-evaluative privilege may be useful, especially as it might apply to the work of third-party auditors such as awards commissioners or accredited bodies under international customary agreements. More traditional tools for encouraging compliance also exist, with mixed results depending upon the prestige available in its proper context, for example, activity regarding recognition of compliance activities through the use of an award for excellence also empowers corporate compliance practitioners, such as technical experts in auditing, safety engineering, design, and industrial hygiene, and provides a place to discuss conditions on-site with impunity for people who might otherwise be unlikely to provide information [1]. Well-planned legislation that addresses risks, while balancing the development of commerce, is the hallmark of twentieth-century scientific precedents; emerging laws for nanotechnology can ensure that every stakeholder (manufacturer, researcher, consumer, or worker) will have the very best information at their fingertips when making decisions about nanotechnology use and its implications for the future.

Society's need for health at work is as ubiquitous and perennial as civilization and the DNA of life itself. Amid modern complexity, there remains constant underlying basic human needs: work, health, and survival have been inextricably linked throughout the history of human civilizations. Without work, society cannot survive, and no work can perpetuate society without health. No society has survived without producing things, without work. We enjoy the fruits of many past civilizations today as we draw upon their architecture such as the Pyramids, the Parthenon, and the Great Wall.

Civilizations can be brought to a halt in times of plague and pestilence; and even the most impressive of collective efforts can be stopped when injuries overtake any individual's ability to work. Society therefore needs both, working people and healthy people, for civilization to survive. But these classifications are not dichotomous or mutually exclusive. The fluid categories of sickness and health, which fluctuate within individual abilities and deficits and within the life of any given individual across time, hold implications for every worker in every job description, ranging from dignitaries in the highest offices of leadership, celebrities, and heads of state in North America or Europe to the laborers tearing apart old ships in the shipyards of Asia and from the boardroom to the mailroom. Philosophies and values embedded in ancient cultures can touch our daily lives even today. Indeed, the remnants that survive from ancient cultures are found in architecture, statues, and pottery artifacts of the skilled crafts and creative labors of lost societies. There has always been work as long as there has been human society. Without the work of architects, builders, and the slaves, who were driven by underpaid overseers and who once in a great while died in riots or revolutions, we would not have the wonderful artifacts of history and past cultures that form the foundations of our society today.

Without the hard work of steel workers and construction workers and oil and chemical workers, no one would have the marvelous modern buildings and simple structures that serve our society today. Since the time when people chipped stone to fashion tools, occupational diseases, as in this example, the respiratory ailment derived from crystalline free-silica dust (silicosis), threatened public health, although silica is still the subject of debate regarding its regulation. Every civilization has left records of occupational deaths in hunting, healing, construction, agriculture, and industry. Although still in use in the twentieth century without much thought given to its major hazards, lead was among the first metals known to the early Egyptians, Hebrews, and Phoenicians: lead colic and paralysis were mentioned by Dioscorides in the first century AD. People may debate whether one type of work is more important than another; some people will argue that such values are socially determined by the economic worth of a particular job; a job that is worth the minimum wage is valued less in society than a highly paid job, and therefore society must have a greater need for the highly paid job.

None of these types of work, not the great monuments, not the writings or the arts, could exist without a modicum of human health. Work is both the key to sustaining daily life and civil society's legacy for posterity, as seen in Figure 12.1, a scene from the video "Lessons Learned from Three Centuries of Occupational Health Laws." In this scene, actors recreate the work of indentured servants in colonial life, cooking and serving in the pre-revolutionary United States. The video was made thanks to the Indian King Tavern Museum in Haddonfield, New Jersey, and is archived by Digital 2,000 Productions. The museum and its historic local environment provide the backdrop for discussing the importance of early health and sanitary regulations regarding trade and commerce in food in eighteenth-century United States. These visions of the role of work may change across time due to social changes in society, and the types of jobs that are available may change because of nanotechnology, but the link will remain between work, health, and the survival of society.

Figure 12.1: Indentured servants cooking in eighteenth-century United States, from the video "Lessons Learned from Three Centuries of Occupational Health Laws".
Copyright: Dr. Ilise L. Feitshans.

12.3.1 What is health?

Answering the ancient question "what is health?" is the greatest challenge for the new generation of epidemiology and the occupational, environmental, and public health sciences in light of the developments in nanomedicine. The notion that social conventions and public health policy should seek equity and fairness instead of equality is an important contribution of disability law to international health jurisprudence. The simple textbook answer to this ancient question is, "Health is a state of complete physical, mental and social well-being and not merely the absence of disease and infirmity."[2]

More than a generation after the writing and ratification of these paradigm-defining words in the World Health Organization (WHO) Constitution, this phrase will have renewed meaning in the case of nanotechnology because nanomedicine will redefine old concepts such as disability and health. This definition will change dramatically because of nanotechnology, because early detection and presymptomatic treatments will change the meaning of "absence of disease or infirmity." Covid19 is just one example. No one predicted the specific viral attack that caused trillions of dollars of economic waste and lockdown of billions of people, even among people who studied pandemic preparedness. Yet no one disputed the WHO's role to prevent spread of the disease, even if the critical role was inadequate to protect civil society. The disease has also produced long-term effects, understood among the so-called long haulers.[3] New treatments will cause a paradigm shift in the stage of illness that confirms diagnoses, and consequently, people who are viewed as "healthy" in the greater society will be "treated" as if they are already ill. This may produce excellent health outcomes, but there also looms at large the specter of uninsured populations increasing in size and new forms of discrimination against a class of people who were not previously viewed as disabled. Convergence of new genetic technologies and the nanomedicines that will transport the products of our understanding of genetic processes across cell walls will reshape our societal concepts of sick or healthy populations. Nanomedicine may signal the demise of "one size fits all" regulatory standards, replacing them with personalized assessments based on flexible regulatory criteria that will redefine populations on both sides of the border: people who will soon be easily "cured" and people whose illness is so small it seems to be invisible.[4] These concerns must be addressed not only without bankrupting health-care systems or saddling unsuspecting third parties with liability but also without creating an underclass of people who lose their employability due to stigma, discrimination, potential future injury, or lack of access to good medical care. For these reasons, convergence of new genetic technologies and nanomedicines may redefine our collective understanding of "safety," "health," or "disability" and may challenge both fundamental legal principles of equity and fairness and also the scientific underpinning of existing standards. The twentieth-century heritage of one-size-fits-all standards will give way to different standards for special needs and vulnerable populations as nanotechnologies enable precision medicine.

---

2 World Health Organization Constitution.
3 Ilise Feitshans Nanotechnology Revolutionizing Public Health in the COvid19 Era Nanotechnology and Nanomedicine Open Access Journal August 7 2020.
4 Ilise Feitshans "Global Health Impacts of Nanotechnology Law" Harvard University T Chan School of Public Health NanoLecture Dec 5 2019.

Nanotechnology will therefore force a redefinition of health and disability, in daily life and under law. Using the promising techniques of nanomedicine, it will be possible to treat diseases sooner, possibly even while the people who are expected to become ill are still apparently healthy. If so, nanotechnology provides people around the world with the exciting opportunity to bring together health sciences, policymaking, and law to address prophylactic treatment on a mass scale as never before. It is not certain, however, whether nanomedicine will be useful for everybody; if an individual's health status reveals a problem that is not clearly covered by existing law, then society must decide whether insurance law and societal protections against discrimination will bend or stretch to accommodate the patient. This is particularly important as nanotechnology and nano-enabled therapeutics successfully provide vaccines and provide the underpinning for the artificial intelligence that will determine which people get vaccinated and which people will be unprotected against Covid19.

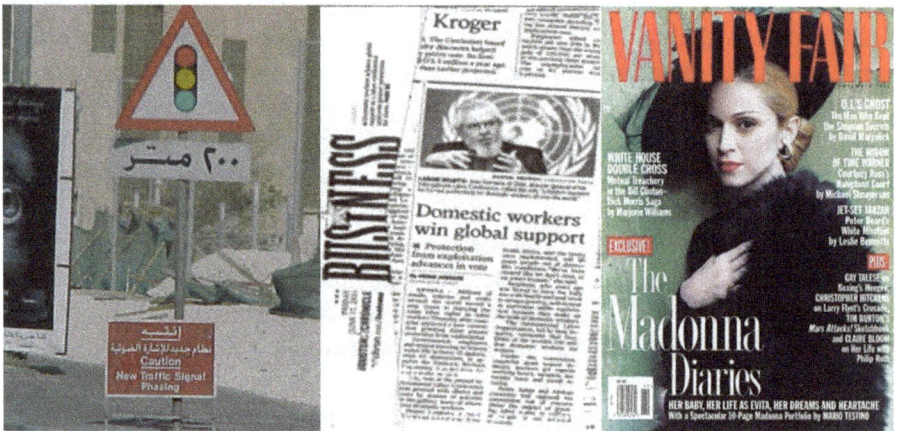

Figure 12.2: Construction warnings in Bahrein, President Juan Somavia announces the international convention protecting the rights of domestic workers, and Madonna describes working as an actor.

12.3.2 What is work?

There is a simple answer to the question "what is work?" "Work is anything you don't want to do, but that you do anyway." Oscar-winning movie star and popular singer Madonna scandalized her peers by describing her workplace on the movie set during the filming of *Evita* [2]. Madonna's diaries document her discomfort with makeup contact lenses and lapses of security that made her both fear and confront threatening crowds of adoring fans. "On the drive from the airport I twice saw graffiti painted on the walls that said, 'Evita Lives, Get Out Madonna.' How's that for a welcome? . . . Today was the day from hell, sort of. First I slept like shit. Shakespeare this was not. And why should they sleep? Everyone is unemployed; no one has to get up and go to work in the morning. The only people making any money are the press, and they will go to any extreme to get a picture or any information about me."

This concept that work is a hardship can be tested in the daily lives of people who are not movie stars, enduring the highly paid hardships of working on location: just ask a teenager to clean someone else's room or his or her own bedroom. Getting that teenager motivated to actually perform the task is work; the teenager is likely to storm out of the household angrily rather than completing the assigned tasks. Yet, work is also a liberating and invigorating source of achievement, pride, and reward, which boosts self-esteem. Although work can be isolating from family and friends, it is also potentially the greatest source of intimate contact with the rest of humanity. Work takes us away from personal life, and yet, it is the embodiment of all our hopes and aspirations, which are, in turn, transferred to our children in their work in the next generations. And thus the paradox of "working for a living" exists, despite the hardships, the risk, the sacrifice of leisure time, even if working with close family and friends. The chance that one might die or become injured while toiling to make life better for one's family, friends, or oneself is at the same time a reality as there is no such thing as zero risk; the goal therefore is to minimize risk and to mitigate its impact through new approaches to risk management, some as simple as roadway construction warnings in Figure 12.2. This well-established notion that healthy working conditions are linked to the survival of civil society will return to the policy arena for nanotechnology.

12.3.3 Is work paid?

There is another component that is also overlooked about work: not all work is paid. Work may require payment under some labor laws. That definition is incomplete from the standpoint of legislative drafting to prevent or reduce risk for two reasons: first, a definition that is based on payment leaves out unpaid labor, especially slaves. Second, the definition excludes volunteers who are unpaid. Yet, even when domestic servants are hired, the need to perform unpaid household tasks remains [3]. Long ago, people recorded causal relations between exposures to dangers and injury in the workplace, even though they did not have the investigative and regulatory tools that exist now to correct these problems. Some of these notions survive from ancient times, along with other artifacts of the work of the ancient civilizations. For example, Pliny the Elder noted that workers in dusty environments should wear fish skins to protect them, an early form of respirators. The impact of sleep deprivation was also understood by moms and the medical community centuries ago.

The unexplored notion that many tasks are work, regardless of whether they are paid or unpaid, leads to inevitable questions of voluntariness or a fallacious choice about risk-based decisions in our culture, whether the risk is encountered on personal business or paid work. Such questions cut across social classes, reaching not merely the lowest paying jobs and the least lauded of jobs held by workers, but even the highest social strata of society – the presidents, movie stars, prime ministers, princesses, and kings. This is an especially important reality, long ignored in all occupational health literature that has come to the fore swiftly in the Covid19 times. Last but not least, there is the question of the impact of these untended problems. Do they hurt only the so-called workers, or do they impact non-working people, including but not limited to the spouses, parents, and unborn generations of workers and their families. This requires looking closer at heretofore unspoken questions: who are the people in contemporary society that we call workers? What is work? Is it merely a construct of performed and assigned tasks, executed for someone's benefit, or is it necessary to be paid, not a volunteer, not nurturing for love, not a slave to some overlord or master, not preparing work for profit, only to find that the purchaser has failed to pay – in order for the tasks performed to be considered as work from the standpoint of preventive health agendas or to trigger protections under law?

Adam Smith wrote in the eighteenth century, in a manner that remains valid,

> Workmen [sic] . . . when paid by the piece are very apt to overwork themselves and to ruin their health and constitution in a few years. A carpenter in London is not supposed to last in his vigor above eight years, and something of the same kind appears in many other trades [4].

Smith's comment was part of an economic argument against slavery; he argued that slavery cost far too much for maintenance and the risk of loss of property compared to hiring overworked and underpaid free men. By the nineteenth century, legislators understood that factories and coal mines ought to be inspected to prevent dangerous conditions because even if one or two people were successful in avoiding the risks of such work by choice, in the aggregate, society needed hundreds of thousands of people to work in the coal mines and factories. The struggle for replacement of people who died working in these dangerous but essential professions was the hallmark of the industrial revolution; the overall disease burden also took its toll, costing money to nonworkers, such as the owners and operators of coal companies, as well as the greater society. In response, therefore, laws were written in which common law theories were modified, not only to protect the harmed workers and their dependent families, but also to ensure the smooth flow of commerce throughout society.

Occupational health literature, which so minutely details the peaks and valleys of exposures to so many toxins in the workplace, surprisingly fails to discuss society's willingness to reconcile the presence of potential harm, injury, or risk while working for a living. Nor does the literature discuss the meaning of these exposures in life with family or life in the home from the standpoint of their impact on other family stressors and the overall well-being that shapes individual health.

Some of society's most important work, such as child rearing and other forms of intrafamilial nurturing, is unpaid work. There is no salary for performing such tasks as cooking a special meal for one's lover or oneself, but it is nonetheless a necessary job. Family caretaking is a necessary job, whether paid or unpaid, fraught with risks. For example, much of the labor performed by housewives, who use harsh chemicals to clean their house, are exposed to the risk of disease as caretakers for the sick or who are sleep deprived are also underpaid or undervalued by not being paid at all [5], with attendant risks: caring for a sick child exposes adults to communicable disease; driving a friend to the mall for clothes or groceries introduces the hazards of driving and perhaps other hazards in the transport of commodities and contact with the general public. So, too, new technologies must be taken into account [6, 7]. As nanomedicine opens the door of employment opportunities for women as surrogate mothers with implanted eggs from different countries and at various stages of fetal development, the questions must be asked whether the terms of the new convention protecting domestic workers will apply to these trades, too. Even if such workers are sometimes outside of their own household, or are required by contract to stay in baby farms where their activities, medical status, and fetal growth can be carefully monitored, there remains a concern that this form of household servitude also has an impact on the life and health of the unborn child because of transplacental transfer of nanoparticles. The exposures to nanotechnology will embrace all parts of society. To be effective, the concept of "work" must be redefined under law as broadly as possible in order to capture the range of risks. Nanotechnology is redefining work by creating new approaches to ancient tasks and new tasks that never existed before, and therefore will force society to revisit these perennial issues under existing law and by drafting new laws.

12.3.4 Voluntariness or choice in major life decisions

Fundamental issues of justice, fairness, and equality that exist in science law and policy are often rooted in a belief system about "choice." As Smith remarked centuries ago, poor working conditions and low wages for long hours may be viewed as a matter of free choice because workers are not obligated to accept those terms or conditions. Modern studies about hidden costs of disease, revealed in the calculations of the global burden of disease (GBD) by WHO, offer evidence that the harm from unchecked freedoms in contract hurts all of society by wasting money and increasing costly health-care burdens upon society as a whole. It is in this context, too, that the fabric is woven to produce a safety net of *corporate social responsibility* designed to foster sustainability and innovation, both simultaneously needed by society. Efforts toward responsible development of nanotechnology reflect both the learning from modern tools of risk management and the free-choice heritage.

Perhaps the most overused word in law and ethics vocabulary is the word "choice." Fundamental issues of justice, fairness, and equality that exist in eugenics, just beneath the glitzy research and "big science" used to map the human genome, have outcomes that seem to depend upon choice or the quest to attain a wider array of choices. Choice may be relevant when standing in a department store, facing three different dresses that all fit perfectly at a reasonable price, but is an inapposite term when characterizing medical decisions. People may select a medicine or decide to continue a pregnancy to term, but the random underpinning of having such an opportunity is undervalued by the use of the word "choice." Genetic studies revealed that societal decisions about what constitutes good or desirable traits suggest that humanity is moving along a spectrum from chance to choice [8], but little attention has been given to what choice really means in terms of baseline information that humans can neither control nor change. It is like fighting gravity – sometimes there is no point in pretending that the outcomes of decisions are the product of anything more than a very narrow spectrum of choice.

Although some people may advocate an oversimplistic view that justice is merely a distribution problem, whether fairness is a question of distribution, a fundamental value, or an elusive social goal, "equal opportunity" has a benefit to the greater civil society that ancient paradigms rooted in racism, sexism, classes, and stigma for disability have been deprived. The line between health and disease will be redrawn by society, but the question must be asked whether people who refuse treatment will have had a genuine choice if, for example, their decision is based on economic constraints and whether failure to comply with the new accepted social construct for treatment will be punished, either indirectly by social stigma or directly by a subsequent refusal of the health-care system to provide obviously needed treatment when the illness becomes manifested. And, people who decide to do something ambitious even though it is difficult should not be punished or chastised for having made a "choice," if in the end all posterity will benefit because they have pursued their difficult goal. Nanotechnology cannot answer these ancient riddles, but the transformation new technology brings to society enables people to consider rethinking the old problems and redefining the correct approaches to choice when measuring outcomes and shaping societal goals.

12.4 What is nonhealth or disability?

Everyone has a disability. Everyone has a gift. Your job is to find the gift
and remove the obstacles of disability

– Sylvia Feelus Levy, 1974 [9]

An emerging issue for policymakers that will gain increasing significance with the successes of nanomedicine concerns the life of so-called healthy disabled people. In every society, there have always been people who are considered disabled: individuals who qualify for national insurance or social assistance due to a specific severe handicap or cluster of impairments with attendant comorbidities but who are not expected to share in the productivity of the general population at large. Due to illness, these populations are often excluded from clinical trials and research efforts because it is presumed that they will not be able to withstand experimental treatments and that their experience is not representative of the larger, seemingly healthy population. Within this group, however, many people may conform to traditional parameters for the social construct of being considered healthy [10]. For example, someone with palsy who cannot walk may have normal-range blood pressure and not be at risk for negative health impacts of diabetes, obesity, or coronary failure. Nanotechnology applied to medical care and nanomedicine will enhance the importance of the healthy disabled population for a great spectrum of research, including the influence of social expectations following diagnosis of illness on psychological status, social well-being, and longevity in relation to an expected survival curve. If applied with forethought when rethinking these vital social values, two sets of benefits can be realized by civilization at the same time, not as competing interests, but as one invaluable set of societal change: the miraculous developments that sound like science fiction to those people who eagerly anticipate these medical products, combined with the new social dimension of protecting rights of people with disabilities, will reshape all of civil society. Thus an unprecedented opportunity exists to benefit from the simultaneous nanotechnology revolution and the revolutionary social change that recognizes individual human potential by promoting the equal opportunity for people with disabilities under law.

12.4.1 Redefining choice in maternal and child health laws

Protecting reproductive health for all requires a new paradigm. Jurisprudence surrounding reproductive health has been carved into small portions by courts and legislatures, with only piecemeal protections for pregnant workers and their offspring. International human rights law does not take into account the special needs of women during pregnancy nor the randomness of events following the decision to carry a pregnancy to term, erroneously labeled as a "choice." Conversely, abortion jurisprudence relies at its root upon a woman's right not to choose to have children and therefore analytically stops at the termination decision. Consequently, important opportunities to protect health and ensure economic well-being of mothers and their children have been lost to humanity [11]. On the microscale, these issues become intensely evidence-specific and problematic [12]. Yet, when viewed from the broad perspective of occupational accidents, death from fire, or other types of disasters that can be avoided through adequate planning [13], and from greater appreciation of the basic societal need to have a new generation populated by the healthy offspring of the previous working generation, the notion of choice becomes an artificial construct for refusing to grapple with social issues that are the product of larger problems such as sexism and embedded discrimination [14]. Since not all mothers have consistently adequate protection of health or job security, many consequences that follow cannot be blamed upon individuals or mothers as a matter of "choice."

These issues cry out for a new approach to solve these ancient conundrums [15]. Piecemeal approaches have been taken for this problem in the past [16], but nano-technology offers new avenues for legal resolutions to the important policy questions surrounding the impact of working conditions on the health of the next generation as it potentially supports or undermines human life. Responding to these issues properly requires rethinking about these problems from a perspective that is maternalistic, rooted in the needs of working women who cannot control the destiny of their pregnancy [17].

A maternally driven model, starting with the working assumption that whether or not they are paid, all mothers work, can better address the uniquely individual but paradoxically universal need for health protection. The appropriate model for asking these questions, and for assessing these answers, employs a new paradigm: taking into account the special needs of women during pregnancy without the unfortunate tendency of society to blame mothers for social ills that may impact their children. From a perspective that is maternalistic, advocates, researchers, and policymakers can use risk governance in emerging nanotechnology frameworks such as EC4safe-nano to address infrastructural determinants, and health and illness for childbearers and their offspring. A new maternalistic model can reflect a deeper understanding of the inextricable link between health at work and the health of posterity. Nanotechnology and the breakthroughs in nanomedicine that rely on transplacental drug delivery offer the opportunity to create a maternally driven model for addressing the need for reproductive health protections. The new model for reproductive health can replace myths and prejudices that blame mothers or pretend that high-level professionals are not workers, replacing these errors from the past with a universal approach to occupational health that can become the vehicle for more efficient protection for health, life, and posterity of civil society.

12.4.2 Bringing health to work for people with disabilities

Society has been radically changed, hopefully irretrievably for the better, by the creation of laws that promote the rights of people with disabilities and prohibit discrimination against them. Disability poses profound challenges to international human rights laws [18]. Due to the revolutionary change in discrimination laws that require hiring, promoting, and protecting people with disabilities as a vital part of the workforce [19], several fundamental aspects of workplace design, implementation of industrial hygiene protections, training for occupational health programs, and "the way we do business" will fundamentally change. The arrival of nanotechnology in the workplace provides an outstanding opportunity to implement such change because the changing demands of work that will come about through the application of new technology also require redesigning the workplace. It is possible, therefore, with forethought, to create opportunities that maximize the benefits of both the social change in disability laws, combined with demographic changes, by rethinking old values and reshaping social constructs, in light of nanotechnologies that will transform disability into health.

At last, an unprecedented opportunity exists to benefit from both the nanotechnology revolution and the revolutionary social change that recognizes individual human potential under international laws preventing discrimination against people with disabilities. The arrival of nanotechnology, praised and heralded as a welcome revolution reshaping industry, provides the perfect opportunity for rethinking workplace design and blending into the weave of the workplace fabric antidiscrimination goals, thereby folding into the fabric of the typical workforce people whose special needs may have previously placed them at the margins. The maturity into the workforce and reproductive age of a generation of people with disabilities who have been raised with an understanding of their rights is an important social change that cannot be ignored. Despite medical disadvantages, such students become professionals, parents, and workers, but they do not lose their consciousness of their rights as people with disabilities.

For the first time under international health laws, there exists a cohort comprising an entire generation of people with identified disabilities who would have been living in institutions a generation or two ago but, instead, are armed with rights as well as a renewed sense of self-worth. Demographic change based on the new social constructs for the role of people with disabilities under law, combined with new avenues for treatment using nanomedicine, will require a different methodology to tease apart the cause and effect between workplace or environmental exposure and health outcomes. The new workforce that is implicitly different compared to the totally "able bodied" workforce of the previous generation holds implications for social theories of aging that will change the target population for treatments using nanomedicine. The presence of disabled workers using nano-enabled assistive technologies and nanomedicine treatments before they experience commonly visible symptoms will change the nature of many job descriptions, as only the "essential functions" of the job will be necessary under law. These combined developments have attendant implications for job design, specialized methods of risk communication, and job hazard analysis. It should not be surprising therefore that the abundance of new methods for accommodations for the young generation will also have intergenerational impacts for the working health of older people, who may wish to remain in the workforce even while undergoing treatment using nanomedicines.

Around the world, law now requires equal opportunity for people with disabilities. Implementing laws promoting the rights of people with disabilities has expanded the definition of disability and the obligations of everyone in society to create opportunities for them, regardless of cause of injury. New methods to measure the cause and effect between workplace or environmental exposure and health outcomes, including the use of big data combining several datasets, will be developed using nanotechnologies in order to understand the role of new exposures in combination with existing disability. The new disability paradigm under law combined with nanomedicine will change the daily lives of chronically ill populations by redefining societal notions of health. This is not an abstract problem because there are very stiff fines and penalties for failure to comply with disability laws preventing discrimination, and nano-enabled methods will allow analysis of individual cases much more precisely. The rightful presence of an identified disabled population within the workforce will change the nature of many job descriptions because only essential functions of the job will be necessary, as mentioned earlier. Jobs will then be custom-tailored to accommodate deficits and to maximize individual productivity. These new members of the vibrant workforce are empowered by refined tools of self-advocacy under law, who possess the assistive technology as well as the regulatory mechanism to operationalize their rights.

Simultaneously, job analysis and job descriptions will be required to take into account inexplicable but nonetheless a common coupling of some diseases together, called comorbidity. Commonly, when a parent or caretaker is asked to assume some responsibility for a person with an identified disability, the caretaker is briefed about the panoply of expected problems that come with the main disability. For example, many people with invisible disabilities may also have an obsessive compulsive disorder. Although nanosciences may not provide clear explanations of why some diseases or illnesses are often found together, the ability to use nano-enabled technologies to bring disabled people to the workforce means that comorbidities will also be taken into account when fashioning job descriptions, even if this health impact may be diminished due to applications of nanomedicine.

Consequently, nanomedicine will accelerate society's increasing interest in providing custom-tailored job analysis and attendant health-care supporting work, in a manner that is similar to the controversial techniques that are proposed by pharmaceutical companies to apply "personalized medicine" using nanotechnology. For these reasons, convergence of new genetic technologies and nanomedicines may redefine our collective understanding of "safety," "health," or "disability" and may challenge both fundamental legal principles of equity and fairness and also scientific underpinnings of existing standards. Antiquated prejudices embedded in existing paradigms for addressing these issues will be discarded in order to construct valid working assumptions. This population shift requires deciding who will pay for treatments, when will they become considered standard without a consent required, when will individual patients be allowed to exercise their right to refusal, and when undergoing treatment for presymptomatic illness will there be a major wave of hard policy choices to be made regarding which people to be considered disabled with all the legal protections available to people with disabilities. Additionally, long-standing but incorrect working assumptions regarding safety and health and the format of preventive programs will be challenged by the presence of an entire new cohort of people with disabilities. The juridical heritage of one-size-fits-all standards will give way to an individualized approach, applying standards that attempt to achieve performance-based outcomes. And the sea change represented by modern disability laws, operationalized by applying nanotechnologies, will acquire the force of a tsunami.

12.5 Integrating nanotechnology into international laws protecting health

12.5.1 An abundance of international laws: resisting the fad of nanoregulation

> The protection and promotion of the health and welfare of its citizens is considered to be one of the most important functions of the modern state [20].
> – George Rosen

A vast and vibrant corpus of laws protects health. This concept of government obligation to protect its citizens is as old as the Great Wall of China, which was built thousands of years ago to keep out invaders and preserve the integrity of an empire. And this concept of government responsibility is met in the actions of thousands of diplomats and government workers around the world; civil servants meet to plan and implement health policy and protect rights, as seen in Figure 12.3. International laws reflect, and do not ignore, the societal need for a legislative response to hazards that exist in daily life. Legal tools exist for promoting the implementation of precautionary principles without civil society seeking to reinvent them. Strong international consensus among laws across a majority of nations and a parallel system of codified international norms demonstrate a universal desire to protect consumers, protect the environment throughout the life cycle of product use, and enhance occupational health protections for all societies.

Precautionary activities are among the fundamental responsibilities of governments, such as primary care [6]. Protecting public safety, defense, and national security and controlling toxic or hazardous substances are reflected among national laws

Figure 12.3: Swiss representative to the United Nations Human Rights Council prepares to address a public session in the Palais des Nations Geneva, Switzerland.
Photo by Dr. Ilise Feitshans.

and intergovernmental agreements designed to promote those goals. International laws governing health and safety for consumers, workers, and the general public therefore provide an important backdrop against which the efficiency of national laws can be compared and measured. International laws provide more than a yardstick for measuring compliance. The large array of national, local, and international laws protecting health underscores the fundamental character of public health protections, which seem to be universal across all societies. Embedded with precautionary principles, many legal systems in the world universally provide the terms and conditions for promoting future development in society. Many laws clearly articulate and codify shared social values that safety and health protections (indirectly following scientific precautionary principles) are invaluable to preserving society.

The dilemma for policymakers concerns context: nanosilver that is wonderful for destroying bacteria and HIV in condoms and killing rodents that eat fabric may also harm children; nanogold that makes rapid communication for electronics and cell phones may bring a message that needlessly hurts someone; titanium dioxide that makes fluffy cream to place in a bowl so that a young child will be attracted to taking his or her needed medicine may also be the source of harm.

Addressing major unforeseen problems before the fact, therefore, requires balancing an admixture of quantifiable and unquantifiable variables in order to give their approval to major programs, sometimes making hard choices before the risks are known. As Figure 12.4 illustrates, policy inputs for information can be an admixture of views from all constituencies and all ages; it is widely accepted that policy can be most responsive when it functions without artificial limits regarding the type of stakeholders involved. Policy choices must be made: sometimes hard choices dictate a response to the will of the people with a law that is popular but unsound, sometimes championing the best practices that meet their long-standing responsibility for protecting people, while, at the same time, balancing the urgent need to foster and develop new industries to stimulate a broken economy.

Figure 12.4: Emalyn Levy Feitshans speaks at the US Capitol.
Photo by Dr. Ilise L. Feitshans.

From an economic standpoint, there can be little doubt that the development of nanotechnology is a shot of adrenalin that can stimulate new commerce and new jobs for an interdependent global economy and therefore a dynamic instrument of social change that can transform many aspects of daily life in civil society, beyond nanotechnology's financial implications. The size, shape, and availability of commercial centers and the contents they sell, including electronic transfers of information, are enhanced and accessible, thanks to nanotechnologies. This brings more jobs and more commercial activity every day, globally. Yet, diffuse use of sovereign power by having too many different systems governing nanotechnology can lead to confusion and conflicts of laws and ultimately block the flow of commerce that had brought the hope of nanotechnology's promises to civil society.

12.5.2 Health protections under the United Nations charter and its human rights framework

Governmental functions protecting health under international law are as old as the treaty law of the WHO Constitution and the soft law of the Universal Declaration of Human Rights (UDHR) [21] treaties such as the International Covenant on Economic, Social and Cultural Rights (ICESCR) [1] and, perhaps, as old as civilization itself. Precautionary principles form the foundation of international rules about toxic and hazardous substance exposure. Few member states of the UN will state aloud that they violate such principles; if anything, rhetoric from every member state can be found claiming that these vital principles are upheld by their law and the actions that implement them when addressing in the general assembly, as seen in Figure 12.5. Even though this simple consensus-based notion may seem to be old fashioned, the

Figure 12.5: Dignitary address the United Nations General Assembly, New York.
Source: US government.

precautionary principles as applied to nanotechnologies in the context of human exposure to nanomaterials are alive and well.

Precautionary principles and government responsibilities protecting health are not new, and therefore these notions are deeply embedded in many laws at every level of governance. These concepts are operationalized in national, state, and local laws, major international legislation, such as the Globally Harmonized System of Classification and Labeling of Chemicals (GHS), and international nongovernmental programs such as the International Organization for Standardization (ISO) work in groups regarding nanotechnology [2] work in close partnership with international governments and treaty organizations such as the Chemicals Committee and the Working Party on Chemicals, Pesticides and Biotechnology, Organisation for Economic Co-operation and Development (OECD). All of these organizations include in their rationale for their findings references to the bedrock notions of applying precautionary principles – even though the actual text of these principles is elusive and never codified.

Figure 12.6: Main entrance, Palais des Nations of the United Nations, Geneva, Switzerland, home of expert meetings about precautionary principles and the law of health.
Photo by Dr. Ilise L. Feitshans.

According to UN Charter Article 13 [3], "The contracting parties state their desire to promote: 'economic and social advancement' and 'better standards of life'." Putting these powerful words into programs and strategic plans has been the work of hundreds of thousands of people who have passed through the gates of the UN in Geneva,

Switzerland, shown in Figure 12.6. As a result of these deliberations, not every policy has been effective or inexpensive or equitable, but many important global health decisions have been successfully implemented at the regional and international levels, with the hope that national laws will be consistent with the best points in the international model. Big risks have been successfully addressed by governments and their legislatures in the past, enabling industry and commerce to flourish by promoting technology, while limiting the scope of liability by creating regulatory barriers to activities that cause avoidable harms. These laws use flexible frameworks for oversight with placeholders in final legislation for new methods. And historically, society has won great benefits by gambling with regulated risk.

By contrast to the slow development at an evolutionary pace for laws about health and international governance infrastructures, nanotechnology laws are sprouting like mushrooms in every nation! Governmental structures at all levels of society presently face a situation in which there is potential risk to public health, but insufficient data exists about actual risk in order to make key policy judgments. Consequently, the regulatory picture of the legal landscape presently looks patchy and disorganized – large gaps in the law where there is uncertainty about the magnitude of risk, and many different sources of law are clustered around tangible, established practices for toxic or hazardous materials. None of these laws question the juridical basis for their enforcement of the state's power to protect health, even when mandating costly engineering controls, medical surveillance through employer-based occupational health services, or global sharing of chemical hazard information.

Recalling the Internal Labour Organization's (ILO) constitutional mandate in 1919 in the Treaty of Versailles, UN Charter Article 55 specifically notes the link between "creation of conditions of stability and well-being" for peace and "higher standards of living" and "universal respect for, and observance of, human rights and fundamental freedoms." During the half century that followed the codification of these words, an elaborate apparatus of capacity building for health care created an infrastructure for implementing these values worldwide, at national, multinational, and international levels [4]. Subsequently, protection for the right to health was written into the fundamental constitutional principles of many nations [5]. This rhetoric equates "adequate" health with related basic rights. But it is difficult to patch together any text explaining "better standards of life" in detail. Like other international instruments that address health, its vague descriptions of protections for life, security of the person, without a benchmark for "well-being," have been used as the legal basis for thousands of programs and new treaties. For this reason, many theorists challenged the notion that health is among the universal human rights codified under international law. The debate has become dormant, however, once the infrastructure has grown to command an impressive role in controlling disease and ameliorating the quality of life worldwide.

12.5.3 International covenant on economic, social, and cultural rights

Broader questions regarding the international legal right to health protections have been addressed, without detailed guidance, in the UN Charter, in ICESCR Articles 7 and 12 and subsequent standards by UN-based international organizations and various international conventions. Obligations that arise under the ICESCR stem from a principle of progressive implementation of the rights it defines, and the principle of nondiscrimination in the enjoyment of those rights. Article 1 establishes substantive obligations that the signatory states undertake to implement in their home legislatures. Using a unique enforcement system of international monitoring and compliance, the treaty tries to balance risks and preventive strategies for using global resources, through international assistance and technical cooperation [7].

12.5.3.1 Article 7 of the international covenant on economic, social, and cultural rights

Article 7 provides greater insight into the meaning of the right to just and favorable conditions of work discussed in other UN documents. "Favorable conditions of work" include terms of remuneration as well as "safe and healthy working conditions." The use of this phrase within the context of favorable conditions of work lends greater meaning to the UDHR's protections and demonstrates the clear nexus between other human rights principles and protection of health, discussed in ICESCR Article 12.

12.5.3.2 Article 12 right to health: promotion of industrial hygiene

Of all the UN-based international human rights documents, ICESCR Article 12 most clearly and deliberately addresses health. It is the clearest of all human rights instruments regarding the explicit right to protection for "industrial hygiene" and protection against "occupational disease." Further, Article 12's mandate to improve "industrial hygiene" is consistent with Article 7(b) of the ICESCR, regarding safe and healthful working conditions. Yet, even this express guarantee of occupational safety and health protections does not offer a detailed exposition of the meaning of these rights, nor does it list the possible approaches that could be applied for achieving the ICESCR's goals. Consistent with the principles articulated in many other international human rights documents, Article 12 employs WHO's constitutional notions of health. Article 12 states the following:

The States Parties to the present Covenant recognize the right of everyone to the enjoyment of the highest attainable standard of physical and mental health. 2. The steps to be taken by the States Parties to the present Covenant to achieve the full realization of this right shall include those necessary for:. . .
(b) The improvement of all aspects of environmental and industrial hygiene;
The prevention, treatment and control of epidemic, endemic, occupational [9] and other diseases . . . .

The plain meaning of this language in Article 12 treats occupational disease as a vector that has the potential global health impact of diminishing public health and increasing the "global disease burden." Under ICESCR Article 12, the states' parties recognize the right to physical and mental health proclaimed in the WHO Constitution as the basis for their commitment to international health programs. Under these terms, countries commit to four "steps" to be taken along a strategic path to achieve the "full realization" of this right. Significantly, Article 12 also pays direct attention to the impact of occupational disease on health and upon disease burden within society, thereby giving validity to occupational medicine as worthy of human rights protection [10]. The text from the Paris Accord on Climate Change also relies very heavily on this precedent in international regulatory history.

Under ICESCR Article 12, the states' parties recognize the right to physical and mental health coequally with all parameters of the WHO constitutional definition. The language of this article is consistent with the view that "health protection in the workplace is as a matter of great significance, but it is also inextricably linked to the task of protecting the population generally" [8]. Like many laws designed to operationalize precautionary principles, the main subject of the law remains undefined: standards for industrial hygiene and the quality of environmental health remain undefined. It is worth noting that this definition, like its predecessors in international law, avoided a list approach in favor of an open-ended definition that allows any condition to be considered an occupational disease for the purposes of prevention, treatment, compensation, and eradication [11]. According to Grad and Feitshans [10], Paragraph 1 of the Draft Covenant prepared under the auspices of the Commission on Human Rights decided to allow the WHO definition to fill this void. ICESCR drafters urged member states "to develop and strengthen occupational health institutions and to provide measures for preventing hazards in work places." Repeating a theme expressed in many international documents and originating in the WHO Constitution, the ICESCR states, "The right of everyone to the enjoyment of the highest attainable standard of physical and mental health." This goal remains as elusive as it is universal.

12.5.4 The World Health Organization Constitution: codifying precautionary principles

The movement to codify health norms as legal principles had a defining moment at the end of World War II, when the entire world cared about attempting to set written legal limits upon behavior by governments and individuals. UN activity brought codification of international norms regarding the right to health into the positivist, plain language of several key international human rights instruments, with a spirit of hope for all humanity's survival.

Although merely an administrative agency, many view WHO as the paragon of rights-based health programming and respected references for health research and health policy (Figure 12.7). The most widely accepted definition of "health" in the world is written in the preamble to the WHO Constitution [12]. This text has been quoted around the globe in constitutions, international treaties, and public health practical guides [13]. Programs to provide vaccination, preventive strategies, quarantine, and successful risk management of diseases relying on WHO's flexible legal framework have become the bedrock for a jurisprudence about human rights to the health, life, and security of a person, codified in subsequent major international human rights instruments. International agreements following this path of legal analysis facilitate transnational scientific collaboration, national and international inspection, and regulatory cooperation and also offer an opportunity for capacity building to implement national programs designed to improve well-being [14].

Figure 12.7: Logo from the WHO website. This logo, as used here, is only to assist readers to identify WHO and does not constitute or imply any endorsement of content.

The WHO Constitution's two-page definition of health begins with this famous text:

> Health is a state of complete physical, mental and social well-being and not merely the absence of disease and infirmity.

This remarkably broad but flexible definition of health bespeaks the basic human need for health. The WHO constitutional definition is so encompassing, however, that it has been criticized as making virtually any human endeavor a matter of health jurisdiction. Few endeavors have no impact on human health. The drafters of the WHO Constitution envisioned scientific breakthroughs that would give insight into the disease process.

WHO's legislative drafters did not know the names and diagnoses of those diseases that had not yet become pandemic or their methods for treatment, but they accurately anticipated that there would be new health problems. The WHO Constitution offers an elastic framework that can be expanded to include new developments and reduced when a problem has successfully been diminished, as in the case of eradication of smallpox. The drafters of the WHO Constitution also understood the notion of latent disease, which is of increasing significance in the areas of reproductive health and human reproductive toxicology, whereby long-term exposures that may appear to be harmless at the outset may, after many years, take their toll due to a cumulative effect. In those situations, a person may enjoy many years of seemingly good health, while, in fact, the illness or disease is silently, slowly progressing. The structure the drafters created, therefore, is sufficiently precise to offer an irresistible promise of global health and sufficiently vague so that new interventions can be designed and implemented to prevent harm and thereby protect health.

Nanomedicine, for example, promises presymptomatic treatments or cures for conditions that may once have been completely disabling or caused death in humans that might not fit into a classic definition of diagnosis or treatment based on old models of recognized symptoms and harm [15]. For this reason, flexible language concerning health protection "in the absence of disease or infirmity" has a very important jurisdictional effect, for it gives rise to WHO's permission to research and implement protective measures regardless of whether people manifest illness or appear to be sick, and will therefore touch presymptomatic nanomedicine patients in disease-specific populations. This ultimate precautionary principle is fundamentally important for justifying any health programs.

WHO's definitions are constructed on the basis of solid fundamental principles rather than a towering paper list of specific conditions and diseases. Consequently, WHO's definition has endured for nearly a century. For example, this language has been used to enable WHO to pioneer efforts in the research, study, treatment, and avenues of prevention of HIV/AIDS. HIV/AIDS was not a known disease at the time the WHO Constitution was written. But once the disease became internationally recognized, the international medical community called upon WHO to create and implement international cooperative research and preventive programs, consistent with its Constitution; no one challenged the basis for WHO's authority to create interventions and inspire research for treatments. Instead, WHO's role as a leader in the global response to the pandemic was encouraged, and the agency had the statutory authority to go forward. The principle of the rights of sovereigns to engage in international relations embraces their obligation to use international relations as a tool to protect the health and well-being of their people. Quarantine, by prohibiting people with specific diseases to cross borders in prior centuries, is the first crude example of this right and duty.

Instituting even-handed needs-based quarantine was an early task of WHO in the late twentieth century. Vaccination as a preemptory strike against pandemics that lead to quarantine provides another example of international collaborative strategies for health protection, facilitated by WHO's constitutional text. To be effective, vaccination must occur to seemingly healthy or disease-free human beings [16]. WHO's efforts have also consistently applied to occupational exposures and related environmental health risks that may be compounded by synergy with human factors or contaminants. Scientists try to determine the precise quantities of toxins that cause cancer, acute toxicity, or other harms in the workplace, compared to the general environment, personal lifestyle, or genetic predispositions. The confluence of these factors may be very different among individuals or populations, but the patterns of disease transmission and illness know no borders.

Recently, WHO reinterpreted key definitions by emphasizing the importance of social context when defining disability [17]. WHO's revised global legal framework includes a commitment from the agency and from each WHO member state to improve capacity building for disease prevention, detection, and response. It provides standards for addressing national public health threats that have the potential to become global emergencies. The rationale for such activity is linked to preventive principles taught in public health schools from the first day: "Investment in these elements will strengthen not only global public health security but also the infrastructure needed to help broaden access to healthcare services and improve individual health outcomes, which would help break the cycles of poverty and political instability . . . " The advantage of this approach is its applicability to existing threats as well as to those that are new and unforeseen, such as natural disasters, industrial or chemical accidents, and other environmental changes, which might cross international borders [22]. WHO Constitutional language has strong links to the right to life and security of person, codified in major international human rights instruments such as the UN Charter, and subsequent international agreements [14]. This text can serve equally well to facilitate evaluation, prevention, and regulation of health risks from emerging nanotechnologies. Nanotoxicity and the implications of exposure for human health therefore are within the scope of WHO's mission to protect human health.

12.5.4.1 WHO's International Programme on Chemical Safety: a precursor to the GHS

WHO's International Programme on Chemical Safety (IPCS) Harmonization Project enables governments and others to work toward the achievement of goals first outlined in Agenda 21, Chapter 19, in Rio in 1992 at the United Nations Conference on Environment and Development. According to Rio Declaration Principle 10, "States shall facilitate and encourage public awareness and participation by making information widely available" [18]. As elaborated in recommendations of the Intergovernmental Forum on Chemical Safety Bahia Declaration of 2000, reaffirmed by governments in the 2002 Johannesburg World Summit on Sustainable Development (WSSD) Plan of Implementation [23], the IPCS is consistent with the Rio principles, which were carried forward in the Strategic Approach to International Chemicals Management (SAICM) in Dubai in 2006 [24].

Implementing WHO's IPCS is one component of the very complex endeavors involving over 25 UN agencies, such as the United Nations Institute for Training and Research (UNITAR) [25] and hundreds of national government agencies in the GHS [26]. Regional governments such as the European Union (EU), the ASEAN nations, and trade organizations such as the ISO have teamed with various stakeholder partners to promote a unified system of identifying chemicals with the same set of labels worldwide. This approach ensures that handling and storage information travels with the chemical so that the same level of accurate right-to-know information is accessible to everyone involved in transport, storage, or industrial use of the substance. This approach avoids spills or other accidents and provides midstream commercial users with a layer of protection against liability.

12.5.4.2 WHO Beijing declaration: occupational health for all

The Beijing Declaration, Occupational Health for All (1994), signed at the second meeting of the WHO Collaborative Centers and adopted in 1996, builds upon WHO's HFA2000 Plan for Action by addressing rapid changes in work organizations that impact human health worldwide. WHO, the ILO, UN Development Programme (UNDP), and the nongovernmental organization (NGO) International Commission on Occupational Health (ICOH), in addition to 27 countries, adopted a proposal for action about target goals. Human existence is an admixture of interdependence. Even the highest-ranking leader can face hazards as a worker that jeopardizes his or her right to health. Too often, policymakers in occupational health and the attendant allied health professions act as benevolent caretakers of the health of a small segment of the working population: speaking in the third person about working conditions, an aloof third-party observer of the flaws and strong points in labor or management structures, unattached to risk and rarely noticing their own health and safety. Following this strategy, WHO's Ninth General Programme of Work, for 1996–2001, stated, "Occupational health and safety at work is a fundamental human right and a worldwide social goal as the basis for his view that successful implementation of WHO's global strategies will depend upon: (1) data sharing, and (2) research and collaboration in partnership with industry, and (3) similar activities in partnership with occupational safety and health compliance professions, through professional associations" [27]. Research examining costs of the so-called disease burden in society reveals that it is in everyone's economic interest to be part of the activities that prevent work-related impairment. WHOs global strategy for promoting occupational health has been developed through a network of collaborating centers (CCs) that share "a common vision . . . to mitigate the adverse effects of occupational hazards and to meet emerging problems."

Pursuant to WHO's General Authority Mandating Action to Protect Worker's Reproductive Health: Implications of the WHO Global Strategy for Health for All Plan of Action 1996–2001 [28], the Director-General of WHO was requested to implement an occupational-health-for-all strategy embracing occupational health care, small enterprises, migrant or informal sectors, and women, as part of the high-risk groups with special needs. Key priorities included developing and disseminating evidence-based prevention tools and raising awareness. By 2009, this mandate was expanded to include programs for the elimination of related diseases [19], the creation of toolkits to improve assessment and management of physical risks in workplaces, the prevention of the effects of noise and vibration, assessments of psychosocial risks in the workplace, self-contained learning units, and a project assessing the hazards from nanoparticles and communication of their risks. Specifically addressing occupational health for the very first time among WHO international instruments, this international instrument, although vague, applies concepts from valid international laws and the WHO Constitution in order to craft a platform for implementing precautionary principles protecting health at work. Point 9 of the declaration reaffirms each worker's "right to know the potential hazards in their risks in their work and workplace, including the development and use of appropriate mechanisms . . . in planning and decision-making concerning occupational health and other aspects of their own work. Workers should be empowered to improve working conditions by their own action, should be provided information and education, and should be given all the information, in order to produce an effective occupational health response through their participation, including the right to know information about health hazards from long-term and acute exposures to substances in their workplace." Unfortunately, the scope of hazard information disclosure is not discussed in the declaration, and thus, risk managers must rely on text from other international documents and national laws to fill this void.

WHO hosts several bodies specializing in hazardous chemicals and waste activities, including:
– The International Programme on Chemical Safety (IPCS)
– Inter-Organization Programme for the Sound Management of Chemicals (IOMC)

- The Intergovernmental Forum on Chemical Safety (IFCS)
- The Health and Environment Linkages Initiative (HELI)

WHO joins UN Environment in providing the secretariat for the Strategic Approach to International Chemicals Management (SAICM).

12.5.5 WHO guidelines for "protecting workers from potential risks of manufactured nanomaterials"

The World Health Assembly identified exposure to nanomaterials as a priority action for the Global Plan of Action on Workers Health adopted in 2007, and the World Health Organization (WHO) global network of collaborating centers in occupational health has selected this field as one of key focus of its activity: guidelines for "Protecting Workers from Potential Risks of Manufactured Nanomaterials."

(WHO/NANOH). to facilitate improvements in occupational health and safety of workers

The purpose of WHO guidelines is as follows:

These guidelines aim to facilitate improvements in occupational health and safety of workers potentially exposed to nanomaterials in a broad range of manufacturing and social environments. The guidelines will incorporate elements of risk assessment and risk management and contextual issues. They will provide recommendations to improve occupational safety and protect the health of workers using nanomaterials in all countries and especially in low- and medium-income countries [29].

Nanotechnology provides the perfect opportunity for WHO to pull together its discordant strands of programming in the global health tapestry and to use the coordinating effort in order to correct long-standing systemic problems in the access, public awareness, and delivery of services associated with workplace health.

Under the auspices of the "Global Plan of Action on Workers' Health" adopted in 2007 [30], WHO published its final report, *Guidelines on Protecting Workers from Potential Risks of Manufactured Nanomaterials* [29], on December 12, 2017. Previously, WHO had accepted and reviewed comments on a draft that it published in 2011: WHO's *Proposal for Guidelines on Protecting Workers from Potential Risks of Manufactured Nanomaterials* [31]. The report's clearly stated reliance on precautionary principles throughout its text underscores the inextricable link between work, health, and the economic viability of any employer in society. By using highly accessible tools to implement best practices that already govern bulk materials, WHO in partnership with the private sector and nonprofit employers has created "a golden opportunity to educate the general public – the novice who is thrown a text and told, 'we need risk mitigation, give me a list for risk mitigation and have it on my desk tomorrow'" [32].

12.5.5.1 WHO guidelines: a flexible definition of nanomaterials

The guidelines offer the standard definition, "'Nanomaterials' refers to materials that have at least one dimension (height, width, or length) that is smaller than 100 nm (10^{-7} m)," but also take into account that manufactured nanomaterials (MNMs) may glom together into substances of much larger sizes [29]. The ability to include the larger group of MNMs without reaching into standards for a bulk form of the same substance is a conceptual breakthrough, showing that nanomaterials are not isolated.

12.5.5.2 Main message: disclose possible risks

WHO developed these guidelines for the target audience of workers, policymakers, and professionals in the field of occupational health and safety in making decisions about protection against the potential risks of MNMs, health professionals, and decision-makers at the local, national, or international level, who are responsible for the health and safety of workers exposed to MNMs. Additionally, the guidelines focus on low- and middle-income (LMI) countries where nanotechnology is an important means of economic progress. Countries such as Brazil and South Africa produce MNMs and have research laboratories that produce CNTs and produce nanosilver that is incorporated in milk packs, fabrics, and clothes, and MNMs are also produced for use in pharmaceuticals. The authors noted, "Despite the publication of a large number of scientific articles about nanotechnology by authors from LMI countries, only a few are about the potential toxicity of MNMs and very few report on safety or risk assessment" [33].

As stated at the start of the report, the guidelines project aims to avoid the catastrophic effects of uncontrolled exposure such as the asbestos industry experienced in the time before occupational exposure regulations existed [34]: "Recourse to precaution should be used to reduce or prevent exposure as far as possible. This was seen as an important underlying approach in the interest of protecting workers' health, especially given previous experience with asbestos" [29]. The big news for people who apply the final guidelines is one message: use the GHS methods of classification and labeling of chemicals, use the authorized safety data sheets (SDS), and disclose potential harm from workplace exposure to nanomaterials. The Guidelines Development Group (GDG) stated that its scope and purpose included prioritization and classification of hazards posed by MNMs [35]. "The GDG recommends updating safety data sheets with MNM-specific hazard information or indicating which toxicological end-points did not have adequate testing available including respirable fibres and granular biopersistent particles' groups" [29]. The GDG report calls for worker exposure assessment using similar methods already in use for bulk materials and for the proposed specific occupational exposure limit (OEL) value of the MNMs, noting that the employer's decision about OEL should be at least as protective as a legally mandated OEL for the bulk form. Since MNM safety is evolving rapidly, the GDG proposes to update the guidelines in 2022.

The final report in 2017 stated that it followed precautionary principles because

> while humans have long been exposed to unintentionally produced nanoparticles, such as those from combustion processes, the recent increase in MNM production demands greater investigation into the potential toxicity and adverse health effects of these materials following exposure. Since newly developed MNMs are not tested sufficiently for possible health hazards, it is generally recommended to take a precautionary approach until testing results are available. This means that MNMs should be considered as hazardous unless there is clear proof that they are not.

To underscore this point and operationalize these goals easily, the authors of the guidelines recommend using the GHS for all MNMs, which requires worker training and disclosure of potential hazards in paperwork that accompanies materials throughout their supply chain in global commerce. Offering a specific list of MNMs, the guidelines strongly recommend disclosure on the SDS for carbon nanofibers (CNFs), CNTs, MNMs, Si-based, and titanium dioxide (TiO_2). To achieve maximum precaution, the hierarchy of controls is an important guide. WHO stated that there is scientific consensus that MNMs do pose potential serious risks to human health, even though high-grade scientific evidence to quantify risks does not yet exist. Yet the reality that it is only a matter of time before scientists will have clear evidence of important factors leading to nanotoxicity looms like a shadow across the entire guidelines. Therefore, exposure must be reduced despite uncertainty about specific adverse health effects.

It would be wrong to claim that the report was based on poor-quality data. Rather, the report forecasts the importance of future risk assessments despite the immaturity of the state of the art. These steps are reasonable and can demonstrate due diligence despite weak evidence in the current state of the art because "the toxicity of MNMs may differ for physicochemical properties, including size, shape (i.e. size in a particular dimension), composition, surface characteristics, charge, and rate of dissolution. There is currently a paucity of precise information about human exposure pathways for MNMs, their fate in the human body and their ability to induce unwanted biological effects such as generation of oxidative stress." The WHO guidelines can be used as a tool for establishing due diligence to prevent harm: every employer should have a program to address exposure to MNMs, including training for exposed workers. Due diligence, embodied in the ability to create a paper trail of evidence demonstrating an employer's good faith efforts to engage in reasonable protective efforts, can reduce harm and subsequent liability [36].

12.5.5.3 Role of due diligence for crafting worker protections

Due diligence, as suggested in the comments about the draft guidelines in 2012 from the Work Health and Survival Project (WHS), is the coherent strand that pulls the entire nanomaterials safety mechanism together. Due diligence is implicit in the process of following these guidelines: First, noting the premature, if not primitive, state of the art of understanding nanotoxicity and the responses for nanosafety, employers, workers, and policymakers who apply these guidelines have nonetheless shown a keen awareness of potential hazards that mandate SaferNano programming. Second, the process itself allows employers and policymakers to write compliance programs with a blank check regarding specific methods of assessment and precautionary measures to be implemented. Although this approach in the final report offers more of a mandate for safety training than was available in the draft proposal, the political reality is that the absence of a clear method for training is open to criticism because it leaves unresolved the legal questions of voluntary compliance that have plagued product liability and workplace health for a century. The guidelines state,

> The GDG considers training of workers and worker involvement in health and safety issues to be best practice but cannot recommend one form of training of workers over another, or one form of worker involvement over another, owing to the lack of studies available.

To assess worker exposure, the authors recommend control banding [37] in combination with the same or similar methods used for the proposed specific OEL value of the MNMs listed in its Annex 1, noting that the MNM OEL should be at least as protective as a legally mandated OEL for the bulk form of the material. Since these lists are guides and not law, there are no penalties for exceeding the proposed OEL. For dermal exposure assessment, there was insufficient evidence to recommend one method of dermal exposure assessment over another.

When exposures exceed the OELs, the report suggests a step-by-step approach for inhalation exposure:
1. First, assessment of the potential for exposure
2. Second, a basic exposure assessment
3. Third, a comprehensive exposure assessment following the standards set forth by the Organisation for Economic Co-operation and Development (OECD) or the European Committee for Standardization.

Given the role of several trade unions such as the International Union of Food, Agricultural, Hotel, Restaurant, Catering, Tobacco and Allied Workers' Associations (IUF), the European Trade Union Confederation (ETUC), and the Australian Council of Trade Unions and their literature available to the experts submitted, the final guidelines are surprisingly ambiguous regarding the nuts and bolts of compliance for nanosafety training despite available information cited by the GDG [38]. The guidelines text is silent about suggested methods for worker training and information disclosures that protect workers' health, despite citing an array of information about safety and health training programs [39] and the consensus cited in the report favoring such programs. This inadvertently leaves to chance whether employers who are the target audience for these guidelines will actually engage in programs that deploy due diligence [40]. Nonetheless, more important than the words used in its actual text, the release of these guidelines represents a milestone in operationalizing the role of due diligence and creating effective SaferNano compliance programs. Effective in-house occupational safety and health programs remain the best-available tool for managing best practices to control risk, serving as both as a weapon to protect against liability and as a tool for preventing harm [41].

12.5.5.4 Bioethical concerns about future research of MNMs

Although no long-term adverse health effects in humans have been observed, there may simply have not yet been enough time for harms to appear; also the guidelines' authors inaccurately cite ethical concerns about conducting studies on humans as a roadblock to a better understanding of nanotoxicity processes in humans. The report authors note with frustration that, except for a few materials where human studies are available, health recommendations must be based on extrapolation of the evidence from in vitro, animal, or other studies from fields (such as air pollution) where humans have been exposed to nanoparticles. Their complaint about these limits overlooks remarkable strides in computer modeling, especially using artificial organs made from stem cells using nano-enabled techniques for research and nanomedicine. These new tools created by nano-enabled technologies can provide accurate simulation of human experimentation without jeopardizing the health and lives of real people.

12.5.5.5 Guidelines are not standards

A specific list of MNMs was published in the guidelines. The authors rated the list based on moderate-quality evidence, which means that potential risks are more clearly documented than other aspects of nanomaterials where risks are unknown and therefore evidence of risk is low. This does not mean, however, that the unknown risks are less important than obvious ones; it simply means that further research is required about nanotoxicity before the risks can be prioritized.

Unquantified risks that may be difficult to explain to workers, consumers, and the general public are nonetheless quite real. For example, the guidelines note that data from in vitro, animal, and human MNM inhalation studies are available for only a few MNMs. Therefore, the report authors strongly recommend updating SDS for the GHS by including MNM-specific hazard information or indicating which toxicological end points did not have adequate testing available. WHO lacks the authority to require enforcement of standards. Therefore, no penalties for exceeding these standards or time frames for abatement are included in the guidelines.

12.5.5.6 Transparency to satisfy UN reform

The WHO guidelines also reflect an effort to modernize the public health approach to occupational health, reformed in order to meet the needs of nanosafety. Stakeholder calls for transparency on the heels of UN reform also have influenced the guidelines. Responding to criticisms by the WHS about a lack of transparency in the draft guidelines [42], the revising team took great pains to explain in detail the methods used by the working group and to spell out the required steps in the review process that answered 11 key questions in the scope of the work. Experts were listed in several annexes. Following established WHO procedures, the Interventions for Healthy Environments Unit in the Department of Public Health, Environmental and Social Determinants of Health obtained planning approval in 2010 to develop guidelines and established a WHO Guideline Steering Group and a Guideline Development Group (GDG) [43]. The GDG was composed of leading experts and end users responsible for the process of developing the evidence-based recommendations. (Members of the WHO Guideline Steering Group and the GDG are listed in Tables A.2.1 and A.2.2 of Annex 2.) Funding for meetings and the costs of the methodologist were provided by the WHO Department of Public Health, Environmental and Social Determinants of Health. Experts participated in the GDG on an in-kind basis, and systematic reviews were conducted by volunteer teams. The project of creating workplace exposure guidelines started with a small team of experts, who prepared a background paper on the development of guidelines for protecting workers from potential risks of exposure to MNMs in 2010–2011. Calls for experts were made to join the GDG and the External Review Group and to identify volunteers to carry out systematic reviews using a Delphi process. To incorporate significant research undertaken in the area of MNM health and safety, teams of researchers were identified who could carry out systematic reviews of the pertinent literature [37] according to the process outlined in the *WHO Handbook for Guideline Development*. (The systematic review teams are listed in Table A.2.3 of Annex 2 [44].)

Key questions identified by the GDG
1. Risks of MNMs: Which specific MNMs and groups of MNMs are most relevant with respect to reducing risks to workers, and what should these guidelines now focus on, taking into account toxicological considerations and quantities produced and used?
2. Specific hazard classes: Which hazard class should be assigned to specific MNMs or groups of MNMs and how?
3. Forms and routes of exposure: For the specific MNMs and groups of MNMs identified, what are the forms and routes of exposure that are of concern for the worker protection?

4. Typical exposure situations: What are the typical exposure situations and industrial processes of concern for relevant, specific MNMs or groups of MNMs?
5. Exposure measurement and assessment: How will exposure be assessed, and are there alternatives to current exposure assessment techniques for MNMs that should be recommended in LMI countries?
6. OEL values: Which OEL or reference value should be used for specific MNMs or groups of MNMs?
7. Control banding: Can control banding be useful to ensure adequate controls for safe handling of MNMs?
8. Specific risk mitigation techniques: What risk mitigation techniques should be used for specific MNMs or groups of MNMs in specific exposure situations, and what are the criteria for evaluating the effectiveness of controls?
9. Training for workers to prevent risks from exposure: What training should be provided to workers who are at risk from exposure to specific MNMs or groups of MNMs?
10. Health surveillance to detect and prevent risks from exposure: What health surveillance approaches, if any, should be implemented for workers at risk from exposure to specific MNMs or groups of MNMs?
11. Involvement of workers and their representatives: How will workers and their representatives participate in the workplace risk assessment and management of handling MNMs?

As the WHS pointed out in its comments on the draft guidelines, the first team did not have women [42]. But women experts were represented in the final guidelines. Extensive effort was made also to bridge the gap between nanotechnology's unique and novel attributes and existing methodologies in order to better understand nanotoxicity. The GDG commissioned systematic reviews of the literature according to the process set out in the *WHO Handbook for Guideline Development*. All recommendations were made on the basis of consensus within the GDG. Eleven systematic reviews were generated, applying the Grading of Recommendations, Assessment, Development, and Evaluations (GRADE) systematic review process, which WHO considers valid for the development of any guidelines. As the team itself noted, this was potentially inappropriate, but the GRADE system was customized. The GRADE systematic review process was developed for medical topics, where clinical trials could be conducted, and health topics, such as environmental and occupational health. Thus the development of these guidelines also provided invaluable experience on adapting the GRADE systematic review process to occupational health and other nonmedical topics. Systematic reviews to answer questions about risks of MNMs and worker training were used to inform Section 12.5 of the guidelines (about best practices). Using the systematic review and GRADE, the guidelines recommendations are rated as "strong" or "conditional," depending on the quality of the scientific evidence, values and preferences, and costs related to the recommendation. The GDG met in Johannesburg, South Africa (2013); then Paris, France (2015); then Brussels, Belgium (2015); and Dortmund, Germany (2016). Following procedures outlined in the *WHO Handbook for Guideline Development*, experts were recruited to synthesize research about MNM health and safety in order to create teams of researchers to conduct systematic reviews of the literature. The systematic review teams were listed in the final report, along with their affiliations.

The first step in the evidence search and retrieval procedure was to identify and define the type of evidence required to address the scoping questions. First, the systematic review teams reformulated the key questions posed in Section 12.1.2 so that they could be answered by a systematic review [45]. Then they defined the best-available evidence to provide answers despite the scarcity of experimental studies directly assessing the impact of interventions on occupational health and safety; several distinct areas of evidence were required for each scoping question. Very few existing systematic reviews were found because this type of assessment is new in the fields of toxicology, occupational health, and environmental exposure assessment. Systematic reviews were commissioned for all questions with the aim of locating studies that could answer the pertinent questions. The systematic review process used for each question varied slightly but followed four PICO elements: **p**opulation, **i**ntervention, **c**omparator, and **o**utcome(s), which are used to assess the exposure or the intervention. The PICO approach guarantees that the systematic review process collects the evidence that is needed to answer the question at hand. The searches conducted for the systematic reviews included observational or experimental study of people or workplaces exposed to MNMs. Systematic review conclusions were based on the findings of the included studies. Following this protocol, systematic review teams determined the quality of evidence for each conclusion. GRADE allows the reviewer to systematically and transparently grade the quality of the body of evidence for the effectiveness of medical interventions. The quality of the body of evidence is then graded on the basis of five specific qualifiers, including risk of bias and inconsistency of results. This results in one of four quality ratings: high, moderate, low, or very low quality of evidence. The GDG broke new ground attempting to craft a use of the GRADE approach for environmental and occupational exposure [43, 46]. The rating process ranked a study design as high quality if it was considered the best for the question at hand. The reviewers did not use any qualifiers for upgrading the evidence, as is possible in the GRADE approach for nonrandomized intervention studies [46].

12.5.5.7 WHO general process from evidence to recommendations

After the systematic reviews had been conducted, the GDG developed recommendations based on the expert conclusions. When formulating its recommendations, the GDG used the balance between harms and benefits, values and preferences, and monetary costs and the quality of evidence in order to determine some relative value of the strength or weakness of the conclusions. Qualitative methods were used due to the absence of no numerical values for benefits and harms. Costs of an intervention, or the implementation costs of a recommendation, were considered and based on expert opinions, but cost–benefit or cost-effectiveness analyses were not performed. There is an explanation for each recommendation linked to the evidence. Proposed recommendations were discussed through face-to-face meetings until there was consensus, with no minority report. Thus, the guidelines are an important step toward protecting workers worldwide from these potential risks.

12.5.6 The International Labour Organization

Since its inception in 1919, the ILO has encouraged promotion of better working conditions. Consistent also with this heritage, ILO C155, Article 3(e) [47] offers the definition of health as "in relation to work, indicates not merely the absence of disease or infirmity; it also includes the physical and mental elements affecting health which are directly related to safety and hygiene at work." This is no typographical error, no stray text from elsewhere that was accidently printed because someone left it on the photocopy machine. Here, the ILO has deliberately applied the WHO constitutional language previously discussed, because it is so well recognized. This classic definition from WHO, as modified for occupational health, however, is deceptively simple and comprehensive at the same time: it assumes that there is a consensus about the term "work" – for example, is it always paid, and if so, how does the law separate out protections for volunteers or slaves, or does it treat them the same as paid employees? This definition of work also relies heavily upon a strong assumption about the relationship between health and work. The text does not question the role of work in disease causation but simply states that the relationship exists. Therefore, this definition bespeaks the complex interaction between dangerous workplace exposures, individual lifestyles, and environmental factors that impact the combined effects of working conditions. In addition, this approach is multidimensional because its concern for both physical and mental elements of health and well-being implicitly takes into account

International
Labour
Organization

Figure 12.8: The International Labour Organization Logo, used for identification and, does not imply endorsement by ILO of any of the text herein.

effects of occupational stress. These concepts are important to people throughout society because of the ubiquitous nature of nanotechnology.

The Preamble to the ILO Constitution of 1919 states, "Universal and lasting peace can be established only if it is based upon social justice" [48]. Since the founding of the Committee of Experts on the Application of Standards in 1926, "ILO standards provide the legislative framework and rationale for dynamic but feasible workplace health protection under law." ILO documents describe occupational safety and health as a multidisciplinary field devoted to the anticipation and control of workplace hazards that may impair health. Precautionary principles codified in these documents reflect fundamental tenets of industrial hygiene, codified in ICESCR Article 12 and repeated in the scientific literature about risk management for nanotechnology [49]. The ILO Convention No. 155 (Convention on Occupational Safety and Health; C155) and the ILO Convention No. 161 (Convention on Occupational Health Services; C161) provide a framework for governance infrastructures that can ensure the implementation of a coherent national policy that can generate robust data for training and updating information about injuries, illness, statistics, education and training tools, risk assessment data, and best practices.

Figure 12.9: (Left) Laborer cutting wood in Switzerland (2015) and (right) signing of the Memorandum of Understanding (MoU) between the American Society of Safety Engineers (ASSE) and the International Labour Organization (ILO), San Antonio, 2009.
Photos: Charoy family.

The ILO Constitution of 1919 makes the link between work health and the survival of civil society quite clear:

> Whereas universal and lasting peace can be established only if it is based upon social justice; And whereas conditions of labour exist involving such injustice hardship and privation to large numbers of people as to produce unrest so great that the peace and harmony of the world are imperiled; and an improvement of those conditions is urgently required; as, for example, by the regulation of . . . protection of the worker against . . . disease and injury arising out of his employment [sic].

By contrast to many state, federal, or local labor codes that segregate workers on the basis of their economic sector, the ILO does not draw such distinctions from the standpoint of the scope of its jurisdiction. Laborers or professionals, as shown in Figure 12.9, disabled or able-bodied, from seafarers to domestic help, and a wide range of highly paid staff across many job categories embracing agriculture, civil employment, and the precarious contracts of informal employees can enjoy the protections enshrined in the ILO precepts, regardless of whether they are paid by multinational corporations like the Better Work Partner of the Disney Corporation or small and middle enterprises. The ILO Constitution protects health and the right to information to ensure safety and health at work. Since its founding at the end of World War I, the ILO has offered the opportunity for trade unions, governments, and employers to achieve social justice and fair employment conditions as a fundamental tenet of creating world peace. The ILO founders were men who believed that fair working conditions prevent a race to the bottom that could lead to slavery and that informed free men would make good decisions as voters and consumers. These notions are exemplified in many of the sculptures at the ILO headquarters, such as the bronze laborers in Figure 12.10 and in the reproduction of the Stele of Hamarabi, the first recorded written laws, also on display in the ILO worksite. ILO C161 and ILO Convention No. 170 (Convention concerning safety in the use of chemicals at work, C170) complement the provisions of C155, which requires member states to establish competent governmental institutions that regulate and inspect workplaces. These duties involve surveillance of the work environment to identify, assess, prevent, and control occupational health hazards at the source. ILO C170 [50] states,

> It is essential to prevent or reduce the incidence of chemically induced illnesses and injuries at work by: (a) ensuring that all chemicals are evaluated to determine their hazards; (b) providing employers with a mechanism to obtain from suppliers information about the chemicals used at work so that they can implement effective programmes to protect workers from chemical hazards; (c) providing workers with information about the chemicals at their workplaces, and about appropriate preventive measures so that they can effectively participate in protective programmes; (d) establishing principles for such programmes to ensure that chemicals are used safely.

ILO Convention No. 187 (Promotional Framework for Occupational Safety and Health) reaffirms ILO C155 by providing an infrastructure for managing the Occupational Safety and Health (OSH) Act, establishing a prevention culture and progressively enhancing occupational health services. This flexible regulatory framework was written as a compromise to end heated debates about whether standards should be customized for specific worker populations or "one size fits all." The visionary flexibility of this text means that ILO C155 is consistent with the trend toward "personalized medicine" that will build upon genetic profiles using the application of nanomedicine.

The heart of ILO C155 about national occupational safety and health laws concerns the creation of effective national, regional, and workplace mechanisms for implementation and compliance with other ILO standards. ILO C155 fosters the creation, implementation, and periodic evaluation of occupational safety and health standards among member states of the ILO. For example, Article 4.1 states ILO C155's goal of fostering the development of a "coherent national policy" concerning occupational safety and health protections. To this end, ILO C155 obligates ratifying member states to promote research, statistical monitoring of hazardous exposures (such as medical surveillance measures, not unlike technical standards in member states), and worker education and training. ILO C155 uses broad terminology to provide a regulatory framework. Consultation with representative organizations and employers is required before exemptions will be granted, and any exclusion for categories of workers requires reporting on efforts to achieve "any progress towards wider application" pursuant to Article 2.3. Some have argued that if a broader economic goal supplants the interest in occupational health, then a government could justify reducing occupational health programming under the terms of this convention, but this rhetoric has not gained traction because occupational health programming saves resources and promotes growth by protecting the employer's economic health.

ILO C155 also fosters education for "representative organizations" and worker participation in the development and enforcement of occupational safety and health regulations internally and on regional, national, and international levels. This language foreshadows the role of worker training and the consumer's right to know that was operationalized by the GHS a decade later. Under the terms of ILO C155, "National competent authorities" who are the authorized administrative agencies undertake to ensure that they will create and implement an inspection system to enforce national laws and regulations on occupational safety and health. If nanotechnology is determined to be part of the occupational health protection system, then nanotechnology workplaces will be embraced in this law without anyone writing anything new. Much of ILO C155 reads like a checklist for a sound occupational health compliance program, except that the requirements apply to governments instead of employers. Governments that ratify this convention basically promise to include these components of risk management systems into their laws, or else they are not in compliance with the international labor standards.

ILO C155 is an international template for the so-called right to know that is granted to workers and communities in civil society. This host of rights includes the right to be informed about hazards, safe handling, and use of dangerous materials; access to working safety equipment free of charge; the right to be involved in the management and supervision of OSH measures at the workplace; the right to be organized in a representative group that can select delegates to OSH committees; the right to regularly scheduled updates concerning information and training on hazards/risks associated with their work and the measures to prevent them; the right to complain with impunity about unsafe circumstances; and the right to refuse hazardous work and not be required to return, in case of imminent serious danger to their health and life, without retaliation, with representation. In parallel, responsibilities of workers regarding such information require that workers follow safety and health rules when using protective equipment; participate in safety and health training and awareness-raising activities; cooperate with their employer to implement safety and health measures; and inform their direct supervisors if they withdraw from an imminent and serious danger, stating their reasons. The same core values of the right to know are applied to communities to protect consumers and the environment, as well as individuals whose time and exposure are controlled by their employers within industry.

The ILO/WHO Committee on Occupational Health, created to advance the purposes of ILO C155 and the chemical safety convention, adopted a comprehensive definition of the aim of occupational health programming: "Occupational health and safety should aim at the promotion and maintenance of the highest degree of physical, mental and social well-being of workers in all occupations" [51]. In 2007, the World Health Assembly (WHA) endorsed the WHO Global Plan of Action on Workers' Health (GPA) (2008–2017) to follow up the WHO Global Strategy on Occupational Health for All endorsed by the WHA in 1996. WHO has developed a global work plan in collaboration with WHO's network of CCs on the basis of the objectives of the GPA for 2009–2012. The main objectives of the GPA apply remarkably well for instituting precautionary measures for workplace exposures involving nanotechnology [52]. Once again, international law text applies a definition that is practically a photocopy of the WHO constitutional definition of "health." The committee has oversight authority to review policies and to promote implementation of these goals without regard to age, sex, nationality, occupation, type of employment, and size or location of the workplace.

A key tool for implementation of the goals expressed in ILO conventions is ratification of the conventions by member states under the ILO Constitution. According to Dixon, "Ratification is the process whereby a state finally confirms that it intends to be bound by a treaty that it has previously signed, consent not being effective until such ratification" [53]. Economists might disagree, noting that the economic imperative for having consistent labels and consistent requirements for material safety during handling outweigh ordinary political concerns, and therefore shift the political will of states in favor of ratification. The universal need for valid information in a predictable format is a common thread across all industries and part of the global safety net that ties together corporations across borders and ties employers to occupational safety and health programs despite governmental monitoring and surveillance. Alston views the ILO as an international model for procedural requirements, which, in his opinion, "legitimize the declaration of new norms" (1984) [54]. Such features of ILO procedures include preparation of a preliminary survey of relevant laws among member states, followed by its governing body's decision on whether to place the item on the agenda of the annual International Labour Conference (ILC), followed by a questionnaire from the ILO Secretariat to participating member states. After the draft has been referred to a technical committee, a draft instrument is circulated to member states and the appropriate worker and employer representatives; a revised draft instrument is then prepared and submitted to the technical committee, discussed by plenary and drafting committee, and adopted after voting by the ILC. This approach allows for maximum discussion and communication between regulated entities and governments.

These procedures, initiated in 1926 at the inception of the Committee of Experts on the Application of Conventions and Recommendations, have continued vibrancy in the international system. For example, the ILO's model forms the blueprint in Convention on the Elimination of All Forms of Discrimination against Women (CEDAW), as discussed in detail in Chapter 3: Article 18 sets forth a mandatory reporting mechanism before an international committee also described within the provisions of the convention. Except in the case of complaints, mandatory reports regarding activities toward implementation and compliance are heard by the committee at the end of the first year following ratification and then at least every four years. Additional reporting procedures for monitoring the application of ILO standards and conventions include but are not limited to direct contact missions [55], commissions of inquiry to investigate particular cases of egregious violations of ILO conventions and constitutional provisions, and regularly scheduled periodic oversight through reporting to conference meetings and reporting to the governing body and the administrative tribunal.

Reporting mechanisms are slow but invaluable; these constitute an important component of a much larger process of mobilizing world opinions toward positive change regarding labor issues. In addition to generating rules for international labor standards through its conventions, the ILO offers codes of practice. These expressions of the best thinking of collective experts regarding safety protections have served as the blueprint for occupational safety laws and regulations in such areas as dock work, transfer of technology to developing nations, civil engineering, and heavy industries. These model codes, which are sometimes applied with minor modification as draft legislation, share the values expressed in several ILO conventions pertaining to occupational safety and health: C62 Safety Provisions (Building) (1937), C77 Medical Examination of Young Persons (Industry) (1946), C78 Medical Examination of Young Persons (Non-Industrial Occupations) (1946), C119 Guarding of Machinery (1963), C120 Hygiene (Commerce and Offices) (1964), and C152 Occupational Safety and Health (Dock Work) Convention (1979). Precedent exists, therefore, to create a code of practice for the safe handling of MNMs and nanosafety, even though no such strategy exists in the ILO agenda.

12.5.7 United Nations Environment Programme

The United Nations Environment Programme (UNEP) has a mandate that strives to balance benefits from chemicals and the recognized potential to adversely impact

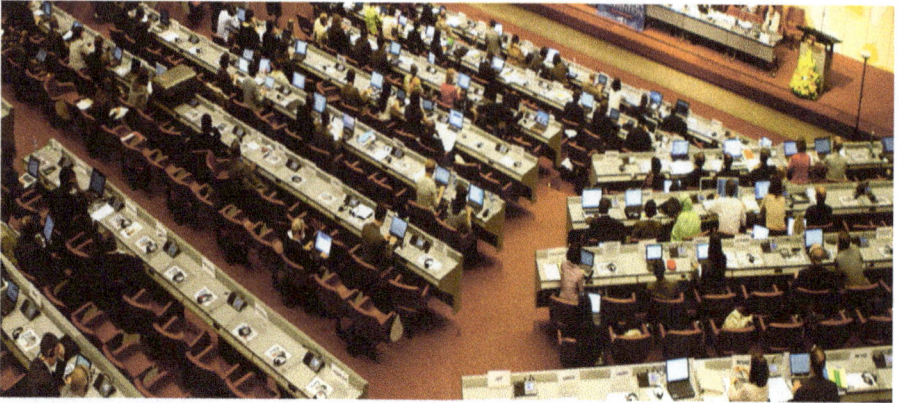

Figure 12.10: A meeting of governments and stakeholders at a conference center to deliberate the UNEP agenda, Geneva, Switzerland. Photo source: United Nations Environment Programme.

human health and the environment if not managed properly. UNEP's mission reflects the international consensus about its working assumption:

> Chemicals are an integral part of everyday life with over 100,000 different substances in use. Industries producing and using these substances have an enormous impact on employment, trade, and economic growth worldwide. There is hardly any industry where chemical substances are not implicated and there is no single economic sector where chemicals do not play an important role [56].

UNEP is, therefore, a repository for data concerning health-related effects ranging from acute poisoning to long-term effects, such as cancers, birth defects, neurological disorders, and hormone disruption. UNEP examines effects on ecosystems, eutrophication of water bodies, and stratospheric ozone depletion. UNEP is a focal point for integration knowledge across disciplines, regardless of whether people are exposed through occupation, activities in daily life through intake of contaminated drinking water, ingestion of contaminated food (e.g. fish contaminated with mercury, dichlorodiphenyltrichloroethane, and/or polychlorinated biphenyls), inhalation of polluted air (outdoor as well as indoor), or direct skin contact. UNEP serves the international community and civil society as an informational resource for many key technical areas of scientific interest that involve data repository and capacity building. UNEP is also an international governmental focal point for coordinating across groups of stakeholders, bridging the channels for communication across governments, NGOs, civil society, and opinion leaders in industry. The Chemicals and Waste subprogramme assists countries and regions in managing, within a life cycle approach, chemical substances and waste that have potential to cause adverse impacts on the environment and human health, including:

- Persistent, bioaccumulative, and toxic substances (PBTs)
- Chemicals that are carcinogens or mutagens or that adversely affect the reproductive, endocrine, immune, or nervous system
- Chemicals that have immediate hazards (acutely toxic, explosives, corrosives)
- Chemicals of global concern, such as persistent organic pollutants (POPs), greenhouse gases, and ozone-depleting substances (ODS)
- Health-care wastes
- E-waste

One key successful conduit for these efforts is Strategic Approach to International Chemicals Management (SICAM), a coalition of several UN agencies, private sector actors, and NGOs that meets regularly in Geneva, Switzerland, to iron out old problems and deliberate strategies for emerging issues. Nanotechnology is included among the emerging issues. The Inter-Organization Programme for the Sound Management of Chemicals (IOMC) (Figure 12.11), established in 1995, has the purpose of promoting coordinated policies and activities among the participating organizations, jointly or separately, to achieve the sound management of chemical safety protecting human health and the environment.

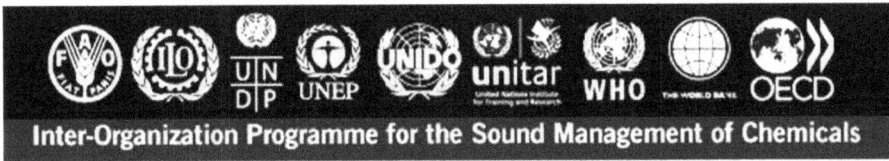

Inter-Organization Programme for the Sound Management of Chemicals

Figure 12.11: The Inter-Organization Programme for the Sound Management of Chemicals (IOMC) logo appears here for information purposes and does not imply any endorsement of this text or its contents.

Given the complexity of the scientific and technical questions to be translated into policy and the remarkably large number of international agencies that have jurisdiction over these issues, the integration of several key UN agencies under the umbrella of one interagency authority to tackle chemical safety issues without duplication is justified. According to UN agreements, the need for the IOMC is underscored by global growth patterns for production, impacting the work and health of future generations. The IOMC, housed in UNEP, is predicated on the precautionary principle that chemicals are major contributors to national economies but also require sound management throughout their life cycle in order to avoid expensive impacts upon human health and the environment.

IOMC participating organizations include the UNEP, ILO, Food and Agriculture Organization (FAO), WHO, United Nations Industrial Development Organization (UNIDO), United Nations Institute for Training and Research (UNITAR), OECD, World Bank, and UNDP. The agencies are tasked under international legal agreements with a division of responsibilities as follows: UNIDO undertakes activities on chemicals through its Energy and Environment Programme; the FAO has activities in the field of pesticides, a major chemicals sector that will be reconfigured as less but more potent pesticide materials will be used by applying nanotechnology; UNITAR provides training for governments and stakeholders; the ILO oversees and provides technical assistance for the application of standards for safe and healthful working conditions and prevention of major chemicals accidents; and the UNDP facilitates amelioration of chemicals management from the start of commercial programming and in tandem with the OECD [56].

In parallel, UNEP and its IOMC partners study the potential risk that hazardous wastes at the end of their life cycles might contaminate otherwise nonhazardous wastes. Chemical processes of concern include sludge from wastewater treatment plants, waste oils, and waste batteries. Addressing this perennial issue requires the legal authority vested in the IOMC to engage governments, industries, civil society, and other stakeholders to seek consensus regarding creative solutions to these intractable problems. UNEP's efforts to improve capacity to manage chemicals and waste soundly throughout their life cycles are realized by developing policy instruments, developing regulatory frameworks, and providing scientific and technical knowledge and tools needed to ensure a successful transition among countries toward sound management of chemicals and waste in order to minimize the impact on the environment and human well-being. The UNEP's Chemicals and Waste Branch is the focal point of UN Environment activities on chemical issues and the main catalytic force in the UN system for concerted global action on the environmentally sound management of hazardous chemicals. UNEP's Chemicals and Waste Branch works directly with countries to build national capacity for the clean production, use, and disposal of chemicals and promotes and disseminates state-of-the-art information on chemical safety. In response to mandates from the UN Environment Governing Council, it facilitates global action, including the development of international policy frameworks, guidelines, and programs.

12.5.8 WTO limitations regarding safety and health for trade agreements

In the WTO, nanotechnology poses a two-edged sword: both as an area of safety and health concern under GATT limitations on trade and as an area of rapidly emerging intellectual property that seeks protection under TRIPS. From the standpoint of emerging risks in nanotechnology, the real challenge for the WTO when striking the balance between these competing values is found in the plain meaning and text in its own treaties and mandates. An invaluable opportunity exists to use the WTO to resolve this fundamental conflict, however, if the WTO proactively serves as a forum for discussion of the emerging risks to human health of applying nanotechnology to products in international trade or commerce, plant safety, or food. An ability to grant patents and then protect the integrity of its patents can turn "big science" into big money. The WTO (Figure 12.12) is a voluntary organization of member states, designed to negotiate global trade policies. The WTO's main functions relate to trade negotiations and the enforcement of negotiated multilateral trade rules (including dispute settlement) regarding the implementation of WTO commitments and multilateral negotiations and trade and tariff data relating to exports. The WTO and the World Intellectual Property Organization (WIPO) have worked together to create an intellectual property regime that prevents fraud and piracy, while, at the same time, promoting research and development of expensive technologies for trade by all nations. The everyday activity of industry and commerce at the WTO was inherited from the General Agreement on Trade and Tariffs (GATT).

WORLD TRADE ORGANIZATION

Figure 12.12: World Trade Organization (WTO). The logo appears here for information purposes and does not imply any endorsement of this text or its contents.
Source: WHO.

On the basis of the work of the Bretton Woods institutions that interact to foster global commerce since the end of World War II, the WTO was founded in 1994 as the result of rethinking its predecessor organization, GATT. GATT had a clear constituency framework and mission: reducing the barriers to global free trade. Although GATT remains famous for its efforts to clarify customs procedures and reduce tariffs [57], its provisions regarding the primacy of safety and health concerns have often been viewed as an interesting loophole in the otherwise unbroken chain of trade rights and responsibilities. Many GATT precedents regarding food quality or safety of chemicals used in industry may apply to nanotechnology. Several OECD projects [58] contribute to global efforts to enhance its regulatory oversight of nanoscale materials health standards for groups of nanomaterials that have a similar chemical composition compared to those covered by existing recommendations [59]. Free trade is not without legal limits. The "public order" exception may be invoked when genuine and sufficiently serious threat is posed to one of the fundamental interests of society. This important exception to trade in favor of protecting human health is reaffirmed in Article 8, Principles ("1. Members may, in formulating or amending their laws and regulations, adopt measures necessary to protect public health and nutrition") and Article 14, General Exceptions ("Subject to the requirement that such measures are not applied in a manner which would constitute a means of arbitrary or unjustifiable discrimination between countries where like conditions prevail, or a disguised restriction on trade in services, nothing in this Agreement shall be construed to prevent the adoption or enforcement by any Member of measures: (a) necessary to protect public morals or to maintain public order [60]; (b) necessary to protect human, animal or plant life or health;" . . . (c) safety").[5] Too little attention has been given to this exception to the overall WTO course of business, but that may change with the ubiquitous use of nanotechnology by industry in a wide variety of consumer products. Unlike previous technology leaps through the development of "big science" in genetics, atomic energy, and astrophysics, many people who are not impressed by new scientific developments will find that their money purse and their shopping carts are nonetheless touched by nanotechnology – in addition to toys and exotic new foods that appear in grocery stores. If so, the nanotechnology "revolution" may herald trade limits based on the role of safety and health, and then, the right to health will take center stage in the WTO. The WTO is also a forum for the views of stakeholders, including but not limited to private firms, business organizations, farmers, consumers, NGOs, competitors, and trading partners, in order to engage in dispute resolution for trade problems.

12.5.9 OECD: nanotechnology protections across industrialized nations

The OECD in Paris, France, was among the first multinational treaty-based governmental organizations to address emerging risks and benefits of MNMs. The OECD has established a widely respected Working Party on Manufactured Nanomaterials (WPMN) that is engaged in a variety of projects to further our understanding of the properties and potential risks of nanomaterials for the following purposes: (i) development of a database on environmental health and safety, (ii) research strategies on MNMs, and (iii) safety testing of a representative set of MNMs. The OECD regularly updates its research into Safety of Manufactured Nanomaterials database, which is a global resource that details research projects that address environmental, human health, and safety issues of MNMs. This database helps identify research gaps and assists researchers in future collaborative efforts. The database also assists the projects of the OECD WPMN as a resource of research information [61].

The OECD WPMN approach also represents a shift in organizational priorities away from trade and economics into the deep inner workings of scientific research and the debates around its development resulting in innovation throughout the supply chain. Additional OECD products include Good Laboratory Practices (GLPs) and

---

5 According to the WTO "Agreement on Trade-Related Aspects of Intellectual Property Rights" (TRIPS), Article 24, International Negotiations; Exceptions: "2. The Council for TRIPS shall keep under review the application of the provisions of this Section. Any matter affecting the compliance with the obligations under these provisions may be drawn to the attention of the Council. . ."

Figure 12.13: Organization for Economic Co-operation and Development (OECD). The logo appears here for information purposes and does not imply any endorsement of this text or its contents. Source: OECD.

ongoing collaboration for information sharing to develop international standards for best practices for nanotechnology. One striking feature of this working party is that it offers a rare mix of stakeholders: it blends voluntary trade and industry-based associations, such as the ISO, with administrative agencies, such as the US Environmental Protection Agency (EPA) and ad hoc groups of scholars and experts from scientific institutions. This group may deal with complex scientific issues, but its existence is a fascinating development from the standpoints of the law. Such partnerships would have been unthinkable in the late twentieth century, when corporations resisted any effort at international regulations. By contrast, such organizations are at the forefront of funding and activity when discussing the revolutionary commercial applications of nanotechnology. The OECD WPMN Safety Testing of a Representative Set of Manufactured Nanomaterials project, in particular, is designed to find and help address important data gaps. The WPMN has identified a representative list of manufactured nanoscale materials for environmental health and safety testing, including fullerenes (C_{60}), single-walled carbon nanotubes (SWCNTs), multi-walled carbon nanotubes (MWCNTs). Additionally, the list for testing includes silver nanoparticles, iron nanoparticles, carbon black, titanium dioxide, aluminum oxide, cerium oxide, zinc oxide, silicon dioxide, polystyrene, dendrimers, and nanoclays. Stakeholders are nonetheless skeptical about the reliability of OECD data as a basis for risk governance and liability conclusions. A study commissioned by the Center for International Environmental Law (CIEL), the European Citizen's Organization for Standardization (ECOS), and the Öko-Institut suggests that much of the information made available by the Sponsorship Testing Programme of the OECD "is of little to no value for the regulatory risk assessment of nanomaterials" [62].

The study was published by the Institute of Occupational Medicine (IOM) based in Singapore. The IOM screened the 11,500 pages of raw data of the OECD dossiers on 11 nanomaterials and analyzed all characterization and toxicity data on 3 specific nanomaterials: fullerenes, SWCNTs, and zinc oxide. "EU policy makers and industry are using the existence of the data to dispel concerns about the potential health and environmental risks of manufactured nanomaterials," said David Azoulay, senior attorney for CIEL. "The fact that data exists about a nanomaterial does not mean that the information is reliable to assess the hazards or risks of the material."

This problem of methodology echoes the issues of scientific certainty that undermined the validity of early "threshold limit values" and scientific consensus standards produced by industrial hygiene organizations in the mid- to late twentieth century. In those cases, the matter of reliability and validation was settled by regulators only after years of protracted litigation [63]. The OECD WPMN published the dossiers in 2015, but the team was unable to draw conclusions on the data quality. The stakeholder groups objecting to the quality and use of the dossiers fear that despite missing analysis, the European Chemicals Agency (ECHA) and the European Commission's Joint Research Centre have presented the dossiers as containing information on nano-specific human health and environmental impacts. *Guidelines on Protecting Workers from Potential Risks of Manufactured Nanomaterials* provides an important contribution to global efforts toward writing clear, reliable nanotechnology regulations and standards that will be consistently reproducible in the future, thereby advancing scientific certainty regarding exposure and its risks. By May 31, 2016, the OECD working party had published 58 reports, which have recently been subjected to scrutiny by nongovernmental organizations in civil society; the Center for International Environmental Law (CIEL), whose name also means "sky" in French; and its partner, the Institute of Occupational Medicine. Industry federations and individual companies have taken this a step further, emphasizing that there is enough information available to discard most concerns about potential health or environmental risks of MNMs. "Our study shows these claims that there is sufficient data available on nanomaterials are not only false, but dangerously so," said Doreen Fedrigo, senior policy officer of ECOS. "The lack of nano-specific information in the dossiers means that the results of the tests cannot be used as evidence of no 'nano-effect' of the tested material. This information is crucial for regulators and producers who need to know the hazard profile of these materials. Analysing the dossiers has shown that legislation detailing nano-specific information requirements is crucial for the regulatory risk assessment of nanomaterials." Yet, these same dossiers were not merely endorsed but relied upon by WHO [64] in its final guidelines for reducing exposure to MNMs.

The report *Analysis of OECD WPMN Dossiers Regarding the Availability of Data to Evaluate and Regulate Risk* [65] provides recommendations for governance of nanomaterials in commerce. The report indirectly represents a call for a new regulatory approach to risk governance, consistent with the revolution that has been heralded about every aspect of nanotechnology in civil society. It therefore remains to be seen whether traditional paradigms for risk management and standard setting, which were inadequate to address many complex risks created by use transport and disposal of toxic substances, can answer this vital need to protect work, health, and survival. "Based on our analysis, serious gaps in current dossiers must be filled in with characterisation information, preparation protocols, and exposure data," said Andreas Hermann of the Öko-Institut. "Using these dossiers as they are and ignoring these recommendations would mean making decisions on the safety of nanomaterials based on faulty and incomplete data. Our health and environment requires more from producers and regulators."

12.5.10 The Council of Europe

With 47 member nations, embracing 800 million people, the Council of Europe (CoE) (Figure 12.14) views its role as the health and human rights vanguard for law governing the right to health and consumer protection throughout Europe. Its human rights court has remained a leading model for jurisprudence throughout the world. In 2012, the CoE Parliamentary Assembly began the first steps toward nanotechnology regulation that would require its member nations to apply safeguards to consumer health and the environment, consistent with its mandate to follow scientific precautionary principles. Pan-European and international reciprocal agreements were offered to the Parliamentary Assembly in the report *Nanotechnology: Balancing Benefits and Risks to Public Health* [2]. The report outlines potential avenues for CoE legislative activity around nanotechnology in commerce such as a study commission, standards, or a binding legal instrument. Rejecting previous calls for moratoria, the CoE would create a multinational regulatory structure that incorporates scientific precautionary principles into real-world technological processes without sacrificing research and innovation, while "protecting the human rights to health and to a healthy environment of everyone."

12.6 Keeping the canary alive: nanoethics and safe work saving society

12.6.1 Inextricable links in work health and survival of civil society

The profound link between health at work and survival of human society is ubiquitous timeless and knows no geographic bounds. Civil society need not write new international laws, if we courageously recognize existing laws. Work health and survival have been inextricably linked throughout the history of human civilizations. Without work, society cannot survive, and no work can perpetuate society without health. The fluid categories of sickness and health, which fluctuate within individual abilities and deficits and within the life of any given individual across time, hold implications for every worker in every job description: ranging from dignitaries in the highest offices of leadership, celebrities and heads of state in North America or Europe, to the laborers

Figure 12.14: Inside and outside the headquarters of the Council of Europe (CoE).
Source: Council of Europe website (www.coe.int).

tearing apart old ships in the shipyards of Asia; from the boardroom to the Uber delivery bicycle. Illness and health are inherent in the human condition. Therefore, occupational health systems and the laws that institute and govern them are essential to
preserving health. Regardless of how health may be defined, accomplishing our mission of preserving health requires reducing risks and providing the opportunity to
work for people who can perform most, if not all, of the traditional functions within a
given job description. But, no one is either: self-sufficient or completely dependent on
others. Human existence is a confluence of interdependence, which makes even the
highest ranking leader a worker with occupational hazards that could be ameliorated
with a sound job hazard analysis. From this standpoint, the link between work, health
and survival becomes inescapable: perpetuating civilization depends on having
healthy new generations who enjoy the fruits, and replenish the labors, of their
ancestors. For this reason, there is probably also great merit in the old wives' tale

that work is part of promoting health. This is only possible, however, because health at work is a precondition for society's ongoing existence.

Rights to health protections are ageless, and therefore changing paradigms for patient choices and "informed consent" in light of "personalized medicine," which applies nanotechnology techniques to pre-existing genetic and proteomic information about the individual patient will also require a redefinition of access to care and the right to refuse treatment culturally and under international laws. Traditional models for health must be revised in order to respond to the needs of a transcultural workforce comprised of older workers, women who were not previously represented in the workforce in great numbers, and workers who work for several employers if not at the same time then certainly during their lifetime.

12.6.2 Nanotechnology confronts millennial questions with new ways to protect public health

In occupational health throughout the twentieth century, there was an ongoing battle to make better, more sensitive canaries who are immune from other causes but more vulnerable, more sensitive to the gases in the mines. We have alarms that make noise, rather than falling silent when we expect disaster. We have nanosensors for on-going monitoring that charts and graphs the peaks and valleys in exposure to various substances and railway cars and cameras to speed people through and record evacuations. And we don't have many women crawling around naked in the depths of the mines, even in third world nations where there may still be canaries in the mines. But we have a host of silent killers that emit from the workplace and those harms impact nonetheless the most vulnerable populations. Preconceptual exposures for parents may bring effects in offspring from exposure to toxic substances; fetoexposures to noise, teratogens, contagious or infectious disease or perhaps even beryllium may cause unreportable or undetected harm – remote and far-fetched, abstract and insignificant, seemingly remote – but linked by a fine thread of investigative reasoning all the same.

Sometimes there is unexpectedly great import in something that at first seems insignificant and far away, removed from our reality. A long time ago, before electricity was used in homes, people mined coal by crawling around the inner bowels of the earth, perhaps with pick axes or other implements, or perhaps using their bare hands. Ivy Pinchbeck, in a brilliant but unsung treatise about the life and work of female miners in eighteenth- and nineteenth-century England and France (yes two or three hundred years ago, women mined coal and other ores in England and France) described how women went into parts of the mines where men who were too highly paid to be risked and horses, who were too large and who cost too dear to replace, would not go. She described women working naked, on their knees and crawling in the veins of coal, working long hours, descending before daybreak and emerging when it was again the night without ever seeing the sun. Pinchbeck described their working conditions. I have seen engravings of such women working, which were used to illustrate a point about the contributions of a doctor in England at a conference on the history of occupational health. But there are no illustrations in Pinchbeck's book, and the engravings I saw are not in any public document that could be reproduced without copyright infringement. But the engravings exist, and they support Pinchbeck's statements. So I believe that her description of these women's toils are historically accurate. I believe too, that they had very difficult working conditions and many women died working in the coal mines.

The nineteenth-century French political novelist, Emile Zola in L'Assemoir also paints a bleak landscape of the coal mining villages in Southern France, which suggests that there is merit to Pinchbeck's statements. He described the life of all the coal miners in the south of France in the early nineteenth century. Where there was famine and little to eat and threats of strikes and threats on the person of the mine owners, it was the outraged female workers who had the hardest, dirtiest, and least paid jobs and were also the workers who lead the call to arms which resulted in strikes, civil unrest and eventually, union contracts with better wages and better working conditions.

In those days too, there were few emergency rescue systems in the coal mines. If there was a cave-in or other disaster few, if any, of the workers could get out alive. The coal that warmed houses in Paris and London and warmed food for urban families of wealth came at a high price in human life, that revolutionaries at the outset of the twentieth century well understood. But sometimes there were emergencies where people could get out – there was not only the hazard of cave-ins and fires, but also the possibility of leaking gas as coal emerged from the innermost layers of the earth. And to protect human life in the event of gas escape, as feeble or inadequate as that attempt may have seemed to be with twentieth-century hindsight, there was the canary in the mine. Subsequently, the canary in the mine has entered popular lexicons in the form of a phrase that means pay heed to this warning, but it is rooted in actual events and the collective experience of those who warmed our society's houses and created the energy sources for the industrial revolution, sometimes by giving their own lives. Genetically selected for a mysterious genetic trait that was observed but at that time not well understood, canaries were posted at key locations in the mines because they were known to be more sensitive to the change in air quality that could mean the loss of oxygen and the loss of precious life. The canaries in the mines that sang or chirped meant all was fine.

But their silence meant death in their hypersusceptible pulmonary system, and it meant that it was time to sound the warning to evacuate the mines. Gases such as methane can escape, replacing life giving oxygen before an explosion leading to a cave-in, and among cautious mine owners there were always the sensitive, the vulnerable, canaries in the mines. While not all could get out in time, the death of the canary meant get out of the mines. And in fact, there were actual debates in accident investigations regarding the time lag between the death of the canaries and the actual command to evacuate the mines. Was there time to do more work before announcing the canary had died, before evacuating the mines? Would every dead canary give cause for evacuation? Couldn't a canary just die of old age on the job, or of some other intervening cause? Would a protesting worker take the oxygen from them, in order to call a strike, to cause a disruption, or to prevent some new technology from being used in the mines? Or by accident? How many canaries had to die before a mine owner was certain it was necessary to evacuate the mines? It cost money to stop work, and it cost time and broken-pace to stop production in an era when coal production was reaching unprecedented demand. And there was unprecedented economic competition between two international superpowers of the time, France and England. And most importantly and most troublesome of all, the accident investigations discussed by Pinchbeck and alluded to dramatically by Zola ask the unanswered eternal question, once you believe the canary is dead, truly dead from gas and not dead from sabotage or other intervening causes, how much time?

How much time does the mine-owner have between the death of the canaries, the genetically selected vulnerable population, until the inevitable destruction of everyone who cannot get out of the mine? No one ever really knows. Gas fluctuates in its ability to travel based on other factors, weather, wind, production quotas the size of the working population inside the caverns, the size and level of development of the work spaces in the mines. There is not much decision time and the mistaken evacuation is expensive, economically costly. But even when there is consensus that the canary is dead, no one really knows the exact evacuation time that remains.

12.6.3 Occupational health is a social good, requisite for and society's survival

Jack Levy Esq. was a lawyer, and before he died he moved my admission to the US Supreme Court Bar. But when he was in college, he was not a lawyer. His summer job, as he referred to it sometimes, was serving on government business in Europe, serving in the army as a solider in World War II. In fact, he was in the Normandy invasion seen in movies like *Saving Private Ryan* and *The Longest Day*. People forget that at the end of the movie, there was still a war going on. The soldiers in those stories woke up after the greatest battle of the twentieth century and went to war, fighting all over again and perhaps getting wounded or killed in much less famous battle that was still part of that same war.

Jack Levy was wounded in another battle days later. His congressional Bronze Star Medal hangs in his granddaughter's bedroom until she moved away for college. The Bronze Star is a tangible reminder of his exposure to risk, his acceptance of the consequences of that risk, and his society's acceptance of the risk in a special place and time.

Warfare is a hazardous situation, and the infantry has a dangerous job if ever there was one.

The risk was high, although socially accepted, whether one was drafted into the armed forces or volunteered. The bullets, my father often said, did not know the difference between the drafted armed forces and the volunteers.

Jack Levy said precious little about The War, but he impressed upon listeners that the thrill of the risk of war was so great that there was no other adrenalin like it – that it was irrational in how it overtook people to fight in a war. It was for that reason, by the way, that addictive quality about the risk inherent in the work of war that my father stated many times in articles and interviews "there was no such thing as a good war" because he believed the irrational overtook people in war so that they lost all sense of perspective. The risk, he believed, was addictive and filled their senses, so that risk overcame their logic, their rational behaviors, indeed, their free will to refuse risk, their ability to choose against risk voluntarily, and ultimately impacted their ability to survive. In his later paid work as a self-employed lawyer, he dedicated the rest of his life to the fine art of reasoning using the rule of law. But even that work in his day, and the work of people who follow in his footsteps, is fraught with hazards.

Travel, stress, long hours, little sleep, and the perennial unpaid toils for clients who fail in their commitment to pay for our work are common hazards. What can we do about it? What are the basic notions for protecting workers, including ourselves, articulated in national and international laws across centuries? And what must be done to rethink our values in order to implement those laws meaningfully? What can we do about it? What are the basic notions for protecting workers, including ourselves, articulated in national and international laws across centuries? And what must be done to rethink our values in order to implement those laws meaningfully?

Survival of our disciplines: occupational medicine, occupational health law and regulations, environmental health. Survival can only occur by rethinking our values, by applying the existing international human rights laws governing the human right to occupational health, and the international human right to reproductive health. Mysterious, baffling, an eternal riddle. No peoples, no individual can survive without health, and no society can survive without reproductive health. This notion is underscored, but not unprecedented, in the discussions about recovery from the Covid19 pandemic of 2020.

Therefore, fundamental universal notion articulated under international human rights law states:

Health is needed by every society and every individual in order to survive.

Work and health are both fundamental for society, and therefore essential for society to survive.

Under the laws discussed here, this notion is called the legal principle of universality in international law. These concepts have erroneously been painted as competing interests, when in fact in order for civil society to survive, they must co-exist. Furthermore, the failure to recognize the importance and the interdependence has significant consequences: we are witnessing unprecedented, perhaps even preventable, illness in vulnerable populations who are not traditionally considered to be part of the worker's health discipline. Older workers, retirees, spouses who do not attend the worksite, and even little children are affected by the harms of unchecked and unnoticed hazards at work.

In response to these factors, civil society can cry and grieve, or complain that change is too unexpected, and that costs of retooling or rethinking are too economically expensive; we can recalculate our prices for premiums for workers' compensation, social insurance, and even employer-based private insurance plans. But all this is not enough if we fail to confront and ultimately address the root causes of these harms. For this, we must begin the painful but deliberate tasks of rethinking our values surrounding work and health and their interaction in contemporary society. No, that is not enough.

We must be unafraid to ask the hard questions that the field of occupational health has shied away from in the last century. Questions about the inadequacies of workers' compensation systems, the perverse incentives that encourage underreporting and under-recognition of disease; the failure of society to provide meaningful health care, even in nations with nationalized health insurance that would clarify the benefits of increased vigilance against occupational health hazards. When we look at hazards in the workplace, wherever that may be, even if it is within the privacy of households or the boardrooms of the captains of industry, we must then ask about our society's notions of health and redefine health in a manner that is consistent with protecting against the hazards of our work. On the one hand, as science and technology uncovers more latent defects and genetic causes of potential illness that are not yet manifested, new technologies offer treatments, preventive measures, and perhaps even cures to old diseases; on the other hand, there may a paradigm shift that will inevitably change the social definition and then the legal definitions of disability and health. What will happen then to the fundamental criteria for injury and illness, codified in workers' compensation statutes throughout many nations, which are the flagstones for understanding occupational health?

Beyond civil society's failure to answer questions, but more importantly the failure to ask the right questions is the essence of our contemporary occupational health problems. By ignoring the last of these questions, the effect of occupational health problems on an ever-expanding circle of people who may never enter the worksite, much less encounter the most dangerous hazards that impact their own health, we have ignored those facets of occupational health problems that would otherwise create a social control to prevent hazards. According to the most recent public health statistics, if we intend to do something to save other people, to ensure society's survival and have a healthy next generation, the starting point of this analysis, is rethinking values about health. There is a crisis in health and there are some possible solutions to these problems that require rethinking our values – looking at occupational health in general and reproductive health in particular from a whole new perspective compared to the approach taken in the twentieth century. The only thing we don't know for those who wish to strategize a healthier future for our society is: How much time? Long ago, people recorded causal relations between exposures to dangers and injury in the workplace. These notions survive, along with other artifacts of the work of the ancient civilizations. By the nineteenth century, legislators understood that factories and coal mines ought to be inspected to prevent dangerous conditions because even if one or two people were successful in avoiding the risks of such work by choice, in the aggregate, society needed hundreds of thousands of people to work in the coal mines and factories that were the hallmark of the industrial revolution, and which still forms the underpinnings of our service-oriented technology-based society. In essence, the ability or inability of any one person to make decisions about risk eventually was overtaken by the fact that the work to be performed, inherent with risks, was needed by society. So laws were written and common law theories were modified to protect working people in order to curb some of the death and illness among workers that disrupted social progress.

Universal norms outlast lifetimes and reach across geographic borders and ethnicities and nationalities.

If everyone must work, however, and if society needs work in order to record its own history, fight its wars, grow its crops, and govern its people, the notion that undertaking risk for work is a matter of personal choice has very little validity. Yet, the present level of occupational health hazards is, by public health accounts, out of control. Sadly, many of these problems are preventable and what is even more disturbing is they are not really about economic issues, if one looks closely and thoughtfully at these problems. It is for this reason that the uncontrolled exposures to occupational health problems threaten our very survival. And rethinking our traditional values and deconstructing our traditional approaches to these old questions are our only hope for survival.

References

[1]   http://www.iso.org discussing ISO TC 229, a series of consensus documents regarding the methods for measuring and examining nanomaterials used in commerce.
[2]   UN Charter. signed June 26, 1945, entered into force October 24, 1945 Center for the Study of Human Rights. *Twenty Five Human Rights Documents*. New York, Columbia University, 1994.
[3]   Feitshans I, United Nations procedures protecting international human rights, Lawline.com. Continuing Legal Education course 2012.
[4]   Constitutional rights to health protections exist in Canada, France, and many nations.
[5]   Garibaldi O. Obligations arising from the International Covenant on Economic Social and Cultural Rights. In: Hannum H, Fischer D eds, *United States Ratification of the International Covenants on Human Rights American Society of International Law*. Transnational, Washington, D.C., Irvington-on-Hudson, New York, 1993, 164.
[6]   Universal Declaration of Human Rights (1948): "Everyone has the right to life, right to work, to free choice of employment, to just and favourable conditions of work and to protection against unemployment."

[7] Parmeggianni L ed., UN ILO *Encyclopaedia of Occupational Safety and Health*, 1488, Occupational diseases "cover all pathological conditions induced by prolonged work, e.g., by excessive exertion, or exposure to harmful factors inherent in aterials, equipment or the working environment."

[8] Occupational Safety and Health Series 74. List of Occupational Diseases 2010. Geneva, Switzerland, Vol. 7, 2010, According to the ILO Protocol of 2002, "Occupational disease covers any disease contracted as a result of an exposure to risk factors arising from work activity." Note the ILO list definition is narrower than the encyclopedia because "conditions induced by prolonged work" may include illnesses that are not recognized as a 'disease'." This difference makes the ILO list a worrisome touchstone inviting litigation, once newly discovered diseases or new negative reactions harming health status are associated with new technologies.

[9] Grad FP, Feitshans I. Article 12: Right to Health. In: Hannum H, Fischer D eds, *United States Ratification of the International Covenants on Human Rights American Society of International Law.*

[10] Ibid., 224.

[11] WHO Constitution basic documents, in several languages, 1948.

[12] Bustreo F, Hunt P. *Women's and Children's Health: Evidence of Impact of Human Rights*. WHO Press, Geneva, Switzerland, May 2013.

[13] The ICPD. Cairo, 1994, adapts language from the WHO Constitution for its definition of "reproductive health."

[14] Hunziker P. *Nanomedicine: The Use of Nano-Scale Science for the Benefit of the Patient Nanomedicine*, European Foundation for Clinical Nanomedicine, 2010.

[15] Greenberg G. Occ_Env_Med_L April 2, 1999, 48, 12, 241–43, citing to the *Mortality and Morbidity Weekly Report* (MMWR) from the Centers for Disease Control and Prevention (CDC) list of "Ten Great Public Health Achievements: United States; 1900–1999," which states, "Vaccination, which has resulted in the eradication of smallpox; elimination of poliomyelitis in the Americas; and control of measles, rubella, tetanus, diphtheria, Haemophilus influenzae type b, and other infectious diseases in the United States and other parts of the world.".

[16] WHO. *World Report on Disability*. Geneva, Switzerland, WHO, June 2012.

[17] Allison LG, James MH, David LH, and Guénaël R. *Global Public Health Security Policy Review* October 2007, 13, 10.

[18] Facts on aligning the Hazard Communication Standard to the GHS International Programme on Chemical Safety Harmonization Project Strategic Plan 2005–2009: harmonization of approaches to the assessment of risk from exposure to chemicals, www.who.int/entity/ipcs/methods/harmonization/strategic_plan_rev.pdf. Last access date: June 2, 2021.

[19] WHO. *Guidelines on Protecting Workers from Potential Risks of Manufactured Nanomaterials*. Geneva, World Health Organization Department of Public Health, Environmental and Social Determinants of Health, 2017, Cluster of Climate and Other Determinants of Health, https://www.who.int/publications/i/item/9789241550048, license: CC BY-NC-SA 3.0 IGO. Last access date: June 2, 2021.

[20] World Health Organization Constitution. The WHO constitutional language is repeated in the International Convention on Population and Development (IPCD). Cairo, 1994, in several national constitutions in the African (Banjul) Charter. The Alma-Ata Declaration, Article 1, reaffirms that "health . . . is a fundamental human right . . .". Last access date: June 2, 2021.

[21] International Covenant on Economic, Social and Cultural Rights (ICESCR), United Nations, adopted 1966, entered into force 1976, Articles 6, 7, 12: "Recognize the right of everyone to enjoyment of just and favorable conditions of work which ensure in particular, safe and healthy

working conditions; the right to the highest attainable standards of physical and mental health, in particular, the improvement of all aspects of environmental and industrial hygiene; the prevention, treatment and control of epidemic, endemic, occupational and other diseases.", https://www.ohchr.org/en/professionalinterest/pages/cescr.aspx. Last access date: June 2, 2021.

[22] Fifty-Ninth World Health Assembly, WHA59.15, Agenda 19, May 27, 2006, collaboration within the United Nations system and with other intergovernmental organizations, including the United Nations Strategic Approach to International Chemicals Management. Last access date: June 2, 2021.

[23] WHO, International programme on chemical safety. WHO/IPCS meeting on strengthening global collaboration in chemical risk assessment in conjunction with the 9th meeting of the Harmonization Steering Committee, https://www.who.int/health-topics/chemical-safety#tab=tab_1. Last access date: June 2, 2021.

[24] In 2008–2009, UNITAR supported national GHS implementation and capacity-building projects in Vietnam, Jamaica, and Uruguay. In 2005–2007, UNITAR supported projects in Cambodia, Indonesia, Laos, Nigeria, Senegal, Slovenia, Thailand, Gambia, and the Philippines. Meetings and workshops were supported in Malaysia, Singapore, and the ASEAN OSHNET. Regional activities exist in ASEAN, SADC, ECOWAS, CEE, the eight Arab states, and other subregions. The projects are executed by UNITAR, in the context of the UNITAR/ILO Global GHS Capacity Building Programme, with funding from the Government of Switzerland, the European Union, and others (https://www.unitar.org/). Last access date: June 2, 2021.

[25] The IPCS promotes consistency among hazard and risk assessment products within the global system for classification of hazards, with a view to facilitating the GHS at the national level.

[26] Mikheev MI, *WHOs occupational health programme*, conference papers Session III, "Health a Changing Challenge," Safety and Health at Work Conference, London, England, March 4–6, 1997, citing WHO Global Strategy, Occupational Health for All, Resolution of the Forty-Ninth World Health Assembly, May 25, 1996 (offset document WHA49.12).

[27] General authority mandating action to protect worker reproductive health: Implications of the WHO global strategy for health for all plan of action 1996–2001. WHA 49.12, May 19, 1996, reprinted in *Int J Occup Med Environ Health* 1997, 10, 2, 113–39.

[28] WHO, Facilitating Projects Guide Work Plan 2009–2012, WHO's global network of collaborating centers in occupational health. Last access date: June 2, 2021.

[29] WHO Global Plan of Action on Workers' Health (2008–2017): Baseline for Implementation Global Country Survey 2008/2009 Executive Summary and Survey Findings Geneva, April 2013. Last access date: June 2, 2021.

[30] WHO draft proposal for *Guidelines on Protecting Workers from Potential Risks of Manufactured Nanomaterials* (WHO/NANOH), Background paper 2011, comments accepted until March 31, 2012.

[31] The Work, Health and Survival Project (WHS), including the International Safety Resources Association (ISRA), Fullerton California, Earth Focus Foundation, Geneva, Switzerland, Digital 2000 Productions, Stafford, Texas, USA; Donald H. Ewert, IH, VP Field Services NanoTox, Inc., and Director Field Services AssuredNano, Dr. Gustav Grob, International Sustainable Energy Organization (ISEO) Geneva, Switzerland. Comments presented by Ilise Feitshans, WHO draft guidelines on *Protecting Workers from Potential Risks of Manufactured Nanomaterials*, WHO/NANOH, background paper, 2011, comments until March 31, 2012.

[32] Boccuni F, Gagliardi D, Ferrante R, Rondinone BM, Iavicoli S. Measurement techniques of exposure to nanomaterials in the workplace for low- and medium-income countries: A systematic review. *Int J Hyg Environ Health* 2017, 220, 1089–97, https://doi.org/10.1016/j.ijheh.2017.06.003.

[33] Nagai H, Okazaki Y, Chew SH, Misawa N, Yamashita Y, Akatsuka S. et al.,, Diameter and rigidity of multiwalled carbon nanotubes are critical factors in mesothelial injury and carcinogenesis. *Proceedings of the National Academy of the United States of America* 108, E1330–E1338, 2011, https://doi.org/10.1073/pnas.1110013108.

[34] Scope of the guidelines and key questions identified by the GDG: Which specific MNMs and groups of MNMs are most relevant with respect to reducing risks to workers, and which should these guidelines now focus on, taking into account toxicological considerations and quantities produced and used? Which hazard class should be assigned to specific MNMs or groups of MNMs and how?

[35] Feitshans I. *Designing an Effective OSHA Compliance Program*, Thomson Reuters/ Westlaw. com, 2013.

[36] Eastlake A, Zumwalde R, Geraci C. Can control banding be useful for the safe handling of nanomaterials? A systematic review. *J. Nanopart. Res* 2016, 18, 169, https://doi.org/10.1007/s11051-016-3476-0.

[37] The guidelines cited the following sources: IUF: Small particles, big risks: IUF, international NGOs release recommendation on the use of nanotech in foods (http://www.iuf.org/w/?q=node/4073); ETUC: First resolution on nanotechnologies and nanomaterials, 2008 (https://www.etuc.org/en/document/etuc-resolution-nanotechnologies-and-nanomaterials); Canadian Union of Public Employees: Fact sheet: Nanomaterials, 2016 (https://cupe.ca/fact-sheet-nano materials); Australian Council of Trade Unions (http://www.actu.org.au/media/149927/actu_factsheet_ohs_-nanotech_090409.pdf). Last access date: June 2, 2021.

[38] von Mering Y, Schumacher C. *What Training Should Be Provided to Workers Who Are at Risk from Exposure to the Specific Nanomaterials or Groups of Nanomaterials?*. Geneva, World Health Organization, 2017, Licence: CC BY-NC-SA 3.0 IGO.

[39] Kulinowski K, Lippy B. *Training Workers on Risks of Nanotechnology*. Washington, DC, US Department of Health and Human Services/National Institutes of Health, National Institute of Environmental Health Sciences, 2011.

[40] Lee N, Lim CH, Kim T, Son EK, Chung GS, Rho CJ. et al., *Which Hazard Category Should Specific Nanomaterials or Groups of Nanomaterials Be Assigned to and How?* Geneva, World Health Organization, 2017, Licence: CC BY-NC-SA 3.0 IGO.

[41] Feitshans I on behalf of the International Safety Resources Association (ISRA), several individual stakeholders, and the Earth Focus Foundation, stakeholder comments to NANOH regarding the WHO draft proposal for *Guidelines on Protecting Workers from Potential Risks of Manufactured Nanomaterials*, WHO/NANOH, background paper 2011, filed March 31, 2012.

[42] World Health Organization. *WHO Handbook for Guideline Development*. 2nd ed. Geneva, World Health Organization, 2014.

[43] Individuals and partners involved in guideline development, including the WHO Guideline Steering Group Guideline Development Group (GDG), systematic review teams, External Review Group gave written statements about of conflicts of interest as part of the WHO protocol for guidelines.

[44] Honnert B, Grzebyk M. Manufactured nano-objects: An occupational survey in five industries in France. *Ann Occup Hyg* 2014, 58, 121–35, https://doi.org/ 10.1093/annhyg/met058.

[45] Lewin S, Glenton C, Munthe-Kaas H, Carlsen B, Colvin CJ, Gülmezoglu M. et al., Using qualitative evidence in decision making for health and social interventions: an approach to assess confidence in findings from qualitative evidence syntheses (GRADE-CERQual). *PLoS Med* 12, e1001895, 2015, https://doi.org/0.1371/journal.pmed.1001895.

[46] ILO Constitution, 1919. Declaration of Philadelphia, 1944.

[47] Workers' health: Global plan of action, Sixtieth World Health Asse (Wha60.26), Agenda 12.13, May 23, 2007.

[48] Editorial, *The Synergist*, February 2011.

[49] C170 Chemicals Convention. 1990 Convention concerning Safety in the use of Chemicals at Work (Date of coming into force: 0411:1993.) Convention C170. *International Labour Conventions and Recommendations 1977–1995*, Volume iii. Geneva, Switzerland, International Labour Office, 1996, 337.

[50] C155 and all ILO conventions are available at ilo.org.

[51] Ibid. "Having considered the draft global plan of action on workers' health;1Recalling resolution WHA49.12 which endorsed the global strategy for occupational health for all; Recalling and recognizing the recommendations of the World Summit on Sustainable Development (Johannesburg, South Africa, 2002) on strengthening WHO action on occupational health and linking it to public health;2Recalling the Promotional Framework for Occupational Safety and Health Convention, 2006,and the other international instruments in the area of one program occupational safety and health adopted by the General Conference of ILO;(International Labour Conference, Ninety-fifth Session, Geneva 2006. Provisional Record 20A) Considering that the health of workers is determined not only by occupational hazards, but also by social and individual factors, and access to health services; Mindful that interventions exist for primary prevention of occupational hazards and for developing healthy workplaces; Concerned that there are major gaps between and within countries in the exposure of workers and local communities to occupational hazards and in their access to occupational health services; Stressing that the health of workers is an essential prerequisite for productivity and economic development, … 2. URGES Member States: to devise, in collaboration with (1) workers, employers and their organizations, national policies and plans for implementation of the global plan of action on workers' health . . ."

[52] Dixon M. *Textbook on International Law*. 6th edition. Oxford University Press, 65, 2007.

[53] Alston P. Conjuring up new human rights: a proposal for quality control. *Am J Int Law* 1981, 78, 607–21.

[54] Samson KT. The impact of ILO direct contact missions. *Int Lab Rev* 1984.

[55] www.unep.org.

[56] Archived at Stanford University, California, USA, The purpose of GATT included Article IV, Increasing Participation of Developing Countries: "(a)the strengthening of their domestic services capacity and its efficiency and competitiveness, inter alia through access to technology on a commercial basis; (b)the improvement of their access to distribution channels and information networks; and (c)the liberalization of market access in sectors and modes of supply of export".

[57] OECD Working Party for Manufactured Nanomaterials (WPMN), *OECD Emission Assessment for Identification of Sources of Release of Airborne Manufactured Nanomaterials in the Workplace: Compilation of Existing Guidance*, ENV/JM/MONO (2009) 16, http://www.oecd. org/dataoecd/15/60/43289645.pdf; *OECD Preliminary Analysis of Exposure Measurement and Exposure Mitigation in Occupational Settings: Manufactured Nanomaterials*, ENV/JM/ MONO (2009) 6, http://www.oecd.org/officialdocuments/displaydocument/?doclanguage= en&cote=env/jm/mono (2009) 6; *OECD Comparison of Guidance on Selection of Skin Protective Equipment and Respirators for Use in the Workplace: Manufactured Nanomaterials*, ENV/JM/MONO (2009) 17, http://www.oecd.org/dataoecd/15/56/43289781.pdf. Last access date: June 2, 2021.

[58] OECD WPMN, *List of Manufactured Nanomaterials and List of Endpoints for Phase One of the OECD Testing Programme*, ENV/JM/MONO (2008) 13/REV, http://www.olis.oecd.org/olis/ 2008doc.nsf/LinkTo/NT000034C6/$FILE/JT03248749. Last access date: June 2, 2021.

[59] Feitshans I. *China in the WTO: The Future of Regulation Protecting the Safety and Health of Workers Using Nanotechnology*. Geneva, Switzerland, on file Geneva School of Diplomacy, 2009. Last access date: June 2, 2021.

[60] http://www.oecd.org/document/26/0,3343,en_2649_37015404_42464730_1_1_1_1,00.html. 270 Ilise L. Feitshans. Last access date: June 2, 2021.

[61] Joint press release by ECOS, CIEL, and Öko-Institut: "1,500-page OECD Dossiers on 11 Nanomaterials are of 'Little to No Value' in Assessing Risks," February 23, 2017.

[62] Occupational Safety and Health Administration (OSHA). "PEL Project," under the Reagan administration, discussed in Ilise Feitshans, *Designing an Effective OSHA Compliance Program*, Thomson Reuters / Westlaw.com, 2013.

[63] Harmonized tiered approach to measure and assess the potential exposure to airborne emissions of engineered nano-objects and their agglomerates and aggregates at workplaces. Series on the Safety of Manufactured Nanomaterials No. 55. Environment Directorate Joint Meeting of the Chemicals Committee and the Working Party on Chemicals, Pesticides and Biotechnology. ENV/JM/MONO (2015) 19. Organisation for Economic Co-operation and Development, Paris, 2015, http://www.oecd.org/officialdocuments/publicdisplaydocu mentpdf/?cote=env/jm/mono (2015) 19&doclanguage=en, accessed August 31, 2017; http:// www.oecd.org/science/nanosafety/publications-series-safety-manufactured-nanomaterials. htm; http://www.iso.org/iso/home/store/catalogue_tc/catalogue_tc_browse.htm?commid= 381983&published=on&includesc=true, accessed May 15, 2017. Last accessed date: June 2, 2021.

[64] Riediker M, Ting Y, Aiken R. *Analysis of OECD WPMN Dossiers Regarding the Availability of Data to Evaluate and Regulate Risk*. report 200–00310 Singapore, Institute of Occupational Medicine, December 2016.

[65] CIEL and Partners, (Center for International Environmental Law, American University School of Law Washington DC USA) Analysis of OECD WPMN Dossiers Regarding the Availability of Data to Evaluate and Regulate Risk (Dec 2016).

Armin Grunwald

13 Twenty years of nanoethics: Challenges, clarifications, and achievements

Abstract: While nanotechnology initially, promising fulfilling high expectations, had raised little ethical awareness only, this situation changed radically in the course of possible horror scenarios involving self-organizing nanotechnology. Early ethical consideration of nanotechnology focused on proceeding responsibly in this tension between paradise-like expectations and dystopian stories. Nanoethics quickly developed into an international scholars' movement contributing a lot to public debate. Beyond this more futuristic debate, a second, more down-to-earth risk debate on nanotechnology emerged. It focused on whether synthetic nanomaterials could cause harm to health and the environment with the precautionary principle having been a major issue of ethical debate. Concerned groups put nanotechnology in close proximity to technologies that many see as highly problematic, such as nuclear technology, and drew an analogy with the asbestos story. In the meantime, ambitious research programs on environmental, health, and safety issues of nanotechnology have been conducted, providing much better knowledge of possible unintended side effects. In the meantime, the more speculative branch of nanoethics, considering, for example, human enhancement and immortality, migrated to other fields such as the convergence technologies, synthetic biology, and artificial intelligence (AI). The more "down-to-earth" reflections on nanoparticle risks, equity, and sustainable development became integral parts of existing regimes of deliberation and regulation. In this way, the ethical aspects moved into familiar fields of the ethics of technology and risk ethics and became "normalized." Therefore, it seems fair to say that we have witnessed the rise and fall of nanoethics over the last 20 years. This story provided the ethics of new and emerging technologies and the approach of responsible research and development with a rich legacy.

Keywords: nanobots, precautionary principle, toxicity, responsibility, justice, sustainable development

Armin Grunwald, Karlsruhe Institute of Technology (KIT), Germany, Institute for Technology Assessment and Systems Analysis at KIT (ITAS)

https://doi.org/10.1515/9783110669282-013

13.1 Brief history of nanoethics

In its earliest days in the 1990s, nanotechnology raised little ethical awareness and perception.[1] It was considered a more or less futuristic technology that promised many attractive and visionary developments, thereby motivating huge public funding [1]. This situation changed radically in the year 2000. The previously overwhelmingly positive utopias of nanotechnology were suddenly accompanied by horror scenarios [2]. A heavy public debate about possible apocalyptic risks posed by nanotechnology arouse in many countries. Although most of these risks had already been described in an influential futuristic essay [3], these "dark" visions received broad attention only after warnings of apocalyptic threats from out-of-control nanotechnology [2]. At that time, people all over the world became familiar with terms such as "grey goo," "nanobots," and the dream of cybernetic immortality [4]. Consideration was mostly given to far-ranging, futuristic, and speculative developments such as nano-robots and radical human enhancement, which provoked criticism of "speculative nanoethics" [5, 6].

Beyond such futuristic elements in public debate and independently of it, a second, more down-to-earth risk debate on nanotechnology emerged. It focused on whether synthetic nanomaterials could cause harm to health and the environment [7, 8]. Concerned groups put nanotechnology in close proximity to technologies that many see as highly problematic, such as nuclear technology and biotechnology [9]. Newspapers drew an analogy with the asbestos story [10]. In the meantime, ambitious environment, health, and safety (EHS) research programs have been conducted in many countries to investigate the possible side effects of nanoparticles [11, 12]. The role nanotechnology could play in and for developing countries also attracted early interest [13] and gave rise to several publications [14].

The early phase of identifying and classifying nanotechnology issues of social and ethical relevance was characterized by interdisciplinary approaches involving technology assessment (TA) cp [15] and ethical, legal, and social implications (ELSI) studies. These studies covered issues such as risk, equity, governance, and participation cp [16–18]. This ELSI period of nanoethics (around 2003–2005) can be regarded as an exploratory phase that contributed decisively to the agenda setting and structuring of this field. Building on these early contributions, nanoethics quickly developed into an international scholars' movement with many publications, concentrated in the years 2004–2008 [19–21]. In this time, also the journal *NanoEthics*. Studies of New and Emerging Technologies (https://www.springer.com/journal/11569) was founded. Parallel to this, the anthology on nanoethics [22] was published, triggering a

---

1 I would like to express my deep thanks to Sylke Wintzer, who did an excellent job in checking language, format and style issues.

short but intense debate on whether "nanoethics" should become its own subdiscipline in applied ethics [23–25].

In the meantime, the more speculative branch of nanoethics, considering, for example, human enhancement and immortality, migrated to other fields such as nano-bio-info-cogno convergence, cp. [26, 27], synthetic biology, and AI. In all of these fields, new configurations between life and technology are addressed by crossing formerly strict boundaries [28]. The more "down-to-earth" reflections on nanoparticle risks, equity, and sustainable development became integral parts of existing regimes of deliberation and regulation. In this way, the ethical aspects moved into familiar fields of the ethics of technology and risk ethics and became "normalized" (Section 13.4). Therefore, it seems fair to say that we have witnessed the rise and fall of nanoethics over the last twenty years. "Fall" is not meant in a negative way here, but is rather intended to point to the rich harvest during the exciting years of nanoethics and to the considerable legacy from that time (Section 13.4).

While the main part of this chapter is dedicated to the application fields of nanoethics (Section 13.3), its motivations and self-understanding (Section 13.2) are also introduced. Finally, the legacy of nanoethics for reflecting and deliberating on new and emerging technologies today are outlined (Section 13.4).[2]

13.2 Motivation and characterization of nanoethics

The presumed penetration of nanotechnology and its potential for deep social transformation have often been cited as reasons for the necessity of studying it from an ethical perspective [22]. In view of the revolutionary potential frequently attributed to nanotechnology in the 2000s, the ethical, legal, and social implications have been examined by commissions and study groups [16, 29], exploring the previously uncharted territory and preparing for deeper ethical analysis. Demands have been raised that nanotechnology be developed in a responsible manner, postulating new forms of ethics:

> In this regard, the advocate could maintain that issues are transformative or revolutionary in some particular way and that, whatever other ethical frameworks we have already developed, those frameworks will be ill-equipped to deal with the force that nanotechnology represents. [23], p. 198

First comments on the need to address the social and ethical issues and challenges of nanotechnology were made in the years following the announcement of the American National Nanotechnology Initiative [1]. The predominant arguments put forward for the urgent need for ethical reflection were supporting innovation, preventing apocalyptic development, and considering unintended side effects. These will be introduced briefly cp [30].

---

2 This chapter builds on several articles and books by the author, most notably [30, 38, 61, 66].

At the beginning of the debate on the risks of nanotechnology, it was noted that nanotechnology and ethical reflection developed at two very different speeds [13]. While nanotechnology developed rapidly, there was hardly any interest in ethical issues at that time, as the authors observed. Their concern was that inadequate ethical appraisal could endanger harvesting the expected benefits and opportunities promised by nanotechnology:

> We believe that there is danger of derailing NT [nanotechnology] if serious study of NT's ethical, environmental, economic, legal and social implications [. . .] does not reach the speed of progress in the science. [13], p. R9

Studying ethical issues related to the consequences of nanotechnology was thus *first* deemed necessary to support the introduction of nano-based innovation in modern societies: ". . . either the ethics of NT [nanotechnology] will catch up or the science will slow down" [13], p. R12

> Without an attention to ethics, it would not be possible to ensure efficient and harmonious development, to cooperate between people and organizations, to make the best investment choices, to prevent harm to other people, and to diminish undesirable economic implications. [31], p. xi

The *second* motivation behind the early demands for nanoethics was a reaction to fears of a possible apocalyptic side of nanotechnology [2]. Apocalyptic fears related to nanotechnology [2] included the grey goo scenario, the cyborg scenario, and the loss of control scenario. From today's perspective, the huge wave of serious concerns and fears may look strange and is hard to understand. However, those years were indeed characterized by a dichotomy between high and sometimes even paradisiacal expectations of nanotechnology, on the one hand, and similarly high concerns up to the feared end of humankind, on the other. A deep ambivalence toward far-ranging stories of the future became obvious: "Tremendous transformative potential comes with tremendous anxieties" [32], p. 4; cp [33].

Moor and Weckert [34] expected an ethics of nanotechnology to minimize the risk of runaway robots. Jean-Pierre Dupuy put the apocalyptic dimension of Joy's fears at the center of his conceptualization of an ethics of nanotechnology [35, 36]. He argued that nanotechnology, insofar as it exploits the principles of self-organization and brings self-organizing nanomachines into the world, would ultimately and inevitably lead to an absolute catastrophe as it would get out of control through unlimited self-replication. It was exactly this kind of speculative ethics that motivated criticism of "too speculative" nanoethics [5]. It expressed concerns that ethics could waste resources and miss the pressing issues of nanotechnology while looking at merely speculative fears [6].

Third, the observation that nanotechnology can have unintended consequences like other technology [15] was also one of the motivations for ethical reflection on nanotechnology. Familiar questions in this respect are: which risks can or should be tolerated in view of the positive consequences hoped for, how should cost-benefit assessments, risk-opportunity analyzes, and comparative risk evaluations be carried out, and when should the precautionary principle [37] be applied in view of insufficient knowledge [8, 38]? Even before the ETC Group issued its widely perceived call for a moratorium on nanotechnology [9], toxicologists had drawn attention to unanswered questions relating to nanomaterials, especially synthetic nanoparticles [7]. The history of asbestos, in particular, was repeatedly referred to as a warning [10]. Empirical research into the consequences of nanotechnology, often referred to as EHS studies, was undertaken (Section 13.3.1).

Hence, different and heterogeneous observations and concerns have motivated the emergence of nanoethics. They converge in the diagnosis of the presumed huge impact of nanotechnology on future development. If there is a large transformative potential of nanotechnology and if the uncertainties and risks involved are also large, then it seems probable that ethical challenges will emerge and need to be tackled. However, this mission of nanoethics was not very specific.

The heterogeneity of its motivations had implications for the development of nanoethics. In spite of the wide consensus about its necessity, nanoethics was expected to cover a mixture of extremely diverse ethical issues in different fields of application (Section 13.3). The order of magnitude of its objects – nanometers – was the only common element, but without being specified in ethical terms. According to this situation, the attempts to establish nanoethics as an own subdiscipline of applied ethics [23] failed due to a missing common perspective in ethical respect. Nevertheless, the term "nanoethics" has developed into an umbrella term covering several ethical and social aspects of nanotechnology. For instance, the book *Nanoethics* [22] includes an informative collection of essays on almost every conceivable topic related to social aspects of nanotechnology. It covers, for instance, research programs, the public (US) debate on technological revolutions and their predictability, issues of democratic theory, public involvement, educational reforms, developing countries, and military developments. The idea of elaborating the specifically ethical aspects of these topics was not considered of interest. Rather, the term nanoethics served as a generic term for diverse social aspects of nanotechnology, including political, social, and methodological ones [30].

The challenge for nanoethics in this broad sense is the need to transgress the boundaries of the classical subtypes of ethics and to working dialogue with natural scientists and engineers. In this way, the term "nanoethics" is understood as a terminological and conceptual platform where the individual paths of ethical reflection on nanotechnology can be brought together. These then reflect on the progress of nanotechnology and nano-based products, processes, or systems and their perspectives through the glasses of ethics. The term "nanoethics" denotes the "interdisciplinary, ethically reflective dialogue in which normative uncertainties and challenges associated with nanotechnology and their consequences for society are debated" [30]. This does not require any kind of nanoethics discipline, but only an awareness of the underlying ethical questions of modern nanoscience and nanotechnology in different respects (Section 13.3).

While most philosophical and ethical questions of nanotechnology are not completely new [20], their co-occurrence in nanotechnology is in many cases new or leads to new constellations of ethical reflection. Analogous to nanoscience and nanotechnology, which blur traditional boundaries between physics, chemistry, biology, and the engineering sciences, the ethical questions surrounding nanotechnology bring together various traditional lines of philosophical and ethical reflection, including the philosophy and ethics of engineering, medical ethics, anthropology, bioethics, information ethics, the philosophy of mind, the theory of justice, and the imperative of sustainability.

13.3 Application fields of nanoethics

Nanotechnology was early characterized as a "technoscience" and an "enabling technology." Both terms emphasize, among other aspects, the openness of nanotechnology to a broad variety of applications in various areas of society and the economy, the huge transformative potential, but also the high uncertainty about possible consequences. Ethical issues usually arise from a combination of technologies, innovation, and their social environment. Therefore, the major fields of activity of nanoethics will be presented along main application areas of nanotechnology in different social environments.

13.3.1 Environment, health, and safety

The unintended release of nanomaterials into the environment and their inhalation or ingestion by humans can lead to previously unknown effects. Synthetic nanoparticles are foreign bodies in the biosphere. They can enter the environment through emissions during production, during the everyday use of products, or during disposal. Issues such as mobility, responsiveness, persistence, pulmonary penetration, and solubility have to be taken into consideration to evaluate the potential spread of nanoparticles and their traces and consequences. The production, use, and disposal of products containing nanomaterials may lead to their occurrence in the air, water, soil, or in organisms and humans [7]. Nanoparticles can enter the human body via the lungs or through food, for example, at the workplace or during leisure activities [39]. As an example, a strong concern was that titanium dioxide nanoparticles used in sunscreens could penetrate the human skin and migrate to the human brain. The challenge of acting under conditions of high uncertainty, but with many nanoproducts already on the market, was at the heart of the ethical challenges posed by nanoparticles in the early days of nanotechnology. At that time, classic risk management, based on measuring and weighting quantitative risks, did not work due to lack of knowledge:

> Much of the public discussion about nanotechnology concerns possible risks associated with the future development of that technology. It would therefore seem natural to turn to the established discipline for analyzing technological risks, namely risk analysis, for guidance about nanotechnology. It turns out, however, that risk analysis does not have much to contribute here.
>
> [40], p. 315

Precautionary considerations – such as are foreseen for environmental issues within the context of European framework legislation (von Schomberg, 2005) – were intensively discussed [8, 38, 41]. The emergence of nanotoxicology around the year 2005 contributed to a strong improvement in knowledge about possible risks. In accordance with the different phases in the life cycle of technologies, it is necessary to distinguish between the production (e.g., for protection at the workplace), use (e.g., for consumer protection), and disposal of nanomaterials. Due to the specific difficulties in tracking nanoparticles in the human body and because of their diversity, even the development of appropriate measurement techniques is an extremely complex and time-consuming task.

Questions on the toxicity of nanoparticles to the ecology or humans, on nanomaterial flow, on the behavior of nanoparticles when dispersed in different environmental media, on their rate of degradation or agglomeration, and on their consequences for the various conceivable targets are, however, not solely ethical questions. Empirical scientific disciplines, such as toxicology, medicine, or environmental chemistry, are competent for such questions. They must provide the knowledge to draw practical consequences for working with nanoparticles and for disseminating products based on them. However, as the debate on the environmental standards of chemicals or radiation has shown, the results of empirical research are not decisive in judging the acceptability of risks. Safety and environmental standards – in our case for dealing with nanoparticles – must be based on sound knowledge, but cannot logically be derived from this knowledge [30]. In addition, normative standards, for example, regarding the intended level of protection, the level of public risk acceptance, and other societal and value-laden issues, come into play [37]. Given this situation, it is not surprising that there are frequent conflicts over the *acceptability* of risks. Determining the acceptability and tolerability of the environmental and health risks of nanoparticles was therefore an ethically relevant issue. Today, familiar risk assessment is much more applicable due to a much improved knowledge base.

Ethical reasoning, however, is not limited to the effort to prevent harm but also motivates searching for ways to shape a positive development of nanotechnology with regard to EHS. The application of nanotechnology in products and systems cannot only lead to pollution of the environment but also contribute to a significant reduction of this burden. Saving material resources, reducing the mass of by-products that burden the environment, improving the efficiency of energy transformation, reducing energy consumption, and removing pollutants from the environment are desirable from an ethical point of view [42]. A number of studies on precisely the positive environmental effects of nanotechnology have been published in the meantime [43],

EHS challenges also concern responsibility for future generations. While nanotechnology is expected to enable new technological approaches that reduce the environmental footprints of existing technologies in industrialized countries, this might come at a cost over the entire life cycle, including long-term effects. Shaping nanotechnologies for sustainable development requires anticipatory sustainability assessments in order to permit distinctions to be made between more and less sustainable technologies [42]. They have to cover the entire *life cycle* of the respective technological products or systems, even if these are still under development. Therefore, prospective life cycle assessment is needed [44]. This extends from the primary storage sites to transportation and manufacturing processes to the product's use, finally ending with its disposal. In many areas, however, the data about its life cycle that would be needed for life cycle assessment are far from being available. Empirical research on the persistence, long-term behavior, and fate of nanoparticles in the environment as well as on their respective consequences would be necessary to enable action to be taken in accordance with ethical criteria of long-term responsibility.

13.3.2 Nanotechnology and developing countries

Positive potentials of nanotechnology for developing countries were already highlighted in the first writings on nanotechnology. Drexler [3] expected the solution of almost all problems of developing countries, such as poverty, hunger, and sanitation. In the meantime, however, the initial optimism has disappeared: "The selected issues discussed in this chapter allow drawing mostly pessimistic conclusions on the impact of nanotechnology on developing countries" [14], p. 303. Instead, nanotechnology goes for other goals:

> among the first products with nanotechnological content were: thermal shoes . . ., dust and sweat-repelling mattresses, more flexible and resistant golf clubs, personalized cosmetics . . ., disinfectants and cleaning products for planes, boats, submarines etc., cream that combats muscular pain, dental adhesives that set the tooth crown better . . . [45]

This indicates that demands in the sphere of lifestyle and luxury in rich countries will be fulfilled first, rather than the existential needs in terms of hunger and poverty in developing countries. An example of similar concern is nanomedicine [46]. With regard to *less developed societies*, the particularly dramatic inequalities that already exist between industrialized and developing nations could be further increased [47]. The problems posed by unfair distribution of access to new technologies are often particularly evident in medicine, where inequality often means unfairness [45].

Developing countries will not automatically benefit from nanotechnology [48]. On the contrary, the self-dynamics of the economic system points in the opposite direction of a widening gap between rich and poor. Counteracting this self-dynamics and improving distributive justice requires political measures [49]. One of the early demands in this direction was to involve developing countries in the examination of the ethical aspects of nanotechnology [13], in order not to exclude large portions of the world's population from the benefits of the expected potentials and, in consequence, to make them victims of discrimination. Without a doubt, nanotechnology offers great potential for developing countries, but this potential will remain *mere potential* unless there are political measures for making this potential a reality.

13.3.3 Military applications

The military applications of new technologies are frequently the focus of ethical consideration [50], and in position papers of non-governmental organizations. The use of nanotechnology in armaments opens the door to improved weapons, innovative materials, and new areas of application [51]. One example is the development of a nanoscale powder for use in propellants and explosives in order to increase the energy efficiency and speed of the explosion. Nanoscale electronic, sensory, and electromagnetic components could improve our capacity to steer and control military vehicles and make them more robust. This could further strengthen the current trend toward autonomous systems in the air, at sea, and in space. There are numerous potential applications in military intelligence based on the use of nanotechnology components for sensors, sensory systems, and sensory networks. Even the field of weapons and munitions will be very directly affected by improved sensory capacities and enhanced computer power and storage capacity enabled by nanotechnology. Developments in nanotechnology are likely to have significant consequences for military personnel, even at the level of personal equipment (soldier as a system). In the foregrounds, the effort to equip soldiers with additional functionality without substantially increasing the weight of the equipment [52]. All of these developments can be seen as an increase in efficiency compared to previous military technology.

A politically and ethically relevant secondary consequence of these developments is proliferation, assuming that the usual moral standards for the application of technology continue to apply in war. Not all countries comply with the standards to the same extent, and some dictatorships could be tempted to use the new technologies internally (e.g., for surveillance). These more-efficient weapons could even fall into the hands of terrorists. The smallness of nanotechnology is occasionally cited as a reason for concern that nanotechnological techniques could lead to the construction of substantially smaller bombs and explosives, which would be considerably easier for terrorists to build and use. In view of these secondary effects, it is not to be expected that nanotechnological advances in military technology will raise dramatically new ethical questions, while other fields, such as autonomous drones, are more at the forefront of ethical concern.

13.3.4 Biotechnology and life sciences

Biotechnology is considered as one of the most important key technologies to have arisen for decades. Technologies such as genetic modification are used for many purposes and form the basis of many branches of industry. Nanotechnology, in the form of nano-biotechnology [30, 53, 54], is expanding the technological understanding and use of biological processes to a considerable extent, for industrial use as well as for nanomedicine [46].

Basic life processes take place on the nanoscale because life's essential building blocks (such as proteins) are precisely this size. Nano-biotechnology makes it possible for nanotechnology to better understand, manipulate, and perhaps even control biological processes and, thus, serves as basis for synthetic biology [55]. The molecular "factories" (mitochondria) and transport systems, which play an essential role in cellular metabolism, can be regarded as models for controllable bio-nano-machines. Nanotechnology at this level could enable synthetic biology to engineer cells or even create artificial life forms [55, 56]. An intermeshing of natural biological processes with technical processes is already underway. The classical barrier between technology and life is increasingly being breached or even crossed [28].

The corresponding discussions of risks are structurally similar to the discussion on genetically modified organisms and gene editing. The discussions are about safety standards for the research concerned, the risks of experiments outside the laboratories, and release problems. Another topic of debate has been the danger of misuse, such as the technical modification of viruses to produce new biological weapons, with the debate on genetically modified organisms serving as a background. However, creating life from scratch seems to be much more dramatic than modifying life by means of genetic technology. Classical patterns of argumentation such as biosafety and biosecurity are at the heart of the debate [56]. Other arguments such as human hubris or the fear of "playing God" also play a role. A wide range of future ethical discussions is opening up for shaping new interfaces between life and technology, for which at present there is insufficient practical background to support concrete reflection, but which requires hermeneutical effort to understand the significance of ongoing changes at the life–technology interface [28].

13.3.5 Human enhancement

In combination with biotechnology and medicine, nanotechnology offers perspectives that go beyond the traditional medical tasks of curing diseases, such as "improving" the human body or radically transforming or redesigning it [26]. Enhancement goes beyond curing and beyond the traditional ethos of medicine. For example, new or advanced sensory functions could be implemented by extending the electromagnetic spectrum that an eye can perceive. It is also possible to create new interfaces between man and machine by directly coupling ICT systems to the human brain. Another focus in the debate on human enhancement is the extension of the human life span up to visions of immortality.

By extending these lines of development into the futuristic and speculative dimension, it would be possible to study topics such as the increased integration of technology into the human body, the increasing convergence of humans and technology, and the creation of cyborgs as technically enhanced humans or technology with human traits. These perspectives raise anthropological questions about our image of man and the relationship between man and technology [57] and, at the same time, lead to the ethical question of the degree to which humans *can*, *should*, or *want* to go on transforming the human body. First, ethical analysis must address the semantic and hermeneutic problems associated with the concepts of curing, doping, and enhancing since the latter are factors that play a decisive role in each of the relevant normative parameters [58].

Ethical reflection does not seem premature despite the speculative nature of the subject. In particular, advances in brain science and developments in converging technologies [26] lead to this expectation – and would justify ethical reflection "in advance." In the combination of the emergence of enhancement technologies and a social situation in which competition strongly influences every domain of life, it seems probable that enhancement technologies will be used if available. The pressure to gain an advantage over competitors is high, and the doping problem in sport could be a forerunner to other social areas. Ethics, on the one hand, is prepared to deal with such issues, considering, for example, issues of fairness, equity, and human rights. On the other hand, however, the magnitude of the challenge and the close relations with societal organization at large, with political philosophy, and with the principles of capitalism lead to the conclusion that this is a field where a new type of ethics is needed. An "ethics of enhancement" should guide a situation in which humans start changing their own nature to an extent that was unknown and unimaginable in earlier times [28].

The relations with nanotechnology, however, are only loose. Though the idea of improving human performance was rooted in visions of early nanotechnology [26, 27], the ethical debate gained a momentum of its own and moved to new questions and topics far from nanotechnology, such as post- and transhumanism [59].

13.4 The legacy of nanoethics

The ethical debate on nanotechnology has declined after 2010. But it undoubtedly had many direct effects on public perception of nanotechnology, on the political and social awareness of its potential risks, on methodological issues of prospective ethics, on the emergence of vision assessment and the discovery of the hermeneutic side of responsible research and innovation (RRI) [60], and on technology assessment [15]. Thus, from today's perspective, we can identify considerable impacts as a legacy of nanoethics.

13.4.1 Embedding nanotechnology in society

Several indirect and mediated impacts of nanoethics on the interface between nanotechnology and society can be stated. Social awareness of nanotechnology risks increased without creating an atmosphere of rejection, fear, or fundamentalism. The risk issue naturally belongs to introductions to nanotechnology and public lectures and has become part of the nanotechnology "identity." Nanotechnology researchers are aware of the fact that their research is "under observation" by ethical debates. The nano-lab is no longer separated from society, but nanotechnology advance takes place under the gaze of society. Ethical consideration has influenced the research agenda of sciences such as toxicology and of the social sciences, in particular in the science and technology studies field and in applied ethics. Funding agencies increasingly demand risk issues be explicitly considered in funding applications. Several ambitious national programs on nanotechnology, ethics, risk, and society have been implemented, for example, in Austria, Germany, and the Netherlands, partly involving stakeholders and citizens. Compared to heavy, sometimes even fundamentalist debates on technologies such as nuclear power and green genetic engineering, the nanotechnology story appears to be a positive example of a frank and honest, but also peaceful management of concerns.

13.4.2 Critique of speculative nanoethics

The emergence [13] and rapid thematic development of nanoethics [19, 20] was accompanied by concerns that the ethics' movement toward earlier stages of research and innovation ("upstream") would lead to a concentration on speculative developments, while very little attention would be paid to questions of nanotechnology design and applications that were actually pending:

but a new gap has opened up because most nanoethics is too futuristic, focusing on nano-enabled devices that can read our thoughts, for example, at the expense of ongoing incremental developments that are more ethically significant. [6]

If this diagnosis were true, then large parts of nanoethics would be misguided and concern themselves with irrelevant and purely speculative ideas, while the really important developments would not be taken into consideration. This accusation motivated several methodological reflections on how to conceptualize ethics in the absence of valid knowledge about consequences [60–63]. The familiar consequentiality approach of applied ethics breaks down as soon as valid knowledge about consequences is not available [60]. The more speculative the considerations of the consequences, the less they can serve as orientation for concrete (political) action and decisions; in this respect, the criticisms of speculative nanoethics are correct. For other purposes, however, this is not a problem: conceptual, pre-ethical, heuristic, and hermeneutic issues then become more significant instead. The primary concern is then to clarify what is going on in the speculative developments considered, what is at issue, which rights might be compromised, what images of man, nature, and technology are formed and how they change, what anthropological issues are involved, and what designs for society are implied in the projects for the future. Ultimately, this leads to the recommendation not to speak of "speculative nanoethics" but to consider these forms of reflection as elements of an explorative philosophy of nanotechnology [61].

13.4.3 Normalization of nanotechnology

The ethics and risk debate on nanoparticles has brought nanotechnology "down to earth" in public debate and "normalized" it. Nanotechnology has lost its futuristic character, for example, in mass media coverage, and has become a "neighbor" of new chemicals and fine and ultrafine particles [64]. Society is familiar with this issue: hundreds or thousands of new chemicals are produced every year. Regulations and procedures have been introduced to manage these problems, for example, the European Union's Registration, Evaluation, Authorization, and Restriction on Chemicals system. The mechanisms of risk management and precautionary thinking are surely not directly transferable to the field of nanoparticles without specific adaptations [4], Chapter 5, but the *type of problem* is not new to society but *normal*.

The risk debate on nanoparticles has contributed to a normalization of nanotechnologies also in a second sense. For many years, nanotechnology was perceived as a "clean" and smart technology, as an ideal technology in strict contrast to traditional technologies symbolized by large chemical plants or the coal and steel industry. The early risk debate has destroyed this purely positive perception. Nanotechnology has become a "normal" field of technology with normal problems of possible hazards to human health or the environment, and also with quite normal requirements for ethical reflection and responsibility.

13.4.4 Nanoethics as origin of responsible research and innovation

Responsible innovation has come to the fore in connection with a wide variety of emerging technologies, such as genetically modified organisms, synthetic biology, nanotechnology, information and communication technology, robotics, and geo-engineering. The discussion about responsible innovation and its cognates began to take off along with large-scale national programs for research and development (R&D) in nanotechnology: The US National Nanotechnology Initiative [1] adopted a strategic goal of "responsible development." The UK Engineering and Physical Sciences Research Council initiated pilot study on responsible innovation in nano-technology for carbon capture. The Netherlands organized a "national dialogue" on nanotechnology, concluding that further development in nanotechnology should be "responsible." The purpose of these and many other efforts is to enhance the possibilities that technology will help improving quality of life. These activities served as models for establishing RRI in European research policy and for developing conceptual approaches [65] and a large number of subsequent projects in many areas.

13.5 Conclusions

The history of nanoethics has been impressive over the last decades. Nanoethics was among the first approaches to help understand the emerging field of nanotechnology about 20 years ago, following the US National Nanotechnology Initiative [1] and the debate motivated by Bill Joy [2]. At that time, it was unclear and contested what nanotechnology is, what was really novel, what potential for science and society it might have, and how its impact on human development could be assessed. Early nanoethics took care by providing several important clarifications through several ELSI and TA studies. They contributed to the development of nanoethics but also to the emergence of ethical inquiry in other fields such as synthetic biology, human enhancement, digitalization and brain research. The traces of nanoethics as the first "NEST-ethics" [21] can be found in those other fields of ethics, which can be regarded as "outsourced" by nanoethics (Figure 13.1).

Several lessons can be learnt from nanoethics and its history. The first is that in any new and emerging field of science and technology, ethical reflection cannot be performed in the familiar manner of applied ethics, that is, by applying ethical norms and principles to new technologies, products, and services. Instead, ethics

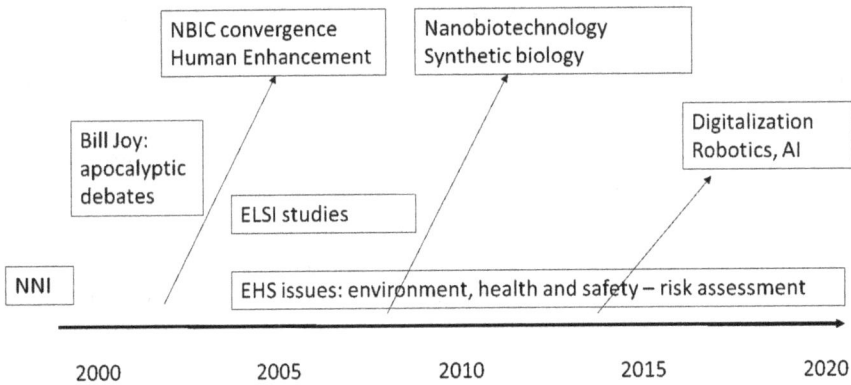

Figure 13.1: The history of nanoethics.

must be a kind of explorative philosophy, seeking to understand the novelties offered by the scientific and technological advance, reflecting the new spaces for human development opened up and taking care of possible challenges, risks, and unintended side effects. Usually, this will be more something like developing a map of possibilities than a strict ethical judgment. In the course of development, however, things become more specific, innovation pathways and application ideas become better visible and can be discussed in ethical regard more in-depth. We could witness this development at the occasion of the history of nanoethics, leading from the early explorative steps to specific risk assessments today, while the more explorative questions migrated to other fields enabled by nanotechnology but with their own dynamics and perception such as synthetic biology, robotics, and AI (Figure 13.1).

Nanoethics has contributed to dealing with nanotechnology risks in a mature and rational manner. While postulates of a moratorium on the use and release of nanoparticles [9] fueled expectations and fears of public protest and rejection, almost nothing has happened in this direction. It seems plausible that ethical deliberation of nanotechnology and related activities have contributed to this "relaxed" development in the following way. In the field of nanotechnology, there have been a lot of activities in recent years which make clear that dealing responsibly with risk is seen as part of scientific advance and its political shaping, funding, and regulation. Risk has not been ignored or denied (especially not by nanoscientists and managers) but has been understood as a challenge for research. In this way, *trust* has been created – and trust is a major issue in avoiding communication disasters. Young nanotech professionals and engineers should learn to stick to "down to Earth" approaches instead of fueling hype cycles, and to be open and willing to participate in public dialogue. Public dialogue needs the voices of scientists and engineers not as advocates of their own developments – this only will create mistrust – but as honest brokers with regard to preliminary knowledge about expected benefits and possible risks in an open-minded communication. In fact, ethics accompanying scientific progress is needed, but at some stages, anthropology and philosophy could be even more important, in particular for the more explorative tasks to be performed in order to create basic understanding of nanotechnology as well as its possible implications on humankind.

Often, there are complains about the divergence of the "two cultures," science and engineering on the one side, humanities and social science, on the other. In many respects, this divergence is unproblematic: nano-researchers shall do excellent nanoscience while social scientists shall provide society with excellent social science. However, again and again, communication and even cooperation between these cultures is needed, in developing orientation with regard to shaping the technological advance and making best use of its outcomes. Obviously, mutual understanding is necessary in those situations, which, sometimes, is difficult to achieve. It is not a good idea to tractate students in natural science and engineering with a huge amount of ethics and social science to prepare them for such challenges. Often, they will feel overburdened with such stuff. It is preferable, however, to teach natural science and engineering in a manner opening their eyes for the ethical dimension of their own activities, often deeply hidden in quasi value-neutral research. Optimal opportunities to bridge the gap between the two cultures is identifying a common research interest among them. A current example is the field of self-driving cars. Making this technology ready for the market needs sound and robust solutions, which can be realized in cooperation between legal science, computer science, engineering and robotics, and ethics. A common problem to be tackled by applying different perspectives is the best opportunity to motivate interdisciplinary cooperation in research and education.

In all of its activities, ethics serves as *advice* to science, engineering, stakeholders, research funding, regulation, public debate, and so on not as a kind of authority. Ethical reflection on nanotechnology needs close cooperation of professional ethics and philosophy with scientists and researchers deeply involved inR&D. Ethical debate must not be a discourse *external* to ongoing scientific and engineering research but should rather be considered an interdisciplinary and inherent part of it.

References

[1]   NNI – National Nanotechnology Initiative. National Nanotechnology Initiative. Washington, DC, USA, US Government, 1999.
[2]   Joy B. Why the Future doesn't need us. Reprinted. In: Allhoff F, Lin P, Moor J, Weckert J eds, Nanoethics. The Ethical and Social Implications of Nanotechnology. Hoboken, NJ, USA, Wiley, 2000, 17–30.
[3]   Drexler KE. Engines of Creation – The Coming Era of Nanotechnology. Oxford, UK, Oxford University Press, 1986.
[4]   Schmid G, Ernst H, Grünwald W. et al., Nanotechnology – Perspectives and Assessment. Berlin, Germany, Springer, 2006.

[5] Nordmann A. If and then: A critique of speculative nanoethics. Nanoethics 2007, 1, 31–46.
[6] Nordmann A, Rip A. Mind the gap revisited. Nature Nanotechnology 2009, 4, 273–74.
[7] Colvin V. Responsible Nanotechnology: Looking Beyond the Good News. Centre for Biological and Environmental Nanotechnology at Rice University, Houston, TX, USA. 2003. http://www.eurekalert.org/. Last accessed date: March 19, 2021.
[8] Haum R, Petschow U, Steinfeldt M, von Gleich A. Nanotechnology and Regulation within the Framework of the Precautionary Principle. Berlin, Germany, Institute for Ecological Economy Research, 2004.
[9] ETC Group. From genomes to atoms. The big down. Atomtech: Technologies converging at the nano-scale. 2003. http://www.etcgroup.org. Last accessed date: March 19, 2021.
[10] Gee D, Greenberg M. Asbestos: From 'magic' to Malevolent Mineral. In: Harremoes P, Gee D, MacGarvin M et al., eds, The Precautionary Principle in the Twentieth Century. Late Lessons from Early Warnings. London, UK, Earthscan, 2002, 49–63.
[11] ENRHES – Engineered nanoparticles: Review of health and environmental safety. Project report, 2010 (https://www.nanowerk.com/nanotechnology, accessed 2020-06-16).
[12] Jahnel J. Addressing the Challenges to the Risk Assessment of Nanomaterials. In: Dolez PI ed, Nanoengineering. Global Approaches to Health and Safety Issues. Amsterdam, The Netherlands, Elsevier, 2015, 485–516.
[13] Mnyusiwalla A, Daar AS, Singer PA. Mind the gap. Science and ethics in nanotechnology. Nanotechnology 2003, 14, R9–R13.
[14] Schummer J. Impact of Nanotechnologies on Developing Countries. In: Allhoff F, Lin P, Moor J, Weckert J eds, Nanoethics. The Ethical and Social Implications of Nanotechnology. Hoboken, NJ,USA, Wiley, 2007, 291–307.
[15] Grunwald A. Technology Assessment in Practice and Theory. Abingdon, NY, Routledge, 2019.
[16] Nano Forum, ed. 2004. Nanotechnology. Benefits, risks, ethical, legal, and social aspects of nanotechnology. https://www.nanowerk.com/nanotechnology-report.php?reportid=3, accessed 2020-06-16.
[17] Paschen H, Coenen C, Fleischer T, Grünwald R, Oertel D, Revermann C. Nanotechnologie. Forschung und Anwendungen. Berlin, Gemany, Springer, 2004.
[18] Royal Society. Nanoscience and nanotechnologies: Opportunities and uncertainties. London, UK, Clyvedon Press, 2004.
[19] Khushf G. The Ethics of Nanotechnology – Visions and Values for a New Generation of Science and Engineering. In: National Academy of Engineering ed, Emerging Technologies and Ethical Issues in Engineering. Washington, DC, USA, The National Academies Press, 2004, 29–55.
[20] Grunwald A. Nanotechnology – A new field of ethical inquiry? Sci Eng Ethics 2005, 11, 187–201.
[21] Rip A, Swierstra T. Nano-ethics as NEST-ethics: Patterns of moral argumentation about new and emerging science and technology. NanoEthics 2007, 1, 3–20.
[22] Allhoff F, Lin P, Moor J, Weckert J eds, Nanoethics. The Ethical and Social Implications of Nanotechnology. Hoboken, NJ, USA, Wiley, 2007.
[23] Allhoff F. On the autonomy and justification of nanoethics. NanoEthics 2007, 1, 185–210.
[24] Lin P, Allhoff F. Nanoscience and Nanoethics: Defining the Disciplines. In: Allhoff F, Lin P, Moor J, Weckert J eds, Nanoethics. The Ethical and Social Implications of Nanotechnology. Hoboken, NJ,USA, Wiley, 2007, 3–16.
[25] Keiper A. Nanoethics as a discipline? The New Atlantis. J Tech Sci 2007, 16, 55–67.
[26] Roco MC, Bainbridge WS eds, Converging Technologies for Improving Human Performance. Arlington, VA, USA, National Science Foundation, 2002.
[27] Wolbring G. Why NBIC? Why human performance enhancement? Eur J Social Sci Res 2008, 21, 25–40.

[28] Grunwald A. Living Technology. Philosophy and Ethics at the Crossroads between Life and Technology. Singapore, Pan Stanford, 2021.

[29] Coffrin T, MacDonald C. Ethical and social issues in nanotechnology. Annotated Bibliography. 2004. http://www.ethicsweb.ca/nanotechnology/bibliography.html. Last accessed date: March 19, 2021.

[30] Grunwald A. Responsible Nanobiotechnology: Philosophy and Ethics. Singapore, Pan Stanford, 2012.

[31] Roco MC. Foreword: Ethical Choices in Nanotechnology Development. In: Allhoff F, Lin P, Moor J, Weckert J eds, Nanoethics. The Ethical and Social Implications of Nanotechnology. Hoboken, NJ, USA, Wiley, 2007, 5–6.

[32] Nordmann A. Converging technologies – Shaping the future of European societies. High Level Expert Group 'Foresighting the New Technology Wave'. Brussels, Belgium, European Commission, 2004.

[33] Grunwald A. Converging technologies: Visions, increased contingencies of the conditio humana, and search for orientation. Futures 2007, 39, 380–92.

[34] Moor J, Weckert J. Nanoethics: Assessing the Nanoscale from an Ethical Point of View. In: Baird D, Nordmann A, Schummer J eds, Discovering the Nanoscale. Amsterdam, the Netherlands, IOS Press, 2004, 301–10.

[35] Dupuy JP The philosophical foundations of nanoethics. Arguments for a method. Lecture at the Nanoethics Conference, University of South Carolina, March 2–5, 2005.

[36] Dupuy JP, Grinbaum A. Living with uncertainty: Toward the ongoing normative assessment of nanotechnology. Techné 2004, 8, 4–25. Reprinted in: Schummer I, Baird D, eds. Nanotechnology Challenges: Implications for Philosophy, Ethics and Society. Singapore, World Scientific Publishing, 2006, 287–314.

[37] von Schomberg R. The Precautionary Principle and its Normative Challenges. In: Fisher E, Jones J, von Schomberg R eds, The Precautionary Principle and Public Policy Decision Making. Cheltenham, UK, Northampton, MA, USA, Edward Elgar, 2005, 141–65.

[38] Grunwald A. Nanoparticles: Risk Management and the Precautionary Principle. In: Jotterand F ed, Emerging Conceptual, Ethical and Policy Issues in Bionanotechnology. Berlin, Germany, Springer, 2008, 85–102.

[39] Seaton A, Tran L, Aitken R, Donaldson K. Nanoparticles, human health hazard and regulation. J R Soc Interface 2010, 7, Suppl 1, 119–29.

[40] Hansson SO. Great Uncertainty about Small Things. In: Schummer J, Baird D eds, Nanotechnology Challenges – Implications for Philosophy, Ethics and Society. Singapore et al., World Scientific Publishing, 2006, 315–25.

[41] Weckert J, Moor J. The Precautionary Principle in Nanotechnology. In: Allhoff F, Lin P, Moor J, Weckert J eds., Nanoethics – The Ethical and Social Implications of Nanotechnology. Hoboken, NJ, USA, Wiley, 2007, 133–46.

[42] Fleischer T, Grunwald A. Making nanotechnology developments sustainable. A role for technology assessment? J Clean Prod 2008, 16, 889–98.

[43] JCP – Journal of Cleaner Production. Sustainable nanotechnology development. Special issue. J Clean Prod 2008, 16, 8–9, 16.

[44] Schepelmann P, Ritthoff M, Jeswani H, Azapagic A, Suomalainen K. Options for Deepening and Broadening LCA. CALCAS – Co-ordination Action for Innovation in life-cycle Analysis for Sustainability. Brussels et al., European Commission, 2009.

[45] Foladori G. Converging Technologies and the Poor. The Case of Nanomedicine and Nanobiotechnology. In: Banse G, Grunwald A, Hronszky I, Nelson G eds, Assessing Societal Implications of Converging Technological Development. Berlin, Germany, edition sigma, 2008, 193–216.

[46] ESF – European Science Foundation. Nanomedicine. An ESF forward look. 2005. http://www. esf.org/publications/forward-looks.html. Last accessed date: March 19, 2021.

[47] Foladori G, Invernizzi N. Nanotechnology and developing countries: Will nanotechnology overcome poverty or widen disparities? Nanotechnol Law Bus 2007, 2, Paper 11.

[48] Court E, Salamance-Buentello F, Singer PA, Daar AS. Nanotechnology and the Developing World. In: Ten Have H ed, Nanotechnologies, Ethics, and Politics. Paris, France, UNESCO Publishing, 2007, 155–80.

[49] Invernizzi N. Nanotechnology for Developing Countries. Asking the Wrong Question. In: Banse G, Grunwald A, Hronszky I, Nelson G eds, Assessing Societal Implications of Converging Technological Development. Berlin, Germany, edition sigma, 2008, 229–39.

[50] Altmann J. Military Nanotechnology: Potential Application and Preventive Arms Control. London, UK, Routledge, 2006.

[51] Lau C. Nanotechnology and the Department of Defense. In: Roco MC, Bainbridge WS eds, Converging Technologies for Improving Human Performance. Arlington, VA, USA, National Science Foundation, 2002, 308–09.

[52] Moore D. Nanotechnology and the Military. In: Allhoff F, Lin P, Moor J, Weckert J eds., Nanoethics. The Ethical and Social Implications of Nanotechnology. Hoboken, NJ, USA, Wiley, 2007, 267–75.

[53] Goodsell DS. Bionanotechnology. Lessons from Nature. Hoboken,NJ, USA, Wiley-Liss, 2004.

[54] Ach JS, Siep L eds, Nano-Bio-Ethics. Ethical and Social Dimensions of Nanobiotechnology. Berlin, Germany, Lit-Verlag, 2006.

[55] Giese B, Pade C, Wigger H, von Gleich A eds, Synthetic Biology. Character and Impact. Heidelberg, Germany, Springer, 2014.

[56] de Vriend H. Constructing Life. Early Social Reflections on the Emerging Field of Synthetic Biology. The Hague, the Netherlands, Rathenau Institute, 2006.

[57] Jotterand F. Beyond therapy and enhancement: The alteration of human nature. Nanoethics 2008, 2, 15–23.

[58] Grunwald A. Converging technologies: Visions, increased contingencies of the conditio humana, and search for orientation. Futures 2007, 39, 380–92, https://doi.org/10.1016/j.fu tures.2006.08.001.

[59] Hurlbut B ed, Perfecting Human Futures: Transhuman Visions and Technological Imaginations. Wiesbaden, Germany, Springer VS, 2016, https://doi.org/10.1007/978-3-658-11044-4_2.

[60] Grunwald A. The hermeneutic side of responsible research and innovation. J Responsible Innov 2014, 1, 274–91, 10.1080/23299460.2014.968437.

[61] Grunwald A. From speculative nanoethics to explorative philosophy of nanotechnology. NanoEthics 2010, 4, 2, 91–101.

[62] Selin C. Expectations and the emergence of nanotechnology. Sci Technol Human Values 2007, 32, 2, 196–220.

[63] Nordmann A. Responsible innovation, the art and craft of future anticipation. J Responsible Innov 2014, 1, 87–98.

[64] Grunwald A, Hocke-Bergler P. The Risk Debate on Nanoparticles: Contribution to a Normalisation of the Science/society Relationship? In: Kaiser M, Kurath M, Maasen S, Rehmann-Sutter C eds, Governing Future Technologies. Nanotechnology and the Rise of an Assessment Regime. Dordrecht et al., Springer, 2010, 157–77.

[65] Owen R, Bessant J, Heintz M. eds, Responsible Innovation: Managing the Responsible Emergence of Science and Innovation in Society. Chichester, UK, Wiley, 2013.

Part V: **Nanotechnology philosophy:
Dilemmas and ethical issues**

Albert Ed. Evrard and Praveen Martis

14 Christian thinking and acting in nanotechnologies: Reflection based on the principles and values of the Social Teaching of the Church based on apps concerning old people

New technology must be researched and produced in accordance with criteria that ensure it truly serves the entire "human family" (Preamble, Univ. Dec. Human Rights), respecting the inherent dignity of each of its members and all natural environments, and taking into account the needs of those who are most vulnerable. The aim is not only to ensure that no one is excluded, but also to expand those areas of freedom that could be threatened by algorithmic conditioning.

Rome Call for AI Ethics (Rome, February 28th, 2020)

Abstract: This chapter explores an ethical conduct of stakeholders within the nano-scene that is in line with the Social Teaching of the Catholic Church. It, comparatively, opens an avenue to a nanotechnology's ethics based on principles and concepts circulating in other areas of the Social Teaching of the Church like Artificial Intelligence, xenotransplantation and transplantation and the teaching about technology and labor. Besides, looking at the needs of the elderly people and applications containing nanomaterial helps to focus on vulnerable people's issues and reflect on how to behave. In addition, the *Spiritual Exercices* out of the Jesuits tradition may contribute to make ethical decisions and take action based on the reflection guided by the context of the aged.

Keywords: applications containing nanomaterial, older people's needs, Social Teaching of the Church, Catholic Church, moral theology, nano-ethics

14.1 Introduction

Ageing with nanoscience and technologies (NST), like other techno-science, is among the crosscutting and global contemporary phenomena [1, 2]. Attached to a growing portion of the population and to a growing number of existing and future applications,

Albert Ed. Evrard, University of Manitoba, Canada
Praveen Martis, Department of Chemistry, St Aloysius College, Mangalore, India

https://doi.org/10.1515/9783110669282-014

these two fields are likely to profoundly and irreversibly transform the environment [1, 3, 4], human beings, ways of thinking, and ways of considering humanity and life.

Because of this impact, an ethical assessment of human action relating to NST seems to be essential nowadays. In this chapter, different from but complementary to other approaches in nano-ethics [5–7], this evaluation will operate from what the Roman Catholic Church says about it in its Social Teaching (STCh), otherwise called the Social Doctrine. Besides, it will combine a way of making decision and taking action that is seminal in understanding decision-making, and specific to its spiritual tradition, the *Spiritual Exercices* of Saint Ignatius of Loyola (SE). For the STCh, it is a question of "forcefully renewing the humanism of life" putting the fraternity and solidarity of individuals and peoples back to the foreground [8]. For the SE, born in humanism, it is a question of freeing the decision from what prevents it from reaching this goal. In short, reorienting ethical action. This responds to the invitation recently issued by Pope Francis on the subject of "emerging technologies". It is necessary to understand and to evaluate this contemporary phenomenon. This is "a very demanding task, given the complexity and uncertainty about possible developments, and requires even more careful discernment than is usually desirable" [4].

Approach and limitations

As this interdisciplinary approach may seem unusual in its field and the complexity of the field of NST and nano-ethics, it is important to set the framework.

14.1.1 At the crossroads of ageing and nanotechnologies

The first particularity of this chapter is that it is located at the crossroads of the two phenomena identified as the unprecedented "silent revolution" in human history [9–11], a crucial and privileged place to raise ethical questions.

In practice, this can be observed in the fields of health, cosmetics, and assistance to the elderly and the management of their daily environment, through examples of technologies that have been developed and which contain NSTs [3, 12]. This emerges from a few reasoned contentions. Firstly, these fields correspond to major expenditure by the public authorities as well as by companies and individuals [13]. Secondly, they make it possible to obtain three degrees of penetration of the nano-metric entity in the individual, an essential point for determining ethical questions and limits to the extent of "intervention in living matter" [1], and therefore, a situation in which NSTs are present outside the human body and one inside it, in order to reflect on ethical action. This is done by pointing out to the direction of a growing attention to the quality of life, to the centrality of and respect for the integrity, freedom, and dignity of the human being, in terms of "individual behaviour" and "the effects of choices and structural arrangements" [1]. It is therefore not a question of making a systematic survey but of raising questions posed by some existing applications.

The point of observation is crucial because the questions are measured by the fragility of the most vulnerable; those that people falsely consider to be the most dependent; those appearing to be far behind new technologies or not very inclined or slower to address them. It guarantees, beyond the prejudices and tenacious assumptions resulting of ageism (discrimination against older people) [3, 12, 14, 15] that has to be fought, a good place to approach ethical issues raised for all, and revealed by NSTs as well as by techno-science, in general [16, 17]. Among other issues: unequal access to medical care, unequal access to technology, inadequate information security/privacy protection, inadequate protections of individual autonomy, inadequate consumer safety information and protection, inadequate resources to develop pro-poor NST, how NST alters long, healthy, and comfortable life, impact memory, sociability, and so on are worthy of consideration. Furthermore, avoiding a general discussion of NST, it defines a precise field, which is necessary if we take for sure that "case-by-case ethical assessment for nanotechnology is crucial" [14, 18].

This point of observation allows everyone to visualize recipients of the activity in NST as well. To the imagination that makes researchers and developers dream of the many possibilities offered by nanotechnologies [19, 20], which sometimes border on science fiction, echoes an imagination anchored in the reality, which makes one envisages possible solutions based on what life is and what human beings are [1, 21], as presented in the Social Teaching of the Church (STCh) or the *Spiritual Exercices* of Ignatius of Loyola (SE) [22, 23]. For both, life remains central in the approach to any question. First, at the individual level: it is not only the constitution of every human being made up of natural nanostructures, fragility, human contingency, and human finiteness, but also the alliance of biological, genetic, and epigenetic mechanisms which ensure an extended homeostasis of the person in relation to his or her environment [24, 25]. Second, at the level of the human family: "the joys of family relationships and social cohabitation", which lead the elderly to believe more in their hopeful "dreams" and the young to have "visions" capable of pushing them to commit themselves courageously to history "without depriving them of the possibility of contributing to the development of society" [1, 10].

The same techno-scientific paradigm underlying NST economics, policy, and development underpins both the Silver Economy and geront'innovation, which many consider as two of the major opportunities of this century [26, 27]. In addition, the commitment to emerging technologies is culturally different in many regions of the world for STCh and nano-ethics; these environments show an underlying common organization of profit and the rapid pace of technological development that puts pressure on ethical reflection and the search for the common good [1, 10].

To sum up, as Pope Francis emphasizes keeping older people in mind helps ask how safe and healthy NSTs, in their effects, are at the service of the "whole human family", including the "most vulnerable", leaving no one behind.

14.1.2 Observing and evaluating from the Social Teaching of the Church

The second particularity of this chapter is that the observation and evaluation of these two phenomena, ageing and NST, are rooted in the Social Teaching of the Universal Catholic Church (STCh), which is off-center with regard to existing nano-ethics, and exploratory as well in matters relating to NST and emerging technologies.

Decentralized, it is. In fact, the STCh that belongs to theology and moral theology in particular [28–30], relies on the real presence of God in the reality of the world and on the human being determining his conduct, with reason, taking into account this reality and presence. whereas Nano-ethics is essentially philosophy-rooted [31–34] and shows the absence of God or, at best, a God reduced to a human idea or hypothesis. Moreover, its places of formation, origin, tradition, goals, criteria and methods, the way of conceiving the human being, and its environment are genuinely different from that of nano-ethics [35], and diverse in the field of religions as well [36]. That said, there is in common the idea of an observation and an evaluation. In this respect, taking into account the state of progress in the examination of emerging techno sciences in STCh, this chapter "cannot be limited to the resolution of questions raised by specific situations of ethical, social or legal conflict" [1] but will explore further a framework of relevant criteria for responsible action.

Moreover, if nano-ethics is largely based on Western culture and more precisely the American and European (even if it seeks, in order to understand itself and to win the battle of competition, to enter into dialogue with Asian and Indian contexts, for example [32, 37]), it is still very fragmented (though NSTs are globalized). On its side, the STCh addresses issues at a universal and global level, including the development of a vision of the human being and the world. However, this encapsulated difference does not prevent a dialogue [33, 38]. In short, STCh and nano-ethics originated from civil society or governmental sources can be associated but not assimilated [39]. The important thing, eventually, could be to situate any contribution before examining it.

Exploratory, it is too. As a matter of fact, the field of NST is not yet the subject of attention for the Church at the universal level. As a result, this chapter is a proposal from the academic world, among others, and not a document with particular authority and relevance to the STCh.

Whatever it is, the work is indispensable, where, at present, there is little agreement on responsibility, the moral standard in research and innovation, funding, production and use in NST [2], as in other areas of techno-science, nor is there any agreement on individual and collective (family, society) moral responsibilities for living with older people. These concerns are differently appreciated in many parts of the world depending on culture, belief, and social organization.

In addition, it is also indispensable because the STCh, like other bodies of thoughts and knowledge [40, 41], does see a powerful techno-scientific paradigm underpinning the globalized environment in which NSTs and existing ethics, including nano-ethics, are developing. An environment that puts ethics under tension, through the inflation of technical information [42], the competition between a few countries or blocs of countries, between investors looking for quick results that can be transformed into lucrative applications and other actors driven by a different pace and objectives [11, 41, 43, 44].

The STCh, therefore, while recognizing existing research, whose dialogue with religions is erroneously often collocated in culture [32], will seek to develop "a different way of looking at things, of thinking about policies, curricula, a lifestyle, and a spirituality." This is supported by an integral conception of the human person, his dignity and sociality [30]. This also assumes objective truth of values and judgment regarding human action and world situations, originated in natural law and divine revelation [39, 45, 46]. This moves away from any libertarian and liberticidal humanist ideology animated by the will to power leading to the division, to isolation and to control of human persons, "with the convinced support of the market and technology" [1]. Finally, this does not respond to a case-by-case ethic, as seems to be implicit in the nano-scene: "ethical principle in research always arises from a particular research context that is grounded in practice, not *a priori* reflection" [2].

Additionally, it should be noted that this approach is convergent with the STCh on older people. Based on the latter, there is an urgent need "to stem the culture of indifference, rampant individualism, competitiveness and utilitarianism which are now threatening all areas of society, and to remove any form of segregation between the generations, a new mentality, a new attitude, a new mode of being, a new culture. A form of prosperity and social justice needs to be pursued that is compatible with the objective of defending the centrality of the human person and his/her dignity" [10, 14], making a choice in relation to an open ethic indicating as: "questionable whether ostracism of ageing is 'good' or 'bad' for the elderly and/or for society as a whole" [47].

Thus, distinguishing itself from existing approaches in nano-ethics, the STCh of NSTs, based on examples related to older people, is developing a "new and universal ethical perspective attentive to the themes of creation and human life" [1]. This ethics in action is opening up an integral humanistic horizon based on God's creative and loving relationship, inspired by fraternity and solidarity. To quote Pope Francis, on a more original note, it seeks conditions that ensure an increase in the areas of freedom for each person [18, 45], his or her rightful place in society, and which promotes life and what is alive in each human being.

14.1.3 To choose the greater good

Finally, the third and last particularity of this chapter: In addition to the orientation of the STCh, there is a "way of proceeding" that can help everyone in the nanoscale to make an ethical action their own, to inhabit their moments of decision. This way of proceeding is contained in the *Spiritual Exercices* of Ignatius of Loyola (SE), belonging to the spiritual tradition of the Church, as well as other forms of methods [22, 23].

Jesuits, of whom Ignatius of Loyola was the founder in the sixteenth century and those who have since experienced the transforming action of the *Spiritual Exercices* in their lives, know how much the tools they contain help in making decisions, in management, education or finance, for example. It is not just a question of having a mentor to talk to about a thorny issue. Crossed by desires and passions and by soliciting memory, intelligence, will, and imagination, the whole being of the exercitant is able to make an informed and true decision. Usually assisted by an external observer, SE offer a fully actual humanistic way, including God, to choose (discernment) between a good and a greater good in every matters. This, after having recognized and rejected evil and having freed oneself from attachments that prevent reaching an authentic decision (indifference).

In short, once the temptation of a moratorium on the development of NSTs and the temptation of regulation solely by the market and by progress linked to unlimited growth have been removed [9, 45], SE invite us to propose a way of proceeding that everyone – students, researchers, technicians, developers, financiers, traders, and consumers – can adopt. This would be evolved by making present to the mind and soul, older people and the situations in which they live, and by allowing oneself to be touched by them.

14.1.4 How to understand nanotechnologies and nano-ethics?

In any case, NSTs and Nano-ethics are sufficiently defined in this book and the questions raised by the complexity of the subject matter have sufficiently been developed to simply refer to them too.

However, for the purpose of this chapter, the NSTs concern two areas. First, the "sciences that study the composition of matter, its characteristics, its intimate properties on the nanometre scale (10–9 m) and develop mathematical tools to model these behaviours" and the technologies "that concern the design, characterisation, production and application of structures, devices and systems through the control of shape and size on the nanometre scale. They require a vast array of instrumentation to fabricate, to measure and possibly to manipulate nanoscale objects" [48].

Second, in the context of this chapter, beyond the multiple components, definition, and issues related to their evolution and regulation [25, 32, 49, 50], NSTs refer, for most, to applications in the environment of the elderly. These are combining technologies (ubiquitous computing, biotechnologies, for example). On the other hand, it relates to applications resulting from research and development activity with nanoparticles (NPs) or nanostructured materials (NSMs) that are not naturally present in the human body or the environment, and their voluntary use [3, 51, 52]. In addition, the stages of a chain of research and activity in NST leading from the nanoscale to the application "in hand" are of interest as well.

As for nano-ethics, this chapter refers to ethical "issues, concepts, basic principles, and critical perspectives associated with nanoscale science and engineering", potentially present in well-established areas of ethics like bioethics, environmental or medical ethics. Moreover, nano-ethics aims to address especially the way "nanotechnology promotes human, environmental, and social goods" [18]. In the realm of this chapter, nano-ethics is *a priori* non-religious and not necessarily seen as an autonomous field of ethics yet. It grows in a context where, depending on the stages of the chain, attention to ethics varies greatly. In addition, nano-ethics, like STCh, assumes that researchers and engineers to be responsible for their research and its impact. Whereas, on the nano-scene, responsibility currently appears moderate. Thus, if the researcher is reputed to be very attentive to his laboratory employees, he is less attentive to social responsibility; whereas the engineer, especially a woman, seems more concerned with social issues, in line with her perception of the risks and benefits of NST [2, 11, 53].

14.1.5 How to apprehend old age and the elderly?

Firstly, from the NST perspective, old age apprehension, which concerns more women than men despite the variety of situations, is based on the idea of compensability-based technique. Incorporating NST, those applications existing in the daily life environment of the old or very old persons aim to support ever-changing loss of capacity or ability (physical and cognitive decline, mobility, isolation) [17]. Thus, like for other studies, the number-side of age does not only intervene directly, nor does the legal framework setting the age, such as withdrawal from the labor market or old age at work. On the other hand, the perspective focuses on: senior citizens, or their individual or institutional entourage. They are seen as consumers of services and products generated by the Silver Economy and the geront'innovation. Besides, consumption here means that products are more rented than bought [3, 27]. Finally, despite the diversity of situations, the perspective does take for granted that senior citizens are generally open to these technologies, though a financial, ethical, psychological, institutional, and regulatory barrier could be a drag on the use of it [27], or a lack of familiarity and support in learning as well.

Moreover, according to the STCh, NSTs, through the selected applications, products, or services, put in tension the needs of older people in terms of health, food [1, 54], housing, and their spiritual needs or the needs of the heart rarely added [55]. The question is how the activity concerned related to the inner self can be authentically encouraged by the technique. In commonly shared values, these basic needs refer to the fundamental rights and freedoms of every human person eventually, and to their requirements [56]. For the STCh, both needs and their declaration's existence are pre-political, not created by governments, and, therefore, have to be fully respected by every person or institution, as referred to in Pope Francis' words highlighted at the head of this chapter.

Hence, for the STCh, the criterion is less of a first glance distinction between essential and non-essential needs. We may think that cosmetics or some of the Elderly-Smart Home technologies ([Assistive or Everyday. It could...] but could be considered essential in certain situations nonetheless) which would address the comfort of healthy and wealthy older people are futile, perhaps. It is true they are not serving the less autonomous and poor older people. At a broader level, the STCh does evaluate the situation based on both the principle of the universal destination of goods and the principle of the preferential option for the poor,[1] aiming at material as well as physical, mental, social, moral, and spiritual poverty and the search for their satisfaction [8, 11, 17, 39, 57, 58]. In other words, a cosmetic might have an important role to support the dignity of someone and an assistive technology might be a burden for the growth of someone else.

Old age, as illuminated by faith, is placed by the STCh within the vast framework of God's providential plan that is love concerning the life of each person. The reality of God associated with the humanity can be understood as a stage of life sustained by the promise of life after death. This broadens the perspective that is only immanent and focusing on material reality, and gives those who seek it the strength to accept old age as a gift and as a task to be responsible for. The STCh, thus, calls for an ethical behavior of responsibility in everyone's life journey coming from and returning to God. It means that it is an individual responsibility. In other words, each has to make his or her own contribution (a more complete vision of life, memory of things, skills, and people, disinterestedness "being vs. having," interdependence, experience that the answers "of science and technology seem to have supplanted"), to be put at the service of others, as well as toward future generations. Within this broad horizon, the STCh does insist that the political society and the intermediary entities between the individual and the State contribute, each according to their dispositions and means (principle of subsidiarity) [8], to making possible the full accomplishment of its responsibilities, for every person at every stage of life [10], including, therefore, old age.

14.2 Social Teaching of the Church and nanotechnologies

Even though Christianity in Protestant area or indeterminate denominational areas or even individual contributions has been expressing itself [19, 36, 38, 59, 60] since the 1980s, there is, to our best knowledge, no document dealing with NST in the

---

1 "164. The principle .... easily [346]". The principle of the common good, to which every aspect of social life must be related if it is to attain its fullest meaning, stems from the dignity, unity and equality of all people. According to its primary and broadly accepted sense, the common good indicates "the sum total of social conditions which allow people, either as groups or as individuals, to reach their fulfilment more fully and more easily". *Compendium de la Doctrine Social de l'Eglise* (2004).

Social Teaching of the Church (STCh). This is despite an urgent and repeated call within Church's circles or beyond to pay particular attention to their social and human impact [1, 34, 61, 62].

However, since Pope John Paul II (biomedicine) and the more recent opening to global bioethics focused less on the beginning and end of life or on a particular technical aspect [41, 63, 64], the expansion of emerging technologies has been taking place in research, education, and communication at the Pontifical Academy for Life and its new statutes (2016) [65]. This platform seems, therefore, to be the one to look at in future to observe further developments on STCh and the emerging technologies, not to mention academic, individual, or collective contributions and Catholic groups initiatives in the area as well.

Moreover, for some authors, the Encyclical Letter *Laudato Si'* (2015) of Pope Francis marks the beginning of the reflection (n° 47) [28, 41]. Although it mentions nanotechnologies (n°102) and the effects of particles (n°21), the Encyclical locates NST mostly in relation to a technical-economical paradigm underlying the techno-science. This paradigm does present itself as a threat to the contemporary world and the future of humanity (chapter III, n°102–136, 198–190) [45].

Next to these contributions at the highest and broadest level of the Catholic Church, let us also mention other regional or local inputs. For instance, at the European level, the Catholic Church in the European Union (COMECE) has proposed some crucial elements. With the help of experts, the gathering of European Bishops has addressed the risks to peace and the future of humanity (military uses) or questioning the capacity to go beyond what is human (trans-humanism, exceeding or overtaking of the human nature) or to limit what is human (eugenics), or evaluating nano-medicine (diagnosis, targeted therapy, regenerative medicine) [66]. Besides, A number of National Conferences of Bishops have also published ethical guidelines in the field of health including the use of emerging technologies, including NSTs [67].

On the other hand, since 1993, the Church started adressing the elderly in the Church and the World in a few number of major documents. In each case, with a biblical or ecclesial approach, the way in which older people live and suffer is addressed based on a location in different circles, a traditional STCh approach. First of all within the family circle, then in wider circles like the neighbourhood, the parish, the city, the country [68–70]. These texts aim at safeguarding their essential dignity, freedom, and equality of every human person [10], while giving little treatment to technological progress and no consideration whatsoever to the possibilities it offers for the life of communities of believers. These texts also state that a purely economic and functional approach to society, judging situations exclusively in terms of "cost-saving benefits", impoverishes the human community by rejecting the wisdom, experience, and enriching presence of older persons [71].

As a way of summing up, let us say that the aim of the STCh is to arrive "at criteria and fundamental ethical parameters capable of providing guidance for ethical problems that occur with the widespread use of technology". Doing so will contribute to develop "an ethic inspired by a vision of the common good, an ethic of freedom, responsibility and fraternity, capable of fostering the full development of people in relation to others and to the whole creation" [8, 72, 73].

14.2.1 Artificial Intelligence (AI)

With the *Rome AI Ethical Appeal* of February 28, 2020 [74], the Church, for the first time outside the bioethics' area, has directly addressed one of the techno-sciences in a universal institutional document. Clearly, through a contextualized analysis and the six principles, its aim is to provide a STCh framework for a digital ethics (algor-ethics); it is also aimed at being foundational reflection for dealing with new technologies, anchored otherwise than in secularized ethics [4, 75]. Innovatively, this document also forms [76] a commitment made and signed by personalities from the high-tech industry. This meets the affirmed need for an inclusive ethics based on the dialogue of stakeholders, a path recommended in nano-ethics [14, 41, 45, 77], and implemented, for example, in the European Union through the platform www. nanocode.eu., or elsewhere, through the platform INSCX™ Global Nanomaterials Exchange and Information Portal www.inscx.com.

The six principles are as follows:

> Transparency: In principle, AI systems must be explicable; Inclusion: the needs of all human beings must be taken into consideration so that everyone can benefit and all individuals can be offered the best possible conditions to express themselves and develop; Responsibility: those who design and deploy the use of AI must proceed with responsibility and transparency; Impartiality: not to create or act according to bias, thus safeguarding fairness and human dignity; reliability: AI systems must be able to work reliably; Security and Privacy: AI systems must work securely and respect the privacy of users.

How could that be relevant for addressing NSTs?

Firstly, it should be noted that these principles are already sought in the ethical field of engineering sciences as well as in ICT ethics or regulation based on the principle of protection of the integrity of the person and privacy [2, 14, 78]. However, if their formulation is the same or similar, the context of reflection differentiates the content.

Secondly, these principles concern AI technologies embedded in the environment of human beings. They, therefore, seem to provide a framework for ethical behavior in the field of assistive technologies and empowerment (ATE) of older people where they live, insofar as they could be subject to the possible transfer and integration of nanoparticles in the human body resulting from the detachment of surfaces, for example [2, 14, 52, 79]. This, as well, insofar as other convergent and ambient technologies combined with NSTs ensuring the miniaturization of components (like AI, technologies for monitoring and perceiving human presence or environmental conditions, GPS or radio frequency identification (RFID) technologies), all together contribute to the development of an "ambient intelligence" and build connectivity forming a "cloud of care." An environment which is more customizable and adaptable, and thus able to best serve the changing circumstances of ageing of the old or older persons (adaptability in the assessment of the cognitive, affective, and physical situation, prevention by anticipation). This, with the aim of walking, being safe. Such an environment is deemed to ensure an independant living by contributing to the prevention of many situations like avoiding and detecting falls, health situations. It is also deemed to promote a sense of security for walking, for indoor or outdoor entertaining and connecting activities, for example [15, 52, 80–83].

With regard to the six principles, worth-noticing elements moving in this direction mainly relate to inclusion. Though timid, some initiatives are worth mentioning. First, involving older people more in the design of the techniques under development [17, 81].

Second, seeing these technologies as a contribution to solve health-related issues like control of blood pressure, stability, calorie consumption, and body temperature [84]. Monitoring health and environmental parameters contributes to maintaining and improving health on a daily basis, and reduce suffering as well, and among its, the one resulting from loneliness and isolation which can lead to a loss of zest for life and, eventually a way to fight marginalization [10, 52, 84], which can be technological in particular [52].

From the IA and STCh, the hereabove observations can be used to establish an ethics of NST. First of all, it invites to place at the heart of any ethical behavior, the preservation of the centrality of the human being, his nature and potential, his environment. There is thus an ethical action of "doing" that can be transposed from AI to NST: (1) to have at the heart of the intention and action, the good of humanity and of every person; (2) to include every human being without discrimination of any kind; (3) to consider the complexity of the ecosystem and its sustainable development. Similarly, there is an ethical action of "do not do": not to exploit human beings especially the most vulnerable.

Thirdly, ethical action in both NST and AI implies a commitment to education for both young people and continuing adult education accessible to all (non-discrimination), based on the principles of solidarity and equity [10, 11].

Fourthly, ethical action in NST involves regulation and legal action in certain cases. According to STCh it would happen both in cases where this is indispensable and in cases where the cooperation of stakeholders in the ethical field cannot succeed and at the appropriate level, by application of the principle of subsidiarity[2].

What is particular to the STCh with regard to law is the insistence on fundamental rights and freedoms. There is a call to embody a renewal of sensitivity and fraternity

---

2 "186 § 3: Subsidiarity, understood in the positive sense as economic, institutional or juridical assistance offered to lesser social entities, entails a corresponding series of negative implications that require the State to refrain from anything that would de facto restrict the existential space of the smaller essential cells of society. Their initiative, freedom and responsibility must not be supplanted [64]". *Compendium de la Doctrine Sociale de l'Église* (2004).

[1], in a contemporary international context where human rights are threatened and the institutional organization of their protection put into question. Besides, for the STCh, the ethics underlying human rights constitute the foundation, a vocabulary, and an angle of approach and understanding of ethical issues: "a central question in the search for universally shareable criteria" [1, 14]. Moreover, through their affirmation and effective protection, these rights give concrete expression to the dignity and a social dimension of the individual [39, 56]. Consequently, for the STCh joined already by some authors in nano-ethics [11], the State has to revisit its regulatory role.

14.2.2 Transplantation

This second layer of reflection situates the STCh mainly in a bioethical environment. Whether it is xenotransplantation (transplantation of organs from animal species into humans) or organ transplantation between humans, both situations deal with a common element in NSTs: the voluntary insertion, into the human body, of an *a priori* non-removable foreign material. This is significant within two areas of NST' developments:

(1) In the field of anti-ageing cosmetics through nano-emulsions, lipid nanoparticles, nano-capsules, nano-accelerator of penetration through the skin, nanomaterials (currently 29) are very widespread both to increase the duration of the effect and its stability and to convey the active product in a targeted manner. This is to satisfy a growing demand to make the appearance younger [85–88];

(2) In the field of older people's health, namely, the developments in areas like cardiovascular, diabetes, osteoporosis, arthritis, or cancer pathologies. Besides, NST incorporated medications also act on suffering whether it is a physical, an emotional or a psychological one (neuro-modulation for instance) that might be highly significant in old age.

In addition, nanomaterials are also used to improve imagery and the monitoring of therapeutic activities in the human body, to build vehicle alike structure able to conduct pharmaceutical substances to the exact and right place not to mention, other developments in biotechnology and biometrics oriented toward predictive and preventive medicine before being curative, through the combined introduction of computer science, biology, high technology, DNA sequencing, cognitive sciences, AI and NST. In other words, the use of already conceived nano-robots, nano-gears, nano-sensors, nano-computers nano-motors, nano-pumps, though still at a conceptual stage could not only improve the health of older people, but of all people, in a more targeted way. [51, 52, 89–93].

The rapprochement of both techniques could show two things that might turn out to be useful for the STCh of NST:

(1) The nanomaterial naturally present in the human body and the environment seems to possess mechanisms causing effects that reach living organisms to a lesser extent. This does not seem to be the case with what human activity can create or

produce [79]. In other words, it is the whole physical, psychic, affective, and spiritual homeostasis that is disturbed and not just some aspects of it, that science and should be focusing on. This is already an invitation to prudence and to the multiplication of impact studies.

(2) Where the presence of a foreign body at the particulate level was the exception (involuntary rarity), by the increase in human activity, the exchanges between the outside and the inside of the human body have intensified. To this presence, which becomes the norm (voluntary and involuntary incorporation), there is no corresponding assessment of the effects on the life and health of human beings, living beings in general and the environment. There is too little work on the degree of adverse impact of foreign bodies or toxicity [14, 94]. The result is an investment in the precautionary principle of nano-ethics through self-regulation (guidelines, social control) or the regulation already mentioned [11, 48]. Knowing that for the STCh: "the law by itself is only a cold standard, a barrier that prevents deviations. The essential thing is the spirit that animates its defenders, the impetus that goes beyond the current perspectives" [95]. Therefore, how does the STCh look at transplantation?

Actually, the STCh sees the entry of an entity into the human body that is initially alien to it as morally lawful. As early as in 1956, Pope Pius XII expressed the hope that research would continue and extend to new therapies that could replace transplantation. He already mentions developments at the molecular (prosthetic) level. In 1982, Pope John Paul II also expressed the hope that techniques for modifying the genetic code would advance as they offer hope for curing genetic or chromosomal diseases. Such an opening, which has now been confirmed, seems to be paving the way for ageing-serving technologies incorporating NSTs among diabetes, cardiovascular diseases, poly-pathology, and neurodegenerative diseases.

This welcome approach assumes that it flows as a part of a culture of life and implements itself in an ethically acceptable manner "the obligation to defend the life and dignity of the recipient and donor; it also indicated the duties of the specialists who perform the transplantation. It is a question of enabling a complex service of life, combining technical progress and ethical rigour, humanising interpersonal relationships and correctly informing the public" [96–99]. As regards the culture of life, the situation of the older people might illustrate tension between eugenic-type legislation regulating the end of life and the promises of increased longevity or surpassed longevity by the help of technology. STCh of NST should provide an opportunity to support the culture of life.

As far as ethics are concerned, at least two criteria common to all types of transplantation might be able to inspire a STCh of NST:

First, the technique must be reliable, possible, and safe, and not expose the recipient to an inordinate risk to his physical and psychological health, based on two characteristics: the level of probability of undesirable future damage and the extent of this probable damage [57, 100].

This question, often debated in nano-ethics, calls for more research in the field, taking note of the number of uncertainties related to risks that are still ill-defined, both for the individual and for the environment. In addition, despite the use of simulation, particularly in engineering sciences, trials would potentially raise the question of the involvement of elderly or very elderly people in the testing process on a human scale. Given the reduced life expectancy and therefore the (wrongly) perceived limited risk, combined with the pressure of the expected benefits of the technologies, and the consent that could seem easy to receive, particularly for people who are alone in life, it is for fear that frail old people could be more easily approached for such experiments.

A second point of consideration is the respect for human nature understood in its wholeness [96] and the respect for the dignity of every human being even at the embryonic stage. At this stage, the question is when moving from an organ to matter, even if it is mobile (nano-organic or composite based) [79], whether the same criteria would be considered in the order of NSTs. In this regard, the following may be considered an appropriate quotation: "the implementation of a foreign organ into a human body finds an ethical limit in the degree of change that it may entail in the identity of the person who receives it" [100]. This question of the degree is a difficult one both objectively and subjectively. It seems anyway that it would reveal a need of nuances in each specific case. Besides, according to the STCh in this area, the use of the technique is not only an individual matter but deep feelings, especially those of the third parties, have to be taken into consideration [99].

In other words, looking at each technical tool, like in AI, the question could be: NSTs have the potential to support (empowerment/enablement) the centrality of the human being, and the development of each individual's potential through restoration (therapeutic techniques) or through enhancement (increase in performance)? *A priori*, the alteration, the modification of human nature combining nano-biotechnology and neuroscience would not comply with STCh' vision on life created by and journeying toward God, and the dignity and integrity deriving from this [51, 73, 101, 102].

Moreover, certain applications in biotechnologies using or not using NST are closely related to life, and consequently capable of bringing about profound modifications and control of human beings by other human beings. Based on the STCh, this raises the question of the attack on both the objective dignity of the unique human being, irreducible to one or another of its components, the free and unsubmissive to foreign powers character of every human being, and the attack on the subjective dignity affecting the respect due to oneself [47, 67]. Where nano-ethics considers ethically questionable human augmentation techniques, anti-ageing rejuvenation, cosmetic surgery, and anti-ageing drugs, to the extent that they cannot deliver what they promise, the STCh adds reasons related to the dignity of the person and the integrity of human nature [14].

14.2.2.1 Xenotransplantation

In addition to these common aspects, further dimensions come from the examination of the two types of transplantation. Firstly, xenotransplantation has since 1956 been the subject of the STCh attention as a solution to the limitation of available human organs for the purpose of transplantation [103, 104]. Results were eventually gained in the treatment of diabetes, Parkinson's, Huntington's, and vascular accidents [100], commonly prevalent in old age.

Xenotransplantation might also raise in STCh three other questions of an anthropological and ethical nature that seem to be relevant for integrating into the field of NST: The first question pertains to the human intervention in the created order. This refers to the "playing God" concept, an activity that is widely debated in the field of nanoethics [41, 43, 105]. In this respect, the STCh is referring to a simple principle: created by God, in his image and resemblance, human beings and the world work and respond to a natural and received order. Therefore, human action cannot go against it.

The second question is related to the use of the animals for the good of man. In terms of NST, this refers to the use of nanomaterials transforming the animals whose organs would eventually be transplanted. It also refers to the possibility of transforming the material or creating new ones from elements unknown in what nature provides. The STCh of NST would here refer to the human being's co-creating responsibility involving "a reasonable use of the power that God has given him" [106].

The third question pertains to the personal identity of the recipient. The question of human nature and the dignity of the individual is uniquely embodied in the personal identity of every human being. At the NST level, the STCh could benefit the identity defined as "the relation of an individual's unrepeatability and essential core to his being a person (ontological level) and feeling that he is a person (psychological level)" [100, 104]. The characteristics expressed in the lived historicity of the person and in the communicability of what he is, so important in xenotransplantation, could help in advancing the question of the insertion of foreign material by NSTs, and the acceptance of the unknowns related to this with which to live.

14.2.2.2 Human organ transplantation

Finally, the transplantation involving a human organ from one human to another human raises at least two other issues relevant to the development of a STCh of NST.

First is the decision to incorporate a technique. For the STCh, the decision leading to transplantation must have authenticity and thus an ethical value. To this end, it has to be offered without compensation a part of one's own body for the health and well-being of someone else, driven by an act of love. According to the STCh, anything that deviates from this is therefore morally unacceptable. Exchange, commercialization, bringing back the use of the body or its parts, as an object, goes against the dignity of the human person [96, 104].

In STCh of NST, this would raise the question of lightness of use. For example, is the entire scope of integration by ingestion (food or structure accompanying the target drug, for example) or cutaneous absorption (cosmetics, cleaning products) of nanomaterials, the effects of their permanence into the human being, their migration, their interactions or possible mutations, their elimination or rejection of these elements, not trivialized? This directly implies that the consent of the donor and the recipient must be an "informed consent" and, therefore, never be subject to an act that is coerced in any way whatsoever. On the other hand, the introduced entity cannot be withdrawn during the recipient's lifetime but only after his death. This aspect might be relevant as well within a STCh of NST [96, 104].

A second issue is about the relationship between the professional and the patient. For the STCh of transplantation requires, among other things, "mutual respect, trust, honesty and appropriate confidentiality" [67], similar conditions apply to the NSTs introduced in imaging, diagnosis, prevention, and treatment.

More precisely, the STCh of NST could have to address at least two issues here. Firstly, the material possibility of the informed consent. Faced with so many uncertainties about the behavior of nanoparticles, in the presence of health risks, about their evacuation in body cells or in the environment, how could possibly an informed consent ever happen? Besides, uncertainties also exist with regard to the consideration of the overall well-being of the person during their use and during the time that this nanomaterial is in them, and especially at the end of their life [66, 67].

The STCh would perhaps suggest this question to be considered in the light of the principle of stewardship rather than ownership of life, which implies that life is received as a gift, a gift to be respected, particularly in terms of the rate at which it leaves the earthly world. Along these lines, there is no way for the NST to develop, hidden or not, nanomaterials with programmed obsolescence or any tools whatsoever that could lead to the death of the subject. Secondly, the notion of intimacy of the person has to be determined. The use of technologies in health and cosmetics invites us to consider the depth at which nanomaterials act and the fact some of their action requires to be controllable (from inside or outside) and could be constantly monitored. Would then the intimacy be changed in an acceptable moral way?

14.2.3 Technology and labor

This reflection situates the STCh more in the area of social justice and the organization of society. As the attitude toward AI and transplantation already shows, far from being "technophobic," the Church is in fact enthusiastic about technology while being theoretically critical in the STCh, which sees it as an essential element of any economy where labor has its own value [107].

14.2.3.1 Technology

Speaking of technology, with NST, like all techno-sciences [1], the STCh should have an overall positive appreciation of "technical progress that enhances good moral actions for the sake of human development but not technological progress that does bad [. . .] not those that hinder life or can be the cause for killing" [41, 45, 108]. Thus, as an exception, technologies integrating NST, that help to suppress life or are based on the idea of weapons, are squarely condemned. Applying the principle of exclusion, it means that others are retained [41], and therefore would include those based on NSTs.

For the rest, the STCh on techno-science emphasizes an ambivalence revealed in the area of NST design or manufacturing jobs, and potentially affecting everyone therefore [109, 110]. On one hand, it enables human beings to develop goodness, justice, beauty in the world (the beauty of a nanometric structure, for example) and to provide a just service to human communities. On the other hand, it generates immense power, especially in the hands of those who have the knowledge and resources to produce the nanomaterials [45].

This is reflected in Chapter III of the Encyclical,*Laudato Si'*, where the approach to environment and to living beings, and among them, the human beings, integrates new technologies and their effects on issues of development and well-being in a coherent and comprehensive manner. Besides, next to the general context, the Encyclical offers an epistemology for developing an ethics of techno-science and *in casu*a STCh of NST. Vulnerable people, including old people (Nos. 21 and 123), are the measure of behavior (No. 10) in an emerging society, underpinned by a hyper-powerful technocratic paradigm due to the rapid and exponential development of techno-science, and in particular NST (No. 16), even if it is dominant (Nos. 103, 109) [45, 111].

Related to the elderly, this technocratic paradigm underpinning the techno-science seems to be linked to the Silver Economy and Geront'Innovation streams. Besides, the question is whether it is reflected in the promise of NST through applications in lines of the "Active Ageing " or "Ageing Well" rhetorical [112, 113]. Both concepts have the plasticity to be used in one way or another, through everyday practices. Consequently, are the NST contributing well to the improvement of the quality of life, however variable in definition or are they threatening every single old person's pace of ageing by instilling dictate about the way to keep healthy as much as possible, for instance? Besides, from a STCh' perspective, it cannot exclude a deep spiritual well-being which is eventually the driven integrating force of all other dimensions [32, 45] that is largely ignored or limited to a mental health issue.

For the STCh, freeing itself from the relationship with God, techniques have, to some extent, wrongfully invaded the sociability of the human being by increasing individualism, contractualizing and monetarizing all relationships. Besides, more recently, with the emerging techniques, it also has tended to invade the human personality. Indeed, some of the new technologies seem to be moving in this direction; while favoring an increased control of sociability and of life and personality through biotechnology for instance [28]. In other words, do both NSTs and techno-sciences contribute to the realization of the greater good or do they reinforce the power of those behind the techno-economic conglomerates supporting a culture of well-being for the few and consumerism for all, based on greed, which would ultimately turn against the human being (No. 4, No. 9, No. 59)?

14.2.3.2 Labor

A second major consideration is the perspective of labor. For the STCh, research and its developments including technology are morally highly valuable and licit. As quoted repeatedly, "it is appropriate that scientists promote research which can enhance and prolong human life and that physicians [and medical and social workers-addition of the redaction], be well informed of the most advanced scientific means available to them in the field of medicine". This is done to prevent, cure. and help people "to bear their suffering with dignity" [114]. This is of utmost importance in the area of NSTs with regard to the field of toxicity [115]. This positive appreciation of science and technology involving the positioning of all workers as human beings in the whole of their environment, allows to claim the idea sustainability "combined with moral values [. . .] is well positioned to offer all scientists, in particular those who profess belief, a powerful impetus for research" [116].

Moreover, human labor, here as elsewhere, whatever its degree of specialization, is seen as a right and a duty, not as something to be done in exchange for something else, or as a simple step in a production chain. Labor responds to the vocation of every human being, and is an activity in constant connection with personal (moral) growth, sanctification (spiritual endowment) and growth in Christian and human freedom [26, 45, 71, 100, 117]. Moreover, for the STCh, labor is carried out with and for others, and, is therefore, oriented toward an action that is considered good (to serve the utility of all men in accordance with the order of creation by God, to be distributed among all, to lead to universal progress).

In STCh of NSTs, this would make it possible to evaluate nano-scene job's practices. Firstly, acting contrary to this cannot be required of the technician, the researcher, any person working or developing technical skills on the chain of discovery, production, and distribution of NST. Secondly, from another angle, the triple T (tradition, time, and technology) may show concerns related to the orientation of applications containing NSTs toward a greater and fairer service of the human community [26, 110].

Labor involving the use and handling of materials is also questioning the relationship to assets. According to the STCh, based on logical and rational processes, the researcher, the engineer, the entrepreneur may get accustomed to think that through their labor, they hold unlimited power over the objects that surround them, that materials are completely open to any sort of manipulation, extraction, or transformation. To this end, the STCh insists on the attitude to be proscribed of considering oneself as the absolute master (ownership) of all things and of all possibilities of action on all things and that rather considering oneself as the guardian of all things of creation (stewardship) to be exercised [45].

Finally, according to the STCh, distortions of the conception of labor and its relationship to goods eventually destroy the individual personal balance between technical skill and human sensitivity that reveals the value of labor, particularly in the area of health and support for older persons [118], which in turn reflects on the lives of older persons themselves.

14.3 Toward an ethics of action in nanotechnologies

Whatever could result from the comparative work and establishment of revisited principles and values of the STCh in order to develop a STCh of NST, taking action according to the Church's vision requires a commitment. A commitment to justice and charity (caritas = love of neighbor) [28] is indeed required from each person according to his or her role, vocation, condition, to the right measure of his or her possibilities and intentions. Practically, for the STCh, everyone involved, from study and research, development, financing, production, distribution, to the consumption of goods and services, as well as the disposal and recycling of NST, has to engage in action in conscience, alone or in dialogue with others [43, 119].

In other words, for everyone, students [120], industrialists, researchers, engineers, consumers, the observation of NST and their evaluation require all the intelligence and memory to accompany their development [45]. However, in terms of action, the example of researchers (USA) showing a generally high degree of separation between private and professional action is suggestive [121]. At the end of the day, the question is not so much one of disagreement or a consent on the content of possible ethical guidelines or principles as on the conditions for the emergence of the individual actualization of the ethical imperative [77], and consequently the decision and the way of appropriating an ethical action through labor.

In fact, the ambivalence that lies at the core of any technique is removed only by a choice. This choice concerns not only in everyone, intelligence and memory, but also the will. The style of acting in accordance with the STCh involves rejecting the reality of evil once identified, and then seeking the good that can be done through using the NSTs. In short, ethical action is based on individual judgement, contrary to what is viewed with suspicion in the research field. It is always based on free will and personal conscience, while encouraging collective deliberation. Moreover, it is an act of good and, ultimately, the pursuit of the greater good among others [2, 122].

Thus, it depends on where the individuals are in their life, the moral and physical resources at their disposal and the time they have to carry out a work. This is a work of logical thinking, informed by the principles and values of STCh, in connection with interiority and spirituality, and whose fruit is to be reflected in the action in search of a just social order, which, for the Christian, is part of the broader perspective of his authentic fulfilment in God [123].

That said, it should be made clear that a profile of commitment of this type cannot be built *a priori* for each person, at each moment of choice and for each possible question. Moreover, would this respect the ongoing formation of conscience? Would it be feasible? The answer is "no". Therefore, both the principles of the STCh and the invitation to practice this work of reflection individually or in groups resist many calls to identify and list the scientific and technological places where ethical questions arise and formulate them. It resists the mirage of a total "catalogue" of answers, even if guidelines, professional and corporate, can help and need to be taken into consideration, like everything else that exists on the nano-scene.

In practice, loaded with knowledge of reality, principles, and concepts evolved from the STCh, each individual or collective group may ask itself the following question: am I a person of the Golden Rule (3.1), the Beatitudes (3.2), or the Exercises (3.3)?

14.3.1 People of the Golden Rule

In the nano-scene, many people perceive issues and act accordingly by putting into practice the values named here by the STCh, without even knowing it. One could say that they are the Golden Rule people: "Don't do to anyone what you wouldn't want to be done to you" [41, 107, 124–125] Unless we seek evil, or intend to act ignorantly, or have gold as the only rule excluding all others, we are all *a priori* Golden Rule people.

At this level, ethical action implies that each actor in the nano-scene asks oneself if his action and his professional responsibility limit itself to scientific-technical knowledge and include a judgment about the older persons and their needs. Everyone might engage according to the personal experience of and knowledge about ageing or by intuition of what is desirable. In line with the Golden Rule way, each actor could also ask oneself if "thoughts are enlightened by the human and humanizing relationship [with the patient] [38, 41, 21]" with old people , the variety of situations in which they find themselves. Finally, to ask one whether or not one has taken the trouble to approach them to understand them, in the reality of lived things rather than in a theoretical and hypothetical way.

14.3.2 People of the Beatitudes

However, the Christian point of view, while taking up this Golden Rule present in the Scriptures [126], goes beyond this measure of the good, which is often limited to a kind of reciprocity. This could start from the commandments recalling duties, which is in a way the other side of the same coin, or through a promise of happiness like the Beatitudes. 'Blessed is he' means happy is "who possesses an inner joy, incapable of being affected by the circumstances around him" (μακάριος'makarios', in the Greek text). Both are originated in compassion raised out of contact with situations involving the weakest, the most disadvantaged [125].

With the Beatitudes, the measure of happiness opens up to a wider measure of action, as if beyond the measure of ourselves, including concern for others: "building a future of freedom requires love of the common good and cooperation, in a spirit of subsidiarity and solidarity" [127, 128]. The true measure of decisions and actions is shown by Christ's way of acting toward the least and even toward his enemy [41, 45]. This forms the basis of an ethic preventing injustice and promoting peace [109].

Within the context of this chapter, the question worth asking, for each stakeholder on the nano-scene, is whether or not: " 'hidden exiles' [. . .] people of all ages, especially the elderly, who, also due to their disability, are sometimes considered a burden, a 'cumbersome presence', and risk being discarded, denied concrete [job] prospects to participate in the creation of their future" [45, 129]. In other words, is the benefit of NST, in the perspective of sustainable development for the elderly and future generations, taken into account or not? and how?

Assistive and empowerment technologies (AET) provide a good example in this respect. From a concern centered perspective on the needs of the individual considered as a customer on the market, research and development seems to be oriented toward the use of technologies in urban areas or in its generalizable dimensions to the whole society [17]. Similarly, developments in the field of health show the combination with mobile devices through apps for the prevention and monitoring of diseases that can affect large populations or areas where the old people are numerous or isolated, or even the blind, deaf, and sick people named in the Beatitudes [52].

14.3.3 People of the *Exercises*

The opening to the least of the Beatitudes through the channel of compassion that pushes for action, can leave one helpless. One must ask oneself whether, in the absence of immediately intelligible steps or crushed by this unknown or the pressure of the world of labor, one does sometimes quickly give up seeking ethical behavior or ways of thinking about ethical conduct [45]. Is it not for the same reasons or just to 'save time', that we take recourse to a pre-edited catalogue to follow of what is to be done or not done in the field of NST? Furthermore, are professional or institutional standards in NST ethics, considered by those who are confronted with ethical question? How do they appropriate the national legal provisions on the subject, not to mention the extensive literature?

Even if they were consulted, the fact still remains that there is a contemporary need to speak about oneself in terms other than "DNA, molecules and genes" terms [22]. The power felt or held by a researcher, an entrepreneur, the intuition of a moral difficulty that may arise or the unknown or irreversible effect that would be created also urge some stakeholders on the nano-scene to think differently, perhaps more deeply and methodically. In one word, there is need to listen to ourselves [23, 32, 107, 130]. That can happen individually or communally [131].

One could say that the people of the *Spiritual Exercices* of Ignatius of Loyola (SE) are thus stakeholders of the nano-scene who, seeking integrity and coherence in their lives, put them in order. They start working on themselves through exercises in the literal sense of the term: "Action or means of exercising or practicing," to train, or "the fact of giving oneself movement", "action of practicing and putting into practice" [132].

Let us start with a comparison present in the very first words of the SE. As for any physical exercise there is warm-up time, postures to be taken: In a similar, way regarding the moral judgment, there is also a question here of exercising one's intelligence, will and memory, in stages, of setting in motion one's sensitivity, one's presence in the world [133]. Addressed to the believers, it is given that the principle that God acts through the Spirit in the whole being and the world, has a part in the exercise, even the greatest one. The important thing is to let him act actually. As a result of it, someone could receive lights that he thinks he would never receive or even take into account during or after the SE. It can be received in terms of ideas, ways of doing things and creativity. For non-believers, it is a matter of debate [134, 135], the SE can be taken as a method (hermeneutical level), integrating the person in all its components and helping in making a decision that does not immediately appear.

In any case, for each actor in the nano-scene, knowing how to navigate the abundance of information, the uncertainties and the complexity of the issues and risks, amounts to organizing a way of proceeding, around two central concepts: discernment (3) and indifference (2), which does not go without preparation (1).

14.3.3.1 Preparation

Firstly, it is essential to prepare for a discernment, whatever is the issue at stake. To do this, it is important to check whether one is open to categories that go beyond the language of mathematics, biology, and technology, beyond what determines one's own perception (linguistic aspects, NST definition, cultural, economic, political, religious, and ethical aspects, for instance) as well as the ethical questioning that exists at the level where one finds oneself [32, 45, 130]. It is also important to check an openness to possible changes in attitude, to the fact that solutions to existing problems may come from elsewhere than NST, for example [23, 45]. It is also important to check one's conception of the human person, for example the elderly, and of nature that is not reductionist or projected into a hypothetical or imaginary future. In terms taken from the language of spirituality but recognizable by all, it is a question of checking whether one enters into the process of a "broad and generous heart," says Ignatius of Loyola, the initiator of the SE [136].

It is important to prepare for discerning. Thus, it is necessary to gather information on the reality of the situation under consideration, on the aspects of the question to be formulated and further then raised. With regard to this, it is also important to notice how the reaction to the elements gathered creates feelings one way or another. Does someone make a note of hope, confidence, fears, worries, indifference? Besides, one must also consider that "The ethical evaluation of a new technology cannot provide definite answers to questions that have partially not even yet been asked" [19, 73].

At this stage, it is important to adequately formulate the ethical question that one intends to answer. Terms must be precise and correspond to a choice to be made. Moreover, the wording must allow for a clear alternative: to do or not to do this, to apply or not to apply for this position, to engage or not to engage in this type of research, to make or not to make use of this type of method, for example. Additionally, the question must be asked in terms that do not reduce the deliberation to come to nothing [137]. Of course, if it turns out that the solution comes immediately with clarity, it might mean that no deliberation is needed because there is no matter for deliberation. It might also mean what experience shows: moving too quickly through this stage makes the exercise more difficult and its success more uncertain.

14.3.3.2 Achieving indifference

Entering the SE, other stages will gradually lead to freeing oneself from any form of attachment. In other words, reaching a degree of indifference. One can define indifference as the inner state of being freed of all what is not God [138]. To help understand this, Ignatius of Loyola speaks of a posture: "as a balance in equilibrium, without leaning to one side or another" [136], about something (material or not) to which one is attached to. In other words, it has nothing to do with not feeling concerned [139] by an ethical action or putting oneself in a position where someone does not feel concerned, but it has to do with being concerned by many good things, and therefore trying to observe within oneself what makes deciding and acting. The purpose is to put order in life, to free oneself from determinants that one perceives, to situate oneself in relation to them, finding oneself more liberated and free, and finally to decide for the greater good: discerning [23, 136, 140].

To give a few examples related to the NST field, we could cite the speed of progress in NST and the control of time that the researcher or entrepreneur thinks they assume they are masters of matter that can be totally transformed indefinitely. It contrasts with the secular rhythm of biological evolution and urges to be guided in the development and use of new discoveries and new applications as well [45, 141]. A determining factor could be, because of the rapidity of things, the anxiety that causes harm to many people, or the specialization in work that prevents the situations and issues raised by NST from being seen in a wider context [45].

Moreover, the context of NSTs used in the perimeter of the elderly puts ethical reflection and STCh in tension and provides other determinants, at least in four aspects:

First, a tension is created between an anti-ageing tendency to prolong longevity or abolish ageing and a tendency to reject old age that can underpin cosmetics and surgery progresses.

Second, a tension is attached between the autonomy and independence of the elderly through AET, bio-geronto-technology, and the interdependence of people through technologies. At the end of the day, the combined use of emergent technologies is requiring monitoring embedded in the human body or external monitoring. In other words, an omnipresent machine–human being link, which is controlled by someone other than the subject himself.

Third, a tension is shown between the abundance of empirical data relating to the conditions of life in old and very old age (biology, psychology, social sciences, medicine, culture, techniques) and a more speculative approach to human nature [38, 142]. Empirical data seems sometimes to be very strong and overruling any kind of other approaches.

Four, more fundamentally, two other tensions must be observed. First, the tension of bringing together or keeping separate, what is God's creation and what is the artificial human realization of living (biomedicine) or the so-called intelligent things. Second, the tension concerning the confusion or keeping separate the fragility inherent in all human life and the promises of overcoming individual humanity and frailty. On this point, as on others, one should observe the important role played by the imagination, particularly in the field of health and the idea of absolute suppression of suffering, for example [89], and observe as well how its promises influence the decision to be taken.

In other words, by noticing the assumptions or prejudices related to the existing state of the nano-scene or to the conditioned intelligence, memory, and will of a person, one distances oneself from it. Consequently, the path of slowing down then opens up to a decision to come by. Besides, it opens up arrival of a true and fair decision on ethical issues without being frustrated or submissive, which is the opposite of helping to expand freedom. Moreover, the fruitfulness of such a decision, in one way or another, or its absence, would later test the decision as well [143]. In this, everyone can recognize the way he or she fulfils or does not fulfil his or her vocation.

14.3.3.3 Discernment

Discernment is defined in relation to the rules ('the how to', a kind of grammar) that allow it to be exercised. It is to say that the "rules for understanding to some extent the different movements produced in the soul and for recognizing those that are good to admit them, and those that are bad to reject them [23, 150, 154], and in exercising them, to allow someone to enter more into indifference. Consequently, the type of standard knowledge as a researcher, as an engineer, for instance, is complemented with knowledge through the exercise of the SE, and by the knowledge resulting from becoming during the SE, eventually [144].

Moreover, throughout the SE, once we have entered them, it is not just the question of going beyond feelings but assuming them, of being able to notice the inner movements, relating to what we are confronted with, while scrutinizing the nano-scene, or by being challenged by an intuition suggesting to act in one direction or another.

In other words, it is a question of "listening to oscillations in life, whether they are intellectual, emotional, or spiritual," taking into account one's temperament, one's physical and psychological inclinations [23]. That said, allowing oneself to expose oneself to meditation and contemplation, which is similar to making oneself present to the mystery of God, makes us experience a real presence beyond ourselves and our understanding, and helps us to move away from ourselves, to see and judge things differently and to find a direction for action eventually. Otherwise formulated, it is God who makes himself present and let his Spirit act, who is "the one who supports, who accompanies so as not to fall, who keeps us solid" [158] when through the exercises, reflection, thoughts, and prayer, we allow ourselves to let him dwelling in us, to be present and active. Such an experience goes beyond any psychological or pure mental experience, though it includes it [23].

It is neither possible in the realm of this chapter, nor is it our purpose, to detail all of these rules. Besides, out of nature, rules framing the experience are progressively unveiled to the exercitant. In addition, rules are in the hands of a coach who makes different use of them, depending on the situation of each person being accompanied, and from where he or she perceives the person accompanied is on this path of experience or the facilities or difficulties he or she encounters. Moreover, through a pre-established flexible path, each exercitant moves in his own pace and experiences a unique passage through the SE.

14.4 Conclusion

If it were immediately clear how man could be born to his authentic human fulfilment with the help of nanotechnology or other technologies, there would be no need to deepen an ethic. On the other hand, "When technology disregards the great ethical principles, it ends up considering any practice whatsoever as licit" [45].

"What can the Christian then say to the one who seeks his own creation?" [34]. For the Social Teaching of the Church and the *Spiritual Exercices*, this human authenticity passes through an examination of the reality of situations, which is including the reality of God present in the heart of man and in the world. This bond gives to the human being his true centrality and humanity. In this regard, the moral lawfulness of acting in the nano-scene is to be evaluated.

One could describe the challenge as following: It is to remain a "part of the nature's biological continuum," signaling the humankind created by God and toward which it is moving, as well as the dignity and individuality of each human being, of every creature and of the creation as a whole [61].

However, in the more or less long term, emerging technologies are capable of being oriented to leave this continuum. It benefits, indeed, from considerable resources, which could be allocated to a rethinking of all ethical reasoning (programmable morality). Whatever the perspective would be, the STCh stands firm that any moral judgment made on any situation must remain a human matter of conscience, which can be exercised by all and cannot be confiscated by some to the detriment of the others. Facing challenges posed by this centrality, the approach of the Catholic Church to emerging technologies is recent and cautious. Moreover, recent documents show that the Church's stated aim is indeed "to intensify the study and confrontation of the effects of this evolution of society in the technological sense in order to establish an anthropological synthesis that is equal to this historical challenge" [1] and so to add a page on the tradition of the Social Teaching of the Church.

With regard to nanotechnologies, as with other emerging technologies, such a mission falls, indeed, to the Catholic Church as well as to other religions: "[. . .] To man, as he is involved in a complex network of relationships within modern societies" [n°76], the Church addresses her social doctrine. As an expert in humanity, "she is able to understand man in his vocation and aspirations, in his limits and misgivings, in his rights and duties, and to speak a word of life that reverberates in the historical and social circumstances of human existence [n°77]" [59].

The attempt made in this chapter indicates that the principles, values, virtues drawn from its heritage, particularly with regard to Artificial Intelligence, transplants, techniques and labor, have the necessary flexibility to open up to reflection in this field that considers the whole human being, the whole of his humanity, and the humankind as a whole. Besides, situations concerning old people epitomized as human frailty, vulnerability, and experienced surviving example make it possible to embody and reveal essential aspects to be taken into account in the evaluation of existing scientific and technical action in NST.

In the final analysis, it is a question of action, indeed, for "The Church does not expect science merely to follow principles of ethics, which are a priceless patrimony of the human race. It expects a positive service that we can call with Saint Paul VI the 'charity of knowledge'" [145]. This service requires commitment and, therefore, prior deliberation. In this sense, it should have a place for the *Spiritual Exercices* of Ignatius of Loyola, as other ways of deliberating and progressing through decision-making process.

So, this approach: 'has been' ideal, utopian, or reflecting a world "of sheep among wolves"? [146]. Admittedly, the path may seem narrow or inappropriately reserved for the believer alone. However, as it concerns principles of social organization, its practicability is open to all. Moreover, the principles contained in the STCh and the rules of discernment and the SE have their weight of validity inscribed in experience even without adhering to the religious or theological context from which they originate. Moreover, depending on the different periods of history, civil societies would borrow them regularly. Certainly they feel that it contributes to the moral imagination subscribed as the "ability to imaginatively discern possibilities for making good choice, and to envision the possible good outcome that might result" [147].

In any case, in the field of nanotechnologies as in others, such an orientation toward the good can only help in advancing in the understanding of the challenges posed to human action in order to direct action to be taken toward any enhancement of hope, justice, and love (caritas – charity), for every person and for the benefit of all humankind.

References

[1] Francis P. Humana Communitas. Letter of his holiness pope Francis to the president of the pontifical academy for life for the 25th anniversary of the establishment of the academy, January 6, 2019 (Accessed March 15, 2021, at http://www.vatican.va/content/francesco/en/letters/2019/documents/papa-francesco_20190106_lettera-accademia-vita.html).

[2] University of Twente (The Netherlands). SATORI Project have your say on ethics assessment in Europe, 2016 (Accessed March 15, 2021, at https://satoriproject.eu/).

[3] García A, Gonzalez M, Frid L, e.a. What Technology Can and Cannot Offer an Ageing Population: Current Situation and Future Approach. In: Biswas P, Duarte C, Langdon P, eds. A Multimodal End-2-End Approach to Accessible Computing. Human–Computer Interaction Series. London, Springer, 2013, 3–22.

[4] Pope Francis. Address prepared by Pope Francis, read by H.E. Archbishop Paglia, President of the Pontifical Academy for Life. Meeting with the participants in the plenary assembly of the Pontifical Academy for Life, 28 February 2020 (Accessed March 15, 2021, at http://www.vatican.va/content/francesco/en/speeches/2020/february/documents/papa-francesco_20200228_accademia-perlavita.html).

[5] Scheiltle Ch, Howard Ecklund E. Scientists and religious leaders compete for cultural authority of science. Public Underst Sci, SAGE Publications, UK 2018, 27, 1, 59–75.

[6] Toumey C. Denominational interpretations of nanotech. In: Nat Nanotechnol Nature, Pub. Group, USA, 2018, 13, 270–71.

[7] Westmore JM. Religion. In: Guston DH, ed. Encyclopedia of Nanoscience and Society. Thousand Oaks, USA, Sage Publications Inc, 2010.

[8]   Kearns AJ. Catholic social teaching as a framework for research ethics. J Acad Ethics, Springer Science+Business Media, The Netherlands, 2014, 12, 145–59.

[9]   Cutcliffe SH, Pense CM, Zvalaren M. Framing the discussion: Nanotechnology and the social construction of technology–What STS scholars are saying. Nanoethics, Springer, USA 2012, 6, 2, 81–99.

[10]   Pontifical Council for the Laity. Documents. The Dignity of Older People and their Mission in the Church and in the World, 1998 (Accessed March 15, 2021, at http://www.vatican.va/roman_ curia/pontifical_councils/laity/documents/rc_pc_laity_doc_05021999_older-people_en.html).

[11]   Vinck V, Hubert H. Nanotechnologies. L'invisible révolution. Au-delà des idées reçues. Paris, France, Editions Le Cavalier bleu, 2017.

[12]   Barska A, Śnihur J. Senior as a challenge for innovative enterprises. Procedia Eng, Elsevier Ltd, UK, 2017, 182, 58–65.

[13]   European Commission, The Silver Economy. Final Report prepared by Technopolis group, Oxford Economics, UK, 2018 (Accessed September 9, 2020, at https://op.europa.eu/en/publi cation-detail/-/publication/a9efa929-3ec7-11e8-b5fe-01aa75ed71a1).

[14]   Mordini M. Ethical Recommendations. In: Mordini E, de Hert P, eds. Ageing and Invisibility. Amsterdam, The Netherlands, IOS Press bv, 2010, 195–218.

[15]   O'Brien M, Rogers W, Fisk A. Understanding age and technology experience differences in use of prior knowledge for everyday technology interactions. ACM J Trans Accessible Comput, Association for Computing Machinery, New York, NY, USA, 2012, 4, 2, Article 9, 1–27.

[16]   Ageism and Intergenerational Tensions in the Covid-19 Period. Press release of the French National Academy of Medicine April 18, 2020 (Accessed March 15, 2021, at http://www.aca demie-medecine.fr/wp-content/uploads/2020/04/Ageism-and-Intergenerational-Tensions-in -the-Covid-19-Period.pdf).

[17]   Meulenbroek L. Innovation and Aging. The Image about the Elderly user in the Smart Homes Sector. Utrecht, The Netherlands, LAP Lambert Academic Publishing, 2011.

[18]   Sandler R. Nano-Ethics. In: Guston DH, ed. Encyclopedia of Nanoscience and Society. Thousand Oaks, USA, Sage Publications Inc, 2010.

[19]   Institut für Kirche und Gesellshaft, Ethical Aspects of Nanotechnology, Statement of the Working Group of Commissioners for Environmental Affairs of the Protestant Church in Germany (EKD), Akzente, Texte-Materialen-Impulse, Iserborn Germany, EKD, 2007, n° 14.

[20]   Toumey C. Early voices for ethics in nanotechnology. Nat Nanotechnol, Nature Pub. Group, USA, 2019, 14, 304–05.

[21]   Papa Francesco Messaggio del Santo Padre Francesco in occasione della Giornata Internazionale delle Infermiere, 12 May 2020 (Accessed March 15, 2021, at http://www.vati can.va/content/francesco/it/messages/pont-messages/2020/documents/papa-francesco _20200512_messaggio-giornata-infermiere.html).

[22]   Modras R. Ignatian Humanism. A Dynamic Spirituality for the twenty-first Century. Chicago, USA, Loyola Press, 2004.

[23]   English J. Spiritual Freedom. From the Experience of the Ignatian Exercises to the Art of Spiritual Guidance. 2d ed. Chicago, USA, Loyola Press, 1995.

[24]   MagninTh, Revol F. Life in terms of Nano-biotechnologies. In: Evers D, Fuller M, Jackelén A, eds. Issues in Science and Theology: What is Life? Cham, Springer International Publishing, Switzerland, 2015, 57–65.

[25]   Jeevanandam J, Barhoum A, Chan YS, e.a. Review on nanoparticles and nanostructured materials: History, sources, toxicity and regulations. Beilstein J Nanotechnol, Germany, 2018, 9, 1050–74.

[26]   Pope Francis. Message of His Holiness Pope Francis to participants in the 108th session of the International Labour Conference, 10 June 2019 (Accessed March 15, 2021, at http://www.

vatican.va/content/francesco/en/messages/pont-messages/2019/documents/papa-francesco_20190610_messaggio-labourconference.html).

[27] Laperche B, Boutillier S, Djellal F, e.a. Innovating for elderly people: The development of geront'innovations in the French silver economy. Technol Anal Strateg Manage, Carfax, London, 2019, 31, 4, 462–76.

[28] Storck T. An Economics of Justice and Charity. Catholic Social Teaching. Its Development and Contemporary Relevance. Kettering OH, USA, Angelico Press, 2017.

[29] Spina A. Reflections on science, technology and risk regulation in Pope Francis' encyclical letter "Laudato Si'. Eur J Risk Regul, Cambridge University Press, UK, 2015, 6, 4, 579–85.

[30] Rahner K, Vorgrimler H. Theological Dictionary. Freiburg/ Montreal, Herder/Palm Publishers, 1965.

[31] Franssen M, Lokhorst G-J, van de Poel I. Philosophy of Technology. In: Zalta Ed, ed. The Stanford Encyclopedia of Philosophy. Standford, USA, Fall 2018. (Accessed April 29, 2021 at https://plato.stanford.edu/archives/fall2018/entries/technology/).

[32] Schummer J. Cultural Diversity in Nanotechnology Ethics. In: Allhoff F, Lin P, eds. Nanotechnology & Society. Dordrecht, The Netherlands, Springer, 2009, 265–80.

[33] Toumey C. Technologies and religions. Nat Nanotechnol, Nature Pub. Group, USA, 2015, 10, 826–27.

[34] de Broca A L'homme du XXIe siècle et l'homme au XXIe siècle. Quels défis pour le théologien chrétien? Revue d'éthique et de théologie morale, Editions du Cerf, France, 2012/HS, 271, 27–39.

[35] Papa Francesco. Al comitato nazionale per la bioetica, 28 gennaio 2016 (Accessed March 15, 2021, at http://www.vatican.va/content/francesco/it/speeches/2016/january/documents/papa-francesco_20160128_comitato-nazionale-bioetica.html).

[36] Toumey C. Seven reactions to nanotehnology. Nanoethics, Springer, New York, USA, 2011, 5, 251–67.

[37] Kte'pi B. Framework Programs. In: Guston DH, eds. Encyclopedia of Nanoscience and Society. Thousand Oaks, USA, Sage Publications Inc, 2010.

[38] Milford R, Wetmore JM, A New Model for Public Engagement: The Dialogue on Nanotechnology and Religion. In: Hays S, Robert J, Miller C, eds. Nanotechnology, the Brain, and the Future. Yearbook of Nanotechnology in Society. Dordrecht, The Netherlands, Springer, 2013, Vol. 3, 97–111.

[39] Hudock B. Faith meets World. The Gift and Challenge of Catholic Social Teaching. Missouri, USA, Liguori Editions, 2013.

[40] Schlag M. Handbook of Catholic Social Teaching : A Guide for Christians in the World Today. Washington DC, USA, Catholic University of America Press, 2017.

[41] Green BP. The Catholic church and technological progress: Past, present, and future. Herzfeld N. Relig New Technol, Special Issue, Religions, MDPI, Switzerland, 2017, 8, 15–31.

[42] Berube DM. How social science should complement scientific discovery: Lessons from nanoscience. J Nanopart Res, Springer, The Netherlands, 2018, 20, 120, 1–12.

[43] Malsch I. Governing Nanotechnology in a multi-stakeholder world. Nanoethics, Springer, USA, 2013, 7, 2, 161–72.

[44] Hongladarom S. Editorial: Nanoethics in the asian context. Nanoethics, Springer, USA, 2012, 6, 117–18.

[45] Pape François. Lettre Encyclique Laudato Si' du Saint-Père François sur la sauvegarde de la maison commune, 24 mai 2015 (Accessed March 15, 2021, at http://www.vatican.va/content/francesco/fr/encyclicals/documents/papa-francesco_20150524_enciclica-laudato-si.html).

[46] Jones S. Catholic Social Teaching: Opening the Chest. Jesuit Centre for Faith and Justice, Catholic Social Teaching in Action. Dublin, Ireland, Blackrock, co, 2004, 86–102.

[47]   General Introduction. In: Mordini M, de Hert P, ed. Ageing and Invisibility. Amsterdam, The Netherlands, IOS Press bv, 2010, 195–218.

[48]   Capolla N. Où se cachent les Nanos ? Démystifier les nanotechnologies. Montréal, Canada, Editions MultiMondes, 2016.

[49]   Toumey C. Later voices on ethics in nanotechnology. Nat Nanotechnol, Nature Pub. Group, USA, 2019, 14, 636–37.

[50]   Guèye T. Les nanotechnologies, par-delà l'"indéfinissabilité". PhilosophiaScientiae, Université de Nancy II, France, 2019, 23, 1, 19–37.

[51]   Margaret Rees M, Moghimi SM. Nanotechnology: From fundamental concepts to clinical applications for healthy aging. Nanomed Nanotech, NBM, Elsevier Inc., USA, 2012, 8, 1, S1–S4.

[52]   Wadhwa K, Wright D. A Survey of Technology for the Elderly. In: Mordini E, de Hert P, ed. Ageing and Invisibility. Amsterdam, The Netherlands, IOS Press bv, 2010, 143–72.

[53]   Corley E, Kim Y, Scheufele D. Scientists' ethical obligations and social responsibility for nanotechnology research. Sci Eng Ethics, Springer, The Netherlands 2016, 22, 111–32.

[54]   Pape Benoît XVI. Lettre Encyclique Caritas in Veritate, 2009 (Accessed March 15, 2021, at http://www.vatican.va/content/benedict-xvi/it/encyclicals/documents/hf_ben-xvi_enc_20090629_caritas-in-veritate.html).

[55]   Pape Jean-Paul Ier. Audience Générale, 6 septembre 1978, (Accessed March 15, 2021, at http://www.vatican.va/content/john-paul-i/fr/audiences/documents/hf_jp-i_aud_06091978.html).

[56]   Glandon MA. Justice and human rights: Reflections on the address of Pope Benedict to the UN. EJIL, European University Institute, Italy, 2008, 19, 5, 925–30.

[57]   Perrot R, Mazibuko Z. The Dynamics of New and Emerging Technologies in Developing Countries and the New Role of the State. The Mapungubwe Institute for Strategic Reflection (MISTRA) Beyond Imagination: The Ethics and Applications of Nanotechnology and Bio-Economics in South Africa. Johannesburg, South Africa, African Books Collective Projet MUSE, 2018, 1–16.

[58]   Drengson A. Shifting paradigms. From technocrat to planetary person1. Anthropol Conscious, American Anthropological Association, USA, 2011, 3, 22, 1, 9–32.

[59]   The Societal FF. Ethical implications of nanotechnology: A christian response. J Technol Stud, Epsilon Pi Tau, USA, 2006, 32, 1/2, 104–14.

[60]   Draper B. Should we fear the future of technology? Canadian Mennonite, April 2019, 12.

[61]   Herzfeld N. Introduction: Religion and the new technologies. Religions, Special Issue, MDPI, Switzerland, 2017, 8, 129, 1–3.

[62]   Toumey C. Religious reactions to new technologies. Nat Nanotechnol, Nature Pub. Group, USA, 2020, 15, 5–6.

[63]   Pontifical Academy for Life. Global Bioethics Working Group, p. 1. (Assessed March 15, 2021 at http://www.academyforlife.va/content/pav/en/projects/global-bioethics.html).

[64]   Pope Francis. Address of His Holiness Pope Francis to participants in the plenary assembly of the Pontifical Academy for Life, 25 June 2018 (Accessed March 15, 2021, http://www.vatican.va/content/francesco/en/speeches/2018/june/documents/papa-francesco_20180625_accademia-provita.html).

[65]   Statutes of the Pontifical Academy for Life. October 18th, 2016, Vatican, 2016 (Accessed March 15, 2021, at http://www.academyforlife.va/content/pav/en/projects/global-bioethics.html).

[66]   Cellule de réflexion bioéthique-réunion du 17 octobre 2006. Avis sur des questions bioéthiques posées par la nanomédecine; Technology at the service of Peace. How can the EU and its Member States address the (mis-)use of force through uncrewed armed systems? Brussels, 2019; Robotization of Life. Ethics in view of new challenges, Brussels, 2019:

Commission des Episcopats de la Communauté Européenne (COMECE) (Accessed March 15, 2021, at http://www.comece.eu/site/fr/home).

[67] Ethical and Religious Directives for Catholic Health Care Services. Washington, United States Conference of Catholic Bishops, 2018, 6th edition, 2018 (Accessed March 15, 2021, at https://healthlaw.org/resource/the-ethical-religious-directives-what-the-2018-update-means-for-catholic-hospital-mergers/).

[68] Pope John Paul II. Letter of His Holiness Pope John Paul II to the Elderly, October 1st, 1999 (Accessed March 15, 2021, at http://www.vatican.va/content/john-paul-ii/en/letters/1993/documents/hf_jp-ii_let_19990110_elderly.html).

[69] Pope Benedict XVI. Address of His Holiness Benedict XVI to participants in the plenary assembly of the Pontifical Council for the Family, 5 April 2008 (Accessed March 15, 2021, at http://www.vatican.va/content/benedict-xvi/en/speeches/2008/april/documents/hf_ben-xvi_spe_20080405_pc-family.pdf).

[70] Pope Benedict XVI. Visit to the Community of Sant'Egidio's home for the Elderly "Viva Gli Anziani" Words of His Holiness Benedict XVI Rome, Monday 12 November 2012 (Accessed March 15, 2021, at http://www.vatican.va/content/benedict-xvi/en/speeches/2012/november/documents/hf_ben-xvi_spe_20121112_viva-anziani.pdf).

[71] Statement by H.E. Archbishop Silvano M. Tomasi Permanent Representative of the Holy See to the United Nations and other international organizations in Geneva on the protection of the Human Rights of Elderly People, 27th ordinary session of the Human Rights Council, Monday, 15 September 2014 (Accessed March 15, 2021, at http://www.vatican.va/roman_curia/secretariat_state/2014/documents/rc-seg-st-20140915_elderly-persons_en.html).

[72] Pope Francis. Address of His Holiness Pope Francis to the participants in the seminar « The Common Good in the digital age » organized by the Dicastery for Promoting Integral Human Development (DPIHD) and the Pontifical Council for Culture (PCC), 27 September 2019 (Accessed March 15, 2021, at http://www.vatican.va/content/francesco/en/speeches/2019/september/documents/papa-francesco_20190927_eradigitale.html).

[73] Kotze M. The theological ethics of human enhancement: Genetic engineering, robotics and nanotechnology. Die Skrifig, South Africa, 2018, 52, 3, a2323.

[74] Appel éthique de Rome AI du 28 février 2020. Workshop international « Le "bon" algorithme? Intelligence artificielle, éthique, droit et santé », Vatican, Académie Pontificale pour la Vie, 26 au 28 février 2020 (Accessed March 15, 2021, at http://www.academyforlife.va/content/dam/pav/documenti%20pdf/2020/CALL%2028%20febbraio/AI%20Rome%20Call%20x%20firma_DEF_DEF_.pdf).

[75] Müller VC. Ethics of Artificial Intelligence and Robotics. In: Zalta Ed, ed. The Stanford Encyclopedia of Philosophy. Stanford, USA, Fall 2020. Edition. (Accessed March 15, 2021, at https://plato.stanford.edu/archives/fall2020/entries/ethics-ai/.

[76] Herzfeld N. Cybernetics and Religion. Oxford, UK, Oxford Research Encyclopedia of Religion, 2015. (Accessed March 15, 2021, at https://oxfordre-com.uml.idm.oclc.org/religion/search?siteToSearch=religion&q=cybernetics&searchBtn=Search&isQuickSearch=true).

[77] Pazgon E. Legal, and Social Implications of Nanotechnology. In: Brar SK, Zhang TCZ, Verma M, e.a. ed. Nanomaterials in the Environment. Reston, VA, USA, American Society of Civil Engineers (ASCE), 2015, 553–61.

[78] Ludescher Imanaka J. Laudato Si', Technologies of power and environmental injustice: Toward an eco-politics guided by contemplation. J Agric Environ Ethics, The Netherlands 2018, 31, 677–701.

[79] Jiménez AS, van Tongeren M. Assessment of Human Exposure to ENMs. In: Tran L, Bañares M, Rallo R, eds. Modelling the Toxicity of Nanoparticles. Advances in Experimental Medicine and Biology, Vol. 947. Cham, Springer International Publishing, Switzerland, 2017, 27–40.

[80]   Schülke A, Plischke H, Kohls N. Ambient Assistive Technologies (AAT): Socio-technology as a powerful tool for facing the inevitable sociodemographic challenge? PEHM, UK, 2010, 5, 8, 1–6.

[81]   Holthe T, Halvorsrud L, Karterud D. e.a. Usability and acceptability of technology for community-dwelling older adults with mild cognitive impairment and dementia: A systematic literature review. Clin Interv Aging, New Zeeland, 2018, 13, 13, 863–86.

[82]   Le Diest F, Latouille M. Acceptability conditions for telemonitoringgerontechnology in the elderly optimising the development and use of this new technology. IRBM, France, 2016, 37, 284–88.

[83]   Fiorini L, Esposito R, Bonaccorsi M. e.a. Enabling personalised medical support for chronic disease management through a hybrid robot-cloud approach. Auton Robots, USA, 2017, 41, 1263–76.

[84]   Al-Shaqi R, Mourshed M, Rezgui Y. Progress in Ambient Assisted Systems for independent living by the elderly. SpringerPlus, Open Access, USA, 2016, 5, 624, 1–20.

[85]   Otlatici G, Yeğen G, Güngör S. e.a. Overview on nanotechnology based cosmeceuticals to prevent skin aging. Istanbul J Pharm, Turkey, 2018, 48, 2, 55–62.

[86]   Sharma B, Sharma A. Future prospect of nanotechnology in development of anti-aging formulations. IJPPS, India, 2012, 4, 3, 57–66.

[87]   Sahu G, Sahu S, Sharma H. e.a. A review of current and novel trends for anti-ageing formulation. IJPCBS, India, 2014, 1, 118–25.

[88]   Fytianos G, Rahdar A, Kyzas G. Nanomaterials in cosmetics: Recent updates. Nanomaterials, Switzerland, 2020, 10, 979, 1–16.

[89]   Dalibert L. Façonnement du corps vieillissant par les technologies. Gérontologie et Société, CNAV, France, 2015, 37, 1, 148, 47–58.

[90]   Michel J-P. The future of geriatric medicine. Eur Geriatr Med, France, 2012, 3, 233–37.

[91]   Kurzweil R. Terry Grossman. Bridge to Life. In: West M, Fahy G, Harris S, eds. The Future of Aging. Pathway to Human Life Extension. Dordrecht, Springer Science-Business Media bv, 2010, 3–22.

[92]   Freitas R. Comprehensive Nanorobotic Control of Human Morbidity and Aging. In: West M, Fahy G, Harris S, eds. The Future of Aging. Pathway to Human Life Extension. Dordrecht, The Netherlands, Springer Science-Business Media bv, 2010, 685–805.

[93]   Allon I, Ben-Yehudah A, Dekel R, e.a. Ethical issues in nanomedecine: Tempest in a teapot? Med Health Care and Philos, EurSoc for Phil of Med and Health Care, The Netherlands, 2017, 20, 3–11.

[94]   Dangers physiques dans les aliments. Corps étrangers, fiche outil. Paris, France, Agence nationale de sécurité sanitaire de l'alimentation, de l'environnement et du travail (ANSES), 2013. (Accessed March 15, 2021, at https://www.anses.fr/fr/system/files/GBPH2013sa0170.pdf).

[95]   Pape Pie XII. Discours du Pape Pie XII à la 127e session du Conseil d'Administration de l'Organisation Internationale du Travail (OIT), 19 novembre 1954 (Accessed March 15, 2021, at http://www.vatican.va/content/pius-xii/fr/speeches/1954/documents/hf_p-xii_spe_19541119_lavoro.html).

[96]   Pope John Paul II. Address of the Holy Father John Paul II to the 18th international congress of the Transplantation Society, 29 August 2000 (Accessed March 15, 2021, at http://www.vatican.va/content/john-paul-ii/en/speeches/2000/jul-sep/documents/hf_jp-ii_spe_20000829_transplants.html).

[97]   John Paul P II. Address of John Paul II to members of the Pontifical Academy of Sciences, 23 October 1982 (Accessed March 15, 2021, at http://www.vatican.va/content/john-paul-ii/en/speeches/1982/october/documents/hf_jp-ii_spe_19821023_pont-accademia-scienze.html).

[98] Pape Jean Paul II. Message aux membres de l'Académie Pontificale des Sciences, 1er février 2005 (Accessed March 15, 2021, at http://www.vatican.va/content/john-paul-ii/fr/speeches/2005/february/documents/hf_jp-ii_spe_20050201_p-acad-sciences.html).

[99] Pape Pie XII Discours du Pape Pie XII à l'association des donneurs de cornée et à l'union italienne des aveugles, 14 mai 1956 (Accessed March 15, 2021, at.http://www.vatican.va/content/pius-xii/fr/speeches/1956/documents/hf_p-xii_spe_19560514_cornea.html).

[100] Prospects for Transplantation Scientific Aspects and Ethical Considerations. Vatican, Pontifical Academy For Life. Not dated (Accessed March 15, 2021, at http://www.pcf.va/roman_curia/pontifical_academies/acdlife/documents/rc_pa_acdlife_doc_20010926_xeno trapianti_en.html

[101] Jotterand F. Enhancement: The alteration of human nature. Nanoethics, The Netherlands, 2008, 2, 15–23.

[102] Balard F. Longévité et immortalité? Aux frontières de la gérontologie et au-delà . . . Gérontologie et Société. CNAV, France, 2016, 38, 3, 151, 9–19.

[103] Pape Jean-Paul II. Message du Pape Jean-Paul II aux participants au congrès promu par l'Académie Pontificale pour la Vie, 2 juillet 2001 (Accessed March 15, 2021, at http://www.vatican.va/content/john-paul-ii/fr/messages/pont_messages/2001/documents/hf_jp-ii _mes_20010702_pc-life.html).

[104] Pape Benoît XVI. Discours aux participants au congrès international sur le thème du don d'organes organisé par l'Académie Pontificale pour la Vie, 7 novembre 2008 (Accessed March 15, 2021, at http://www.vatican.va/content/benedict-xvi/fr/speeches/2008/november/docu ments/hf_ben-xvi_spe_20081107_acdlife.html).

[105] Waters B. Willful Control and Controlling the Will: Technology and Being Human. In: Herzfeld N, ed. Religion and the New Technologies, Special Issue, Religions, MDPI. Switzerland, 2017, 70–76.

[106] Pape François. Le travail est la vocation de l'homme. Homélie. Célébration matinale retransmise en direct depuis la chapelle de la maison Sainte-Marthe, 1er mai 2020 (Accessed March 15, 2021, at http://www.vatican.va/content/francesco/fr/cotidie/2020/documents/papa-francesco-cotidie_20200501_illavoro-primavocazione-delluomo.html).

[107] Pope Francis. Visit to the joint session of the United States Congress. Address of the Holy Father, 24 September 2015 (Accessed March 15, 2021, http://www.vatican.va/content/fran cesco/en/speeches/2015/september/documents/papa-francesco_20150924_usa-uscon gress.html).

[108] de la Rochefoucauld A. Introduction to the Documents of the Holy See. Working Paper. The humanization of Robots and the Robotization of the Human Person. Chambéry, Switzerland, Caritas in Veritate Foundation, 2013, 69–72. (Accessed March 15, 2021, at http://www.fciv.org/downloads/WP9-Introduction-Church-Texts.pdf.)

[109] Pape Paul VI. Audience Générale, 3 mai 1978 (Accessed March 15, 2021, at http://www.vati can.va/content/paul-vi/fr/audiences/1978/documents/hf_p-vi_aud_19780503.html).

[110] Pape Jean Paul II. Discours du Pape Jean-Paul II à des dirigeants syndicaux et à des chefs d'entreprise au lendemain du jubilé mondial des travailleurs, 2 Mai 2000 (Accessed March 15, 2021, at http://www.vatican.va/content/john-paul-ii/fr/speeches/2000/apr-jun/docu ments/hf_jp-ii_spe_20000502_workers-audience.html).

[111] Esmark A. Technocratic Revolutions: From Industrial to Post-industrial Technocracy. In: Esmark A, eds. The New Technocracy. Bristol, UK, Bristol University Press, 2020, 19–52.

[112] Vieillir en restant actif. Cadre d'orientation. Genève, Switzerland, WHO, 2002. (Accesssed March 15, 2021, at https://apps.who.int/iris/bitstream/handle/10665/67758/WHO_NMH_NPH_02.8_fre.pdf;jsessionid=B3FBC47F17FEBBF0EE5C589E6C141577?sequence=1).

[113] Burghardt WJ, SJ. Courage. Absence of fear or grace under pressure. Natl Cathol Report, Kansas City, USA, 2006, 14, 42, 12.

[114] Pope John Paul II. Address of John Paul II to the scientists of the Pontifical Academy of Sciences, 21 October 1985 (Accessed March 15, 2021, at http://www.vatican.va/content/john-paul-ii/en/speeches/1985/october/documents/hf_jp-ii_spe_19851021_pontificia-accademia-scienze.html).

[115] Pope Paul VI. Address of the Holy Father Paul VI to the participants in the international congress of toxicology, 6 September 1970 (Accessed March 15, 2021, at http://www.vatican.va/content/paul-vi/en/speeches/1970/documents/hf_p-vi_spe_19700906_congresso-tossocologia.html).

[116] Pope Francis. Address of His Holiness Pope Francis to participants in the plenary session of the Pontifical Academy of Science, 28 november 2016 (Accessed March 15, 2021, at http://www.vatican.va/content/francesco/en/speeches/2016/november/documents/papa-francesco_20161128_pontificia-accademia-scienze.html).

[117] Constitution dogmatique Lumen Gentium du 21 novembre 1964 (Accessed March 15, 2021, at http://www.vatican.va/archive/hist_councils/ii_vatican_council/documents/vat-ii_const_19641121_lumen-gentium_fr.html).

[118] Pope Francis. Address of His Holiness Pope Francis to Members of the Italian federation of the boards of nursing professions (FNOPI), Saturday, 3 March 2018 (Accessed March 15, 2021, at http://www.vatican.va/content/francesco/en/speeches/2018/march/documents/papafrancesco_20180303_ipasvi.html).

[119] Recommendations for the Ethical Conduct of Quality Improvement, [former] National Ethics Committee (NEC- US), Washington DC, USA, 2002. (Accessed March 15, 2021, at https://www.ethics.va.gov/pubs/necreports.asp).

[120] Punzi VL. A social responsibility guide for engineering students and professionals of all faith traditions: An overview. SciEng Ethics, Springer, The Netherlands, 2018, 24, 4, 1253–77.

[121] Toumey C. God in the lab. Nat Nanotechnol, Nature Pub. Group, USA, 2009, 4, 11, 696–97.

[122] Rialle V. Robotique humanitaire versus robotique suicidaire : ou comment ré-enchanter la Silver économie. Gérontologie et Société, CNAV, France, 2018, 3–4, 17–25.

[123] Barry R. Action for justice. Jesuit Centre for Faith and Justice, Catholic Social Teaching in Action, Blackrock co. Dublin, Ireland, The Columba Press, 2004, 78–82.

[124] Bible, Old Testament, Book of Tobit, 4, 15.

[125] Pope John Paul II. Address of John Paul II to the Representatives of the Christian Churches and Ecclesial Communities and of the World Religions, 27 Octobre 1986 (Accessed April 29, 2021, athttp://www.vatican.va/content/johnpaul-ii/en/speeches/1986/october/documents/hf_jp-ii_spe_19861029_religioni-non-cristiane.html).

[126] Bible, Gospel according to Matthew, chapter 7, verses 12, 22, 39; Gospel according to Luke, chapter 6, verse 31; Letters to the Romans, 13, 8; Letters to the Galatians, 5, 14.

[127] Bible, Gospel according to-Luke, chapter 6, verses 20 to 23, Gospel according to Matthew, chapter 25, verses 31 to 46.

[128] Centre National de Recherche Textuelle et Lexicale; V° « Béatitudes », Nancy, France, 2012 (Accessed March 15, 2021, at https://www.cnrtl.fr/definition/beatitudes).

[129] Pope Francis. Message of the Holy Father Francis for the International Day of Persons with Disabilities, 3 December 2019 (Accessed March 15, 2021, at http://www.vatican.va/content/francesco/en/messages/pont-messages/2019/documents/papa-francesco_20191203_messaggio-disabilita.html).

[130] Pope Francis. Visit to the joint session of the United States Congress. Address of the Holy Father, 24 September 2015 (Accessed April 29, 2021, at http://www.vatican.va/content/fran cesco/en/speeches/2015/september/documents/papa-francesco_20150924_usa-us-con gress.html).

[131] Kočandrle Bauer K, Štěch F, Kušnieriková M. Helpful models of theological, moral, and spiritual discernment in catholicism, protestantism, and orthodoxy. Theologica, Belgium, Catholic University of Louvain, Belgium, 2019 9, 2, 45–66.

[132] Centre National de Recherche Textuelle et Lexicale, V° « Exercice ». Nancy, France, 2012 (Accessed March 15, 2021, at https://www.cnrtl.fr/definition/exercice).

[133] O'Connor D, Myers J. Ignatian values in business and accounting education: Towards the formation of ethical leadership. J Bus Educ Leadersh, San Diego, CA, American Society of Business and Behavioral Sciences, 2018, 7, 1, 24–136.

[134] Beste JE. Integrating christian ethics with ignatian spirituality. Stud Christ Ethics, SAGE Publications, UK, 2020, 33, 1, 61–67.

[135] Oakes K. Spiritual direction for seekers: How the ignatian tradition can help those on the margin of the faith. America, 2019, 221, 12, 26. (Accessed March 15, 2021, at Online: Gale OneFile: CPI.Q. https://link-gale-com.uml.idm.oclc.org/apps/doc/A609856770/CPI?u= winn62981&sid=CPI&xid=2e2a95c0).

[136] de Loyola I. Exercices Spirituels. Paris, France, Desclée de Brouwer/ Bellarmin, 1986, coll. Christus n° 61.

[137] Miller KD. Discernment in management and organizations. J Manag Spiritual Relig, Routledge, UK, 2020, 1–22, (Accessed March 15, 2021, at. https://doi-org.uml.idm.oclc.org/ 10.1080/14766086.2020.1812425).

[138] Communauté de Taizé. Freedom: Is Everything that happens Decided by God in Advance? France, Taizé. No date. (Accessed March 15, 2021 at http://www.taize.fr/en_article3634. html).

[139] Centre National de Recherche Textuelle et Lexicale, V° « indifférent », V° « Indifférence ». Nancy, France, 2012, (Accessed September 8th, 2020, at https://www.cnrtl.fr/definition/in diff%C3%A9rent).

[140] Dyckman K, Garvin M, Liebert E. The Spiritual Exercises Reclaimed. Uncovering Liberating Possibilities for Women. New York, NY, USA, Paulist Press, 2001.

[141] Pape François. Le Maître du Temps. Méditation matinale en la chapelle de la maison Sainte-Marthe, 26 novembre 2013 (Accessed March 15, 2021, at http://www.vatican.va/content/fran cesco/fr/cotidie/2013/documents/papa-francesco-cotidie_20131126.html).

[142] Schweda M, Coors M, Bozzaro C. Introduction : Aging and Human Nature- Perspectives from Philosophical, Theological and Historical Anthropology. In: Schweda M, Coors M, Bozzaro C, eds. Aging and Human Nature. Switzerland, Springer Nature AG, 2020, 1–9.

[143] McDonough P. Ignatian Spirituality. In: McDonough P, Bianchi EC, ed. Ignatian Passionate Uncertainty : Inside the American Jesuits. Berkeley, CA, USA, University of California Press, 2002, 110–31.

[143] Pape François. Célébration matinale retransmise en direct depuis la chapelle de la maison Sainte-Marthe. Homélie du 11 mai 2020 (Accessed March 15, 2021, at http://www.vatican.va/ content/francesco/fr/cotidie/2020/documents/papa-francesco-cotidie_20200511_lospirito-donodidio.html).

[145] Pope Francis. Address of His Holiness Pope Francis to participants in the plenary session of the Pontifical Academy of Sciences, Monday, 12 November 2018 (Accessed March 15, 2021, at http://www.vatican.va/content/francesco/en/speeches/2018/november/documents/papa-francesco_20181112_plenaria-accademia-scienze.html).

[146] Bible, Gospel according to Matthew, chapter 10, verse 16.

[147] Douglass Warner K, Lieberman A, Roussos P. Ignatian pedagogy for social entrepreneurship: Twelve years helping 500 social and environmental entrepreneurs validates the GSBI methodology. J Technol Manag Innov, Universidad Alberto Hurtado Chile, 2016, 11, 1, 80–85.

Michael Vlerick

15 Calibrating the balance: The ethics of regulating the production and use of nanotechnology applications

Abstract: Nanotechnology (henceforth NT) is a rapidly advancing field with the potential of revolutionizing diverse areas such as electronics, health care, transport, and energy production. NT products and applications come with (potential) benefits and (potential) harms. The presence of potential harms calls for regulation. Both underregulation and overregulation – I argue – are morally undesirable. In the case of underregulation, stakeholders fall victim to the harmful effects of the technology. In the case of overregulation, stakeholders are deprived of the benefits of the technology. In this chapter, I identify the biases and factors that lead to underregulation and overregulation and offer solutions in response. More precisely, I argue that a lack of specific regulation, the presence of conflicts of interest and short-term economic incentives could lead to the underregulation of NT products and applications. Conversely, I argue that a negativity bias, harm aversion, the fear of opening "Pandora's box," and the intuition that what is natural is good and what is artificial (human-made) is bad could lead to overregulation. To avoid these pitfalls and the woes of underregulation and overregulation following in their wake, we need to set up a process – which I describe – in which policymakers and independent scientists closely collaborate.

Keywords: nanotechnology, underregulation, overregulation, conflicts of interest, economic incentives, negativity bias, harm aversion, laypeople intuitions

15.1 Introduction

Nanotechnology (henceforth NT) is a rapidly advancing field with the potential of revolutionizing diverse areas such as electronics, health care, transport, and energy production. An increasing number of nanoproducts (e.g., engineered cosmetics and food packaging) and NT applications(e.g., health-care applications for the diagnosis and treatment of diseases) find their way to the market. These products and technologies come with (potential) benefits and (potential) harms. The presence of potential harms

Acknowledgments: I want to thank my colleagues Dr. Stefaan Blancke and Dr. Alfred Archer for their insightful comments on a previous draft.

Michael Vlerick, Tilburg University, The Netherlands; University of Johannesburg, michaelvlerick@gmail.com, m.m.p.vlerick@uvt.nl

https://doi.org/10.1515/9783110669282-015

calls for regulation. At present there is no comprehensive legal framework regulating the production, commercialization, and use of NT applications. There is – as Mnyusi-walla, Daar, and Singer [1] have argued – an important gap between ethical reflection on and regulation of NT products and applications on the one hand and the development, production, and use of these products and applications on the other hand. The gap has only widened since they published their paper in 2003.

This does not mean that we have to start developing regulation from scratch. As Ebbesen and colleagues [2] have pointed out, existing regulations in areas such as biotechnology and genetics provide a good basis for developing the regulatory framework for NT. Moreover, many of the moral considerations developed in these fields will also apply to NT. Nevertheless, we cannot merely apply existing regulations to NT products and applications. NT – like other technologies such as the genetic modification of crops(GMO's) and CRISPR gene editing – is a *sui generis* and potentially disruptive technology [3] with its own set of benefits and risks. This calls for the development of an equally *sui generis* regulation.

In this chapter, I want to contribute to this important task. I will analyze which biases and other factors lead to morally undesirable regulation and suggest ways to overcome these. In Section 15.2, I argue that both underregulation and overregulation of NT products and applications are undesirable. In the case of underregulation, stakeholders fall victim to the harmful effects of the technology. In the case of overregulation, stakeholders are deprived of the benefits of the technology. We should therefore "calibrate the balance" and strive for an optimal trade-off between (potential) benefits and (potential) harms. In Section 15.3, I identify elements that lead to underregulation and solutions to avoid this. In Section 15.4, I do the same for overregulation. In Section 15.5, I conclude.

15.2 What kind of policy is morally desirable?

15.2.1 The problem with erring on the side of caution

When confronted with a novel kind of technology that might cause harm, policymakersare often tempted to err on the side of caution. The influential "precautionary principle" incites policymakers to take precautionary measures if the potential harm for human health and the environment is substantial, even if it is uncertain that the harm would materialize in the absence of precautionary measures [4, 5]. In this context, the EU has implemented a very stringent regulation for researching and producing GMO (genetically modified) crops to protect human health and the environment. Such a prudent approach – not willing to take any environmental or health risk – may seem morally desirable at first glance. After all, doesn't the policymaker have a moral duty to protect the environment and the health and well-being of all stakeholders?

Erring on the side of caution, however, can be morally undesirable. It can lead to overregulation which denies society of the benefits of a technology [6]. We should not only consider the potential harm a technology can cause but also its (potential) benefits. Overregulation is not inherently preferable to underregulation.A human life lost because of precautionary measures barring the use of NT applications (e.g., applications to cure cancer) should not weigh less than a human life lost because of unwanted side effects of using NT applications. In this context, the overly restrictive European GMO regulation was recently – and I believe rightfully – criticized by leading scholars at European institutions. In an open letter, they argue that the current regulation robs European societies of the many benefits of GMO research and production.[1]

In a similar vein, given the many and substantial benefits of NT applications, the regulator should not deny us these benefits by playing it *too* safe. They should trade-off (potential) benefits and (potential) harms with the aim of providing "the greatest amount of good for the greatest number," as prominent "utilitarian" moral philosophers such as Jeremy Bentham [7] and John Stuart Mill [8] have argued. Defending such a utilitarian normative approach against rival approaches – such as Kantian deontology – is beyond the scope of this paper. However, I dare hope that many readers will agree with the moral imperative is to take into consideration both benefits and harms and strive for regulation that serves the interest of all of the stakeholders – human and non-human organisms, present and future generations – as well as possible.

15.2.2 The difficulty of trading-off benefits and harms

Trading-off (potential and established) benefits and harms of new technological applications is by no means an easy feat. Firstly, we do not know the extent of some of the benefits and harms and the probability that they will materialize. A NT cancer treatment, for example, might turn out to be more or less effective than expected. This uncertainty is especially relevant when it comes to harms. Many of the potential harms ascribed to nanoproducts (e.g., their negative impact on the environment and on human health) are not firmly established. They are risks that might or might not materialize. To make matters worse, there are not only the "known unknowns" but also the "unknown unknowns." NT applications might have totally unexpected benefits (e.g., entirely new areas in which they can be used than the ones for which they were developed) and cause completely unexpected harms.

Secondly, even the trade-off between known benefits and harms is not straightforward. Such benefits and harms are not always quantifiable. It might not be possible to

---

1 The initiative for the open letter came from the renowned Max Planck Institute for Developmental Biology and was signed by researchers from 117 research facilities. The letter can be retrieved from: https://www.mpg.de/13748566/position-paper-crispr.pdf – Last accessed on 19th of May 2020.

quantify environmental and health-related benefits and harms. Moreover, even if these benefits and harms can be quantified (or if we can precisely estimate the extent of these benefits and harms), another major problem looms. When trading-off benefits and harms of a certain NT application, we are not only trading off benefits and harms on the same dimension. In many cases, we have to trade-off economic benefits and environmental costs (for example, if some useful applications could be "ecotoxic"); or we have to trade off health benefits (e.g., benefits provided by a NT cancer treatment) and social harms (such as increased inequality because the treatment is particularly expensive and only accessible to the wealthy). This trade-off along different dimensions is not specific to NT, but an inescapable part of most policy decisions.

Given that we are forced to compare apples and oranges (i.e., trade-off costs and benefits on different dimensions), given that we cannot quantify all harms and benefits (or even estimate their magnitude precisely), and given that we can never know (the full extent of) all harms and benefits, trading-off is not a matter of crunching the numbers. The policymaker must make value judgments. She must determine how much importance she accords to these different dimensions (i.e., to environmental integrity, human health, human wealth, social harmony, and so on) and trade-off the expected benefits and harms on these dimensions accordingly. Moreover, the policymaker must make decisions under uncertainty. Given the inherent uncertainty, policy decisions are particularly vulnerable to being biased. A number of biases and factors can lead to underregulation and overregulation, both of which are undesirable as pointed out above. The aim of this chapter is to identify these factors and biases and suggest ways to overcome them (or at least to keep them in check as much as possible).

15.3 Underregulation

15.3.1 Major causes of underregulation

The first cause of underregulation is a *lack of specific regulation*. This is our current predicament. There is no regulatory framework devoted to NT applications in place at the national or supra-national level. This is problematic, as I have pointed out in the introduction. Given the *sui generis*character of NT and the potentially major impact on health, environment, and economic development, the production and use of nano-products and NT applications should be regulated (over and above generic regulations – such as biotechnology or food and drug regulations – under which they now fall).The first step toward avoiding the harm that might arise because of underregulation is to develop specific regulation taking into accounts both the specific risks and benefits associated with NT. For this, we must overcome a "status quo bias" and the inertia that follows in its wake.

A second possible cause of underregulation is *conflict of interest.* When the regulator has a vested economic interest in the commercialization of nanoproducts and NT applications, or is strongly influenced by people who do, she might be tempted to ignore some of the risks and harms in order to bring the product to the market. The same goes for the scientists informing the policymaker (see below on the importance of the involvement of scientists). If they are on the payroll of a company producing NT applications, there is an important conflict of interest and they might downplay the risks and harms. This is nothing new. Scientists employed by companies manufacturing harmful products have acted as "merchants of doubt" to prevent precautionary measures against smoking, pollution, and environmental degradation [9].[2]

A third possible cause of underregulation is *short-term economic incentive.*When trading off economic benefits and environmental harms, nations might be tempted to tip the balance in favor of those economic benefits. The reason for that is that the economic benefits are reaped by the nation producing and exporting NT applications, while the environmental and health cost is often shared globally. Compare it with climate change. Nations have a short-term economic incentive to keep CO_2 emission high since reducing CO_2 emission requires investment in alternative technologies and/ or reducing industrial activities. The environmental cost of CO_2 emission, however, is shared globally.

When benefits are reaped individually and costs are shared, we face a "tragedy of the commons" [10, 11]. Imagine a pasture that can sustain a hundred sheep. It is used at full capacity by ten shepherds who each possess ten sheep. Any shepherd might nevertheless be tempted to add another sheep to his stock since this would mean a ten percent increase in wool revenue, while the cost of adding another sheep is marginal for the shepherd since it is shared by all shepherds (slightly less nutrition for all of the sheep). The fact that the shepherds have a short-term incentive to add more sheep (beyond the pasture's capacity) might ultimately lead to the demise of the pasture to the detriment of all shepherds. Similarly, a cost–benefit analysis in which individual nations trade off economic benefits and environmental costs might lead to the destruction of the common environmental good.

15.3.2 Solutions

In order to avoid underregulation we must first of all *develop specific policy* regulating the production and use of nanoproducts and NT applications. We must – as Mnyusiwalla and colleagues [1] put it – "close the gap" between regulation and

---

2 While conflicts of interest typically lead to underregulation, they might also distort policy in the opposite direction (leading to overregulation). For instance, the stringent European regulation banning meat that contains certain artificial beef growth hormones (which are deemed safe and approved in the US and Canada), might in part be attributable to protectionist economic policy.

rapidly evolving technological innovation. In order to do so, it is of great importance that the policy is drafted by independent parties and informed by independent researchers to avoid conflicts of interest.Those policymakers should be free of undue influence or pressure exerted by lobbying parties (which could be the case, for instance, if they received financial support from industrial players with an economic interest in the products they are regulating).

A good democratic solution to avoid such conflicts of interests is to have a policy drafted by deliberative citizen councils. Such councils are composed of a representative sample of the population (often appointed by sortition). After being thoroughly informed by experts, they deliberate with the aim of reaching a high degree of consensus around certain policy proposals. Not only are the appointed citizens free from pressure by interest groups, they are also free to consider the long term (in contrast to elected politicians who might be reticent to impose measures which are costly in the short term for fear of losing the next elections). Moreover, social experiments with such citizen councils show that they often come up with thoughtful and sensible proposals [12–15].

Finally, to avoid a "tragedy of the commons," there should be multilateral – ideally global – agreement on and coordination of NT regulation. If nations draft and implement regulation independently, they might tip the balance in favor of their short-term economic interests at the detriment of the (shared) environment. Global coordination on policy, however, is by no means an easy feat. Think of the difficulty to reach a global agreement on climate policy [14, 15]. Supra-national and global institutions – such as the United Nations (UN), the World Health Organization (WHO), and the World Trade Organization (WTO) – have a crucial role to play in facilitating such an agreement. While some global institutions are concerned with the ethical implications of NT and have made recommendations – the WHO drafted a document in 2012 on NT and human health reviewing the extant evidence of possible harms and advising a precautionary approach[3] – this involvement should increase. If we leave policymaking to nations, we risk underregulationand the mass production of certain (potentially) harmful applications.

15.4 Overregulation

15.4.1 Major causes of overregulation

A number of important biasesand unfounded intuitionscan lead to overregulation. These biases explain why policymakers sometimes take overly precautionary measures

---

[3] See: https://apps.who.int/iris/bitstream/handle/10665/108626/e96927.pdf;jsessionid= 5F33A5D5177F8C8EE2A78D1E938AA15A?sequence=1 – Last accessed on 19th of May 2020.

when faced with new technologies (e.g., the GMO regulation of the EU) and why a rigid application of the precautionary principle – urging policymakers to eliminate risks at all costs (without considering the potential benefits – see [6, 16, p. 4], but also [17] for a reply) is so influential. A first important bias contributing to overregulation is the *negativity bias*. As Rozin and Royzman [18] explain, expected negative outcomes generally create a stronger affective response and therefore tend to weigh more on our decisions than expected positive outcomes. Rozin and Royzman [18, p. 298] refer to this as "negativity dominance." From an evolutionary perspective this makes sense, since paying attention (and remembering) threats generally increases an organism's chance of survival and reproduction more than paying attention to gains and benefits. This evolved bias explains why policymakers might accord too much weight to potential harms and costs and undervalue benefits when trading-off.

A second psychological feature that contributes to overregulation is *harm aversion*. Most humans are endowed with an aversion to harm others, especially if those others are considered in-group members. Cushman and colleagues [19] found that even "pretend" harmful acts, such as pointing a fake gun at somebody, causes strong physical reactions of aversion (namely increased peripheral vasoconstriction) (see also [20]).The moral aversion we feel for harmful acts is stronger than the aversion we feel for harmful omission (compare throwing somebody in a pond and causing that person to drown with refraining from saving a drowning person even if there's absolutely no personal risk or cost involved in doing so). Whether or not this intuition is justified – many moral philosophers have argued that – it is beyond the scope of this chapter.

Harm aversion leads policymakers to accord more weight to negative effects brought about by allowing the production of NT products or use of NT applications (which constitutes a harmful action) than by negative effects brought about by not allowing the production and use of such products and applications (which constitutes a harmful omission). Moreover, given that the public shares this intuition and that harmful actions are more conspicuous than harmful omissions, the former is typically (much) harder to justify to the public than latter (which often remains unspecified). Therefore, the prospect of people dying from nanoparticle intoxication, for example, is likely to be a stronger concern for policymakers than (the same number of) people dying because a medical NT application was not made available. This too leads to overregulation since it moves the policymakers to accord more importance to (potential) harms – which if ignored or underestimated would cause harmful actions – than to (potential) benefits – which if ignored or underestimated would cause harmful omission.

A third aspect that can lead to overregulation is the fear of "opening Pandora's box." The idea is that once we allow the development and production of disruptive technologies, there is no way back and we open the door to the many harms they might cause. As Davis and Macnaghton [21] point out, this is a prominent lay narrative. It captures the fear we have of unknown risks and harms. To a certain extent this fear is warranted. As indicated above, there are unknown risks, and the policymaker should be aware of this and proceed cautiously. However, she must also take the unknown benefits into account. The fact that we tend to focus on the unknown risks is a product of the first bias we discussed: our evolved negativity bias.

There is however an added dimension to the fear of opening Pandora's box. It also assumes that if we allow the production and distribution of some NT applications that are deemed safe, we irrevocably open the door for other NT applications that could be harmful. This however need not be the case. We can – and should – regulate the production and use of NT applications on a case-by-case basis. Allowing one kind of application deemed safe, does not mean that other applications should be allowed without further consideration. Furthermore, while it is true that, given the presence of unknown risks, the policymaker might mistakenly allow the production of harmful NT applications, this does not mean that we cannot put these applications "back in the box." In fact, we frequently managed to put harmful technological innovations back in the box. We banned the use of toxic asbestos in building materials and CFC (chlorofluorocarbons) – which turned out to damage the ozone layer – in aerosols. There is no reason we would not be able to do so with NT applications that turn out to cause unexpected harm.[4]

A final unfounded but exceptionally tenacious bias which could lead to the overregulation of NT applications is the view that what is "natural is good and what is artificial (human-made) is bad." Nature, according to this view, should not be messed with. If we alter natural substances artificially (e.g., by altering matter at the nano-level) we invite all sorts of harms and hazards upon us. This widespread intuition (among laypeople) has turned the public opinion (and the European regulators) against GMO crops, despite the many scientific studies establishing that GMO crops are safe to consume [22]. While NT has not evoked a popular opposition of this magnitude (yet), the same fear of "messing with nature" applies and has been observed in lay narratives on NT [21]. This might lead to a backlash against NT and move policymakers to overregulate. A proper assessment of the risks of technological innovations (and a proper trade-off between benefits and risks) should be based on evidence not on intuition.

15.4.2 Solutions

In order to prevent overregulationas a result of these psychological biases and unfounded intuitions, we must ensure that regulation is based on the best available

---

4 It is of course possible that the fear of opening Pandora's box is justified. An application could cause strong and irreversible harm if it is produced and distributed (even if it is banned shortly after). If this doom scenario is at all likely, this should of course tip the balance in favor of strong precautionary measures. Remember that the policymaker should not only consider known benefits and harms but also take into account the probability of unidentified benefits and harms materializing when she trades off benefits and harms.

scientific evidence. This in turn requires close collaboration between policymakers and (independent) scientists. Scientific experts can provide the evidence and they are less likely to succumb to some of the unfounded intuitions (especially the view that what is natural is good and what is artificial is bad). We must also adhere to a formal process of decision-making.

Such a *process* would consist of four important steps. First, the (potential) benefits and (potential) harms must be identified. This should be done on a case-by-case basis, since different nanoproducts and NT applications come with different benefits and harms. Second, the positive and negative impacts on different dimensions – such as human health, the environment, and economic development – must be estimated as precisely as possible (although quantifying these impacts will not always be possible, as mentioned above). Third, the policymakers must trade-off (possible) benefits and (possible) harms with the aim of producing the best possible outcome for all of the stakeholders. It is vital that they are aware of the biases (described above) that might distort their judgment and that they should attempt to eliminate these to the best of their ability. Fourth, after the policy has been implemented, its effects must be closely monitored. New insights are bound to arise once the products and applications are distributed and used. These new insights should then inform new measures (e.g., relaxing the regulation if it seems safe to do so or making the regulation more stringent if unexpected harms occur).

In order to make the best trade-off in the light of rapidly evolving scientific insights and new data, we should strive for a *dynamic– not a static – regulation*. Regulation should regularly be updated in the light of advancing knowledge. This should also alleviate worries of opening up Pandora's box. It is not because we allow the production and use of nanoproducts and NT applications based on current and flawed information that we cannot retract this in the future when unexpected harms turn up. Updating regulation regularly protects us against suffering from the woes of underregulation and overregulation for extended periods of time.

Throughout the entire process, the *involvement of scientific experts* is vital. The input of independent scientific experts is indispensable in every step of the process. It is obviously necessary to identify (possible) benefits and (possible) harms of a novel NT application (step 1), to estimate their impact on different dimensions (step 2) – which requires the input of scientific experts in these different relevant fields – and to monitor the effects of the regulation to adapt it when necessary (step 4). But even in step 3, the actual drafting of the regulation by policymakers, the involvement of scientific experts is important.

This does not mean that the regulation should be drafted by scientists – this would amount to technocratic decision-making – only that the democratically appointed policymakers should be closely assisted and informed by scientific experts in all relevant fields (i.e., in all dimensions on which the applications might produce benefits and harms) while making their decisions. This is necessary in order to prevent that the policymakers from falling prey to some of the biases and unfounded intuitions discussed above.

Of course, this will not eradicate all biases from the process. Policymakers might decide to ignore the advice of scientists. Moreover, whereas scientists will be less likely to succumb to some of these biases and unfounded intuitions, they are not immune to biased reasoning. Nobody is. They might, for instance, ignore or underestimate some of the risks in their enthusiasm for the potential of NT applications, especially if they have contributed to developing the application. Conversely, some experts might play it too safe because of a negativity bias or because of harm aversion. Both policymakers and assisting scientific experts should therefore be (made) aware of the biases that might affect their judgment in order to eliminate them to the best of their ability.

15.5 Conclusion

NT is developing at a swift pace. This is good news. Technological innovation often leads to an increase in human well-being. It leads to longer lifespans, living in better conditions. NT is no different. It promises to contribute substantially to better, longer lives and to a better environment for present and future generations. However, the rapid scientific progress in NT also presents us with an important challenge. Given that nanoproducts and NT applications come with risks, their production, distribution, and use should be regulated.

The task that faces the regulator (and the scientists informing and assisting the regulator) is considerable. Not only must they regulate on a case-by-case basis – given that the risk vs benefit profile of different applications can vary widely – they must also frequently revise regulation in the light of the latest findings (and the monitoring of the effects of the policy). In addition to being time-consuming, their task is a difficult one. Given the possibility of unknown benefits and harms materializing and the necessity to trade off benefits and harms on different dimensions, deciding on the optimal trade-off is no exact science. It is therefore especially vulnerable to biased reasoning.This is where the importance of a processand the involvement of experts come in. The purpose of such a process is to ensure a "calibrated balance" on which benefits and harms are weighed. Only with a properly calibrated balance (and the right scientific input) can we avoid the woes of under-regulation and overregulation.

References

[1]   Mnyusiwalla A, Daar AS, Singer PA. Mind the gap: Science and ethics in nanotechnology. Nanotechnology 2003, 14, R9–R13.
[2]   Ebbesen M, Andersen S, Besenbacher F. Bulletin of science. Technol Soc 2006, 26, 6, 451–62.

[3] Lu L, Lin B, Liu J, Yu C. Ethics in nanotechnology: What's being done? What's missing? J Bus Ethics 2012, 109, 4, 583–98.

[4] Sandin P. Dimensions of the precautionary principle. Hum Ecol Risk Assess 1999, 5, 889–907.

[5] Sandin P. Common-Sense Precaution and Varieties of the Precautionary Principle. Lewens T Ed., Risk: Philosophical Perspectives. Routledge, 2007, 99–112.

[6] Sunstein C. Laws of Fear: Beyond the Precautionary Principle. Cambridge (UK), Cambridge University Press, 2005.

[7] Bentham J. An Introduction to the Principles of Morals and Legislation. New York, Prometheus Books, 1789.

[8] Mill JS. Crisp R ed. Utilitarianism. Oxford, Oxford University Press, 1863.

[9] Oreskes N, Conway E. Merchants of Doubt: How a Handful of Scientists Obscured the Truth on Issues from Tobacco Smoke to Global Warming. New York, Bloomsbury Press, 2010.

[10] Hardin G. The tragedy of the commons. Science 1968, 162, 1243–48.

[11] Vlerick M. The evolution of social contracts. J Soc Ontol 2020a, 5, 2, 181–203.

[12] Fishkin J. When the People Speak: Deliberative Democracy and Public Consultation. Oxford, Oxford University Press, 2009.

[13] Steiner J. The Foundations of Deliberative Democracy: Empirical Research and Normative Implications. Cambridge, MA, Cambridge University Press, 2012.

[14] Vlerick M (2019). De tweede vervreemding: Het tijdperk van de wereldwijde samenwerking. Tielt, Belgium, Lannoo.

[15] Vlerick M. Towards global cooperation: The case for a Deliberative Global Citizens' Assembly. Global Policy 2020b, 11, 3, 305–14.

[16] Bodansky D. Scientific uncertainty and the precautionary principle. Environment 1991, 33, 4–5, 43–44.

[17] Sandin P, Peterson M, Hansson S, Ruden C, Juthe A. Five charges against the precautionary principle. J Risk Res 2002, 5, 287–99.

[18] Rozin P, Royzman EB. Negativity bias, negativity dominance, and contagion. Pers Soc Psychol Rev 2001, 5, 4, 296–320.

[19] Cushman F, Gray K, Gaffey A, Mendes W. Simulating murder: The aversion of harmful action. Emotion 2012, 12, 1–7.

[20] Vlerick M. Better than Our Nature? Evolution and Moral Realism, Justification, and Progress. Ruse M, Richards R eds, Handbook of Evolutionary Ethics. Cambridge University Press, 2017, 226–39.

[21] Davies SR, Macnaghten P. Narratives of mastery and resistance: Lay ethics of nanotechnology. Nanoethics 2010, 4, 141–51.

[22] Blancke S, Van Breusegem F, De Jaeger G, Braeckman J, Van Montagu M. Fatal attraction: The intuitive appeal of GMO opposition. Trends Plant Sci 2015, 20, 7, 414–18.

Nobuyuki Haga
16 Importance of social morals in nanotechnology

Abstract: Nanotechnology is a field of advanced science that flourished in the twenty-first century. Although nanotechnology has deeply penetrated into modern human life, it has not been well-discussed about its concern for human safety, its negative impact on ecosystems, and its impact on economic globalization. Setting a global standard for assessing the toxicity of nanomaterials is an urgent task. For that purpose, it is important to clarify the definition of nanomaterials and select model organisms that serve as the evaluation criteria for toxicity. Ethical considerations regarding the manufacture and use of nanomaterials are essential to ensure competition that guarantees coexistence between nations. Nanoethics is responsible for establishing a universal code of conduct for these issues. Nanoethics is an area where we work together in a language common to science and philosophy. Nanoethics will help to enhance our code of conduct by applying complementary relationships between science and philosophy.

Keywords: Nanomaterial, globalization, manufacture, moral, bioassay, cytotoxicity

16.1 Introduction

The nanomaterials considered in this book are carbon nanotubes, titanium oxide, silver particles, gold particles, and so on. If we define nanoethics as the norm of behavior when applying nanotechnology to our lives, how should we assess whether our behavior is right or wrong? We believe that assessing the toxicity of nanomaterials is a top priority for establishing a code of conduct.

We will clarify the subject of nanoethics and examine the usefulness of unicellular eukaryotes, *Paramecium*, as one of the candidate model organisms. We will overview the development of nanotechnology in Japan and extract the requirements necessary for future international cooperation.

16.2 The central dogma of molecular biology

The central dogma of molecular biology provides the basis for a unified understanding of the expression of genetic information and the control of life phenomena. The

Nobuyuki Haga, Department of Biological Sciences, Ishinomaki Senshu University, Miyagi, Japan

https://doi.org/10.1515/9783110669282-016

central dogma is a robust system that enables prediction and evaluation of the characteristics of life that will be inherited by the next generation.

To summarize the central dogma, the genetic information stored in DNA is transferred to RNA by the transcription process and reflected in the amino acid sequence of proteins by the translation process. In addition, genetic information is transferred from RNA to DNA by reverse transcription, and the information relating to the three-dimensional protein structure is transferred between protein molecules of the same type in prion proteins. All these steps are potential targets for nanomaterials to act.

16.3 Three categories in life: material, information, and energy

In modern biology, it is assumed that the characteristics of living organisms are divided into three categories: material, information, and energy. Although these categories are closely related to each other, they are studied in independent analytical ways, so the data obtained from the experiments are considered independently.

16.3.1 Material and information: nucleotides to protein

The relationship between material and information appears between nucleic acids and proteins. The central dogma of molecular biology has guided the formation of a new paradigm in biology as a brief expression of the flow of genetic information by F. Crick [1]. The central dogma is described as follows: genetic information flows in one direction from DNA to RNA to protein. Later, the concept of the central dogma was expanded by the discovery of reverse transcriptase, which revealed the process by which genetic information flows from RNA to DNA [2]. Furthermore, an information transduction pathway without the involvement of nucleic acids was discovered during the formation of the three-dimensional structure of proteins. The protein that performs this action was named "Prion" and was established as a new category in the transduction of genetic information [3, 4].

16.3.2 Information and energy: combined in ATP molecule

The relationship between information and energy is embodied in the function of adenosine triphosphate (ATP). ATP is used as one of the bases encoding the amino acid sequences of proteins in the genetic information system. The base sequence information is converted into amino acid sequence, and the amino acid sequence defines the function of the protein. Therefore, the "Meaning" conveyed by the genetic information is "Protein function." One of the main functions of proteins is energy metabolism.

16.3.3 Energy and material: from glucose to ATP

The relationship between energy and material appears in the covalent bond between atoms in the compounds that compose the living body. The binding energy used for the covalent bond of glucose is released by breaking the covalent bond and converted into a form usable for other chemical reactions. The available energy generated in this process is stored in the ATP molecule. Thus, ATP also plays a central role in the relationship between material and energy.

16.4 Nanomaterials and nanotechnology

As the size of materials decreases to the nanometer level, they change shape and increase the ratio of surface area to mass. This causes changes in the degree of dispersion in the solution and the characteristics of surface chemistry. When these changes act on life phenomena in an inhibitory manner, it is recognized as toxicity.

There are various processes in the life phenomenon that can be targeted by nanomaterials and cause toxicity.
1) Enzyme–substrate interaction
2) Biopolymer formation or degradation
3) Antigen–antibody interaction
4) Antibody production
5) Energy metabolism
6) Sexual cell recognition
7) Gametogenesis
8) Supramolecular assembly
9) Membrane function and transmembrane signal transduction
10) Generation of ions and various radicals

In the *Paramecium* cytotoxicity assay system, we have detected the toxicities of nanomaterials in (2), (5), (7), (9), and (10) categories.

In general, nanotechnology is defined as a technology that includes all three items below.
1. The materials studied are those that have at least one dimension in the range 1–100 nm and are related to the studies of atomic and molecular structure.

2. Create and use structures, devices, and systems with properties and functions due to their nano size.
3. Technology and research related to atomic scale control and operation functions.

16.5 Definition of toxicity

Toxicity is defined as the following description: the degree to which a single chemical substance or a specific mixture of substances can damage an organism.

The central concept of toxicology is that toxic effects are dose-dependent. Dose dependency appears in the form of a continuously changing effect in response to changes in the concentration of the substance. Toxicity generally appears as a characteristic feature of each species. Therefore, the results of laboratory animals used to assess human safety must be carefully considered. A new drug toxicity index (DTI) has recently been proposed [5]. DTI redefines drug toxicity, provides mechanistic insights, and predicts clinical outcome. It also has potential as a screening tool.

16.6 Ethics in nanotechnology

One of the issues that nanotechnology must consider is social morals. This is because nanotechnology has the potential to revolutionize human life, but on the other hand, the toxicity caused by nanomaterials has not been fully verified. The consideration of social morals and ethics derived from the development of nanotechnology will be called nanoethics. The problems handled by nanoethics can be divided into three categories. The first category relates to individual rights in society, the second relates to interactions between states in the world, and the third relates to interactions between humans and other organisms.

Companies that manufacture nanomaterials are subject to significant personal moral issues. Nanomaterials can be freely manufactured at the individual level and can be incorporated into various products either alone or mixed with other materials. Therefore, the initial considerations for side effects such as toxicity and addiction of nanomaterials depend on the individual morals of nanomaterial manufacturers.

Moral agreements discussed between countries impose legal restrictions on trade and pollution. Today, nanotechnology has become one of the key strategic challenges of the national economy. Therefore, there is a need for a moral way of thinking, which is the basis of policymaking between the raw material producing countries of nanomaterials and the processed product producing countries. Moreover, the treatment of environmental pollutants associated with the production of nanomaterials requires appropriate legal regulation between countries.

Human economic activity affects living organisms that share the same environmental factors. It is a great challenge to continue economic activities without compromising the survival conditions of these organisms. There are other situations in which humans have to consider other organisms that should be considered as a matter of relying on the ethics of the individual rather than maintaining morality under domestic laws and regulations. These include food chains in the ecosystem, laboratory animals used to assess human safety, and domestic pets.

From the early 2000s, the concept of "responsible research and innovation" has come to be adopted in the research field. At the present time, it is generally accepted that the unique health and environmental effects of nanomaterials have not been scientifically determined. Scientists are gradually recognizing that there may be no biological mechanism of action specific to nanomaterials. In recent years, there has been growing interest in developing evaluation methods that clarify the characteristics and risks of nanomaterials individually.

Next, we will consider the standardization of rules for evaluating the safety of nanomaterials. Since standards for safety evaluation of nanomaterials have not been established between manufacturers and users, it is not possible to set standards for describing information on nanomaterials used in products. The nanomanufacturing industry should work with academic societies to set standards for safety evaluation methods for nanoproducts and to establish an organization that ensures the safety of nanoproducts. Creating a simple computer program that uses human tolerance and nanoproduct exposure data to calculate the risk of nanoproducts helps to increase nanoethics in the nanomanufacturing industry.

We believe that nanotechnology-related social morals are facilitated by the establishment of a system in which the nanomanufacturing industry cooperates with government authorities to carry out risk assessments and communicates the assessment results to society. The government must create guidelines for nanomanufacturing industries, including items regarding sample safety evaluation requirements, product sales requirements, and in-house research, development and manufacturing handling. The government should work with the International Organization for Standardization to establish safety standardization for nanoproducts.

It is important to establish standardization of scientific research methods to obtain practical and convenient safety data of nanoproducts in public institutions using known toxic substances as models. We believe it is important to establish a public certification system for the data created at each business site and build a database on the safety of nanoproducts. We also believe that an information disclosure system that meets customer requests and the dissemination of safety information in cooperation with the media will improve social ethics. The information display of the nanomaterials contained in the product and the limited condition display of the safety guarantee help users' safe use.

The government needs to set up countermeasures that introduce a preventive perspective, and should strive for early disclosure of legal regulations. In addition, as an important role of the government, it is necessary to collect information on the direction of regulation from the global perspective of nano products and to clarify and present the international regulation range.

16.7 *Paramecium* as a bioassay system to elucidate the safety of nanomaterials

16.7.1 *Paramecium*

Paramecium is a unicellular organism belonging to the eukaryote and was first described by A. Leeuwenhoek of the Netherlands at the end of the seventeenth century. The English name of *Paramecium* is "Slipper animalcule" because of the cell morphology. The cells are spheroids that are lightly twisted, and the head and tail are close to the shape of a triangular pyramid. The lengths are 150 to 180 µm and the width are about 50 µm. There is a depression called the cytopharynx (a tube-like opening in the lateral surface of the cell) near the center of the cell body, and this structure is used to ingest various substances from the outside into the cell [6]. The cilia grow almost uniformly in the whole cell, but they have a special arrangement behind the cytopharynx and the cilia are used to take food into intracellular food vacuoles. The food taken up by the food vacuole is digested, and the useful component is absorbed into the cell (Figure 16.1). The food vacuole travels inside the cell along with cytoplasmic flow, and the undigested substances are released from the cell anus at the back of the cell.

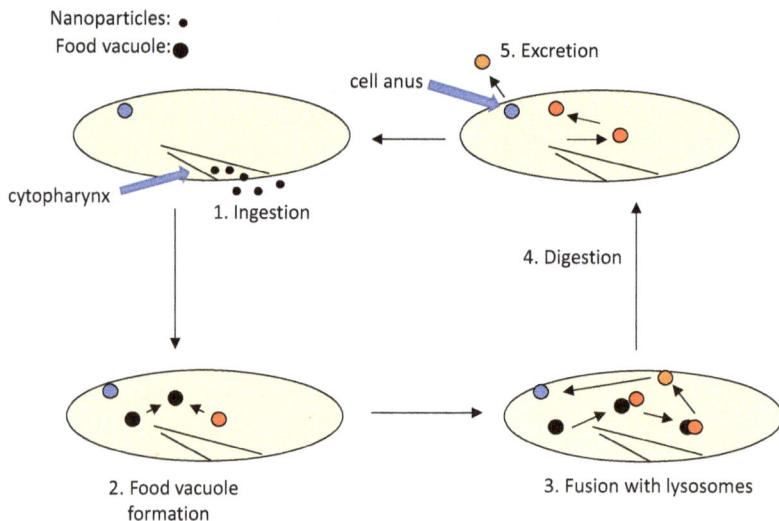

Figure 16.1: Schematic diagram showing the flow from capturing nanoparticles to excretion.

Living *Paramecium* cells can be observed up to magnification 40 times under a stereoscopic microscope and up to 1,000 times under a biological microscope (differential interference microscope, polarization microscope, and fluorescence microscope). In addition, by using a microinjection device, any substance such as nanomaterials, DNA, RNA, and protein can be injected into cells if dissolved or dispersed in water [7–9].

Paramecium can be easily cultured in the laboratory, and many culture methods have been reported so far. Depending on the purpose of the experiment, we can select and culture from the following five culture scales: Erlenmeyer flask (1.5 L culture medium), test tube (20 mL culture medium), handmade Pasteur pipette test tube (2 mL culture medium), slide glass with holes (500 µL culture medium), and glass capillaries (50 µL culture medium). The lettuce juice culture method enables high-density culture of 1×10^4 cells/mL [10]. A genetically homogeneous cell population can be prepared in a few days by isolating and culturing a single cell.

The sexual reproductive process of *Paramecium* is called conjugation (Figure 16.2) [10]. Individual cells have a mating type defined by an allelic pair which have a dominant/recessive relationship and are transmitted to the progeny according to Mendelian law [10]. Mating types are usually designated by the E and O symbols, where E is predominant over O. The conjugation is initiated by sexual contact between E-type and O-type individuals.

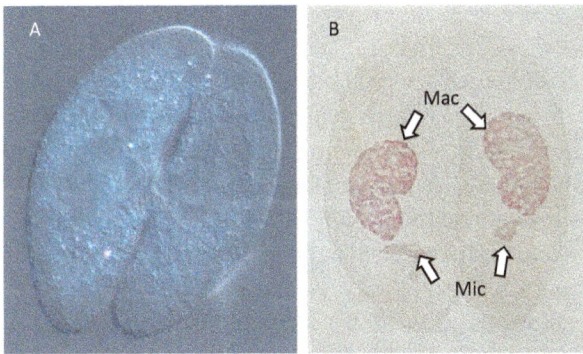

Figure 16.2: *Paramecium* in conjugation. The conjugation performed by *Paramecium* is homologous to sexual reproduction in multicellular animals. In mating cells, the germinal nucleus undergoes meiosis to form two gamete nuclei. They receive one gamete nucleus from the mating partner and fuse it with their own gamete nucleus to create a new generation of nucleus. (A) Mating pair photographed with a differential interference microscope. (B) Nuclei stained by the Orcein acetate method. The large red nucleus is called the macronucleus (Mac), and the genetic information necessary for life is supplied from this nucleus. The small nucleus below the macronucleus is the germ nucleus, called micronucleus (Mic) and is preparing to enter meiosis.

There are two breeding methods in the life history of *Paramecium*, asexual reproduction and sexual reproduction. Asexual reproduction increases the population by cell division and sexual reproduction forms the next generation. The life history of

Paramecium begins with conjugation (Figure 16.3). The progeny cells produced by conjugation increase the number of individuals by repeating asexual cell division, and the sexual function changes from immature to mature to senescence depending on the cumulative number of divisions. The aging phenomenon appears in many cell functions such as sexual activity and cell-division rate [11, 12]. This phenomenon, called clonal aging, follows a process very similar to the life history of multicellular organisms and eventually ends in clonal death.

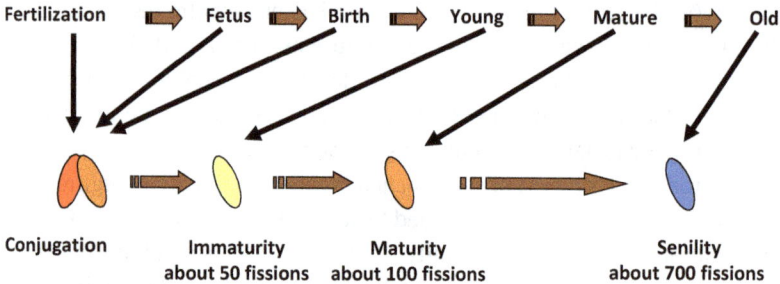

Figure 16.3: Schematic diagram comparing the biologically common features of human and *Paramecium* life history. The upper line shows changes in humans and the lower line shows changes in *Paramecium*. Black and slender arrows correspond to stages in which common functions are expressed.

In sexual reproduction, two individual cells of complementary mating type conjugate to build a new generation of genome. This resets the divisional age of the progeny to zero and restarts life history as a new generation. Therefore, in the life history of *Paramecium*, sexual reproduction creates a new generation of individuals, and asexual reproduction promotes sexual maturation, which in turn leads to an individual's readiness for the next generation. If an individual does not have the chance to mating, it will die due to senescence.

The average number of divisions per day can be calculated by transferring one cell to a glass slide with a hole containing fresh culture medium and counting the cell number after 24 h. By repeating this single-cell culture method every day, the expression pattern of cell functions necessary for sexual reproduction can be positioned in the life history. This method can be used to create an experimental system to monitor the effects of nanomaterials throughout the life history of *Paramecium*.

Paramecium has excitable cell membranes that generate action potentials when subjected to various stimuli [13, 14]. Several types of membrane potential-gated ion channels have been identified electrophysiologically and biochemically. In particular, calcium ion channels localized in the cilia membrane and potassium ion channels localized in the cell body membrane are important. It has been confirmed that the action potential is generated by a temporary increase in intracellular calcium ion concentration. Therefore, in *Paramecium*, calcium ions play the same role as sodium ions in multicellular animals.

Paramecium swims by the movement of about 4,000 cilia arranged regularly on the cell surface. Ciliary movements are linked to the membrane potential of the cell membrane and, as a result, are linked to the action of membrane-gated ion channels. In particular, the reversal direction of effective stroke in ciliary movement is triggered by the action of voltage-gated calcium ion channels [14].

16.7.2 Cytotoxicity assessment of silver nanoparticles

Silver is a metal closely related to human life and is used in, for example, surgical and dental materials, chemical catalysts, electrical devices, and antibacterial substances [15]. *Paramecium* system can be used to assess single-cell cytotoxicity [16]. Furthermore, intracellular injection of nanomaterials can assess toxicity from inside the cell.

Paramecium exhibited a very unique three-step swimming pattern in silver nanoparticle dispersion. In the first stage, the *Paramecium* repeated forward and backward swimming immediately after it was placed in the silver nanoparticle dispersion, and the cell body occasionally rotated like a *"top"* These changes in swimming behavior suggest a disorder of ciliary movement synchronization. In the second stage, the long axis of the cell was shortened and the tip of the tail was deformed to be thin (Figure 16.4). In

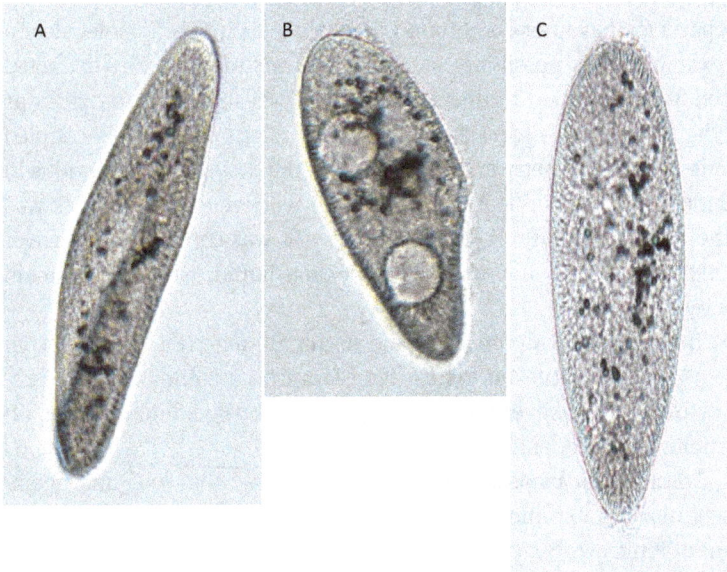

Figure 16.4: Photographs showing a morphological change caused by silver nanoparticles. All three photographs have the head at the top and the tail at the bottom. (A) 1 min after being placed in the silver-nanoparticle dispersion, (B) 10 min after. (C) 10 min after being placed in the gold-particle dispersion. The cytotoxicity of silver particles is only seen on (B) cells. In (B) cells, there is a marked decrease in the posterior area of the cell body. After about 30 min, the entire cell body becomes smaller and death occurs.

the third stage, all cells showed only backward swimming, and after a few minutes, the swimming behavior was stopped, leading to cell death [16].

The changes in a series of swimming behaviors are caused by ciliary movement. Since the direction of effective striking of ciliary movement depends on the intracellular calcium ion concentration, it was suggested that the cytotoxicity of silver nanoparticles is related to the increase in intracellular calcium ion concentration.

The effect of inside application of silver nanoparticles was examined by microinjection experiments. When 1 mg/mL of the 17-nm-silver dispersion was injected about 10% amount of the total *Paramecium* cell volume, a toxic effect was not detected in either swimming behavior or survival (Table 16.1) [16].

Table 16.1: Summary of cytotoxicity of silver ions.

Application site	Effect	Lethal density	Exam period (hour)
Outside	Yes	125–250 ng/mL	24
Inside*	No		24

*The concentration of silver ion used for microinjection is 100 µg/mL.
The injection volume is about 40 pL, which corresponds to about 10% of the cell volume.

Since silver nanoparticles have not been found to produce small toxic molecules in dispersions, we examined this possibility using the dialysis method. The extracted solutions produced at seventh and thirteenth day of dialysis showed cytotoxicity as strong as that of the non-treated silver nanoparticle dispersions. Inductively coupled plasma atomic emission spectrometry analysis indicated the presence of silver ions in the extracted solution of silver nanoparticles, although reactive oxygen species was not detected in the extracted solutions [16]. In this way, it was confirmed that silver ions were generated in the silver nanoparticle dispersion liquid, which is the main cause of cytotoxicity.

We examined human serum albumin, bovine serum albumin (BSA), and casein as antidote proteins against silver-ion toxicity and found that serum albumin proteins reduce the cytotoxicity of silver ions. These proteins showed detoxification effects with the manor of dose dependency (Figure 16.5) [16].

In summary, *Paramecium* bioassay system demonstrated that excitable membranes and intracellular calcium-mediated ciliary movement would be a useful tool for the assessment of ionic cytotoxicity. Some serum albumins perform a useful role for chronic cytotoxicity for silver ions.

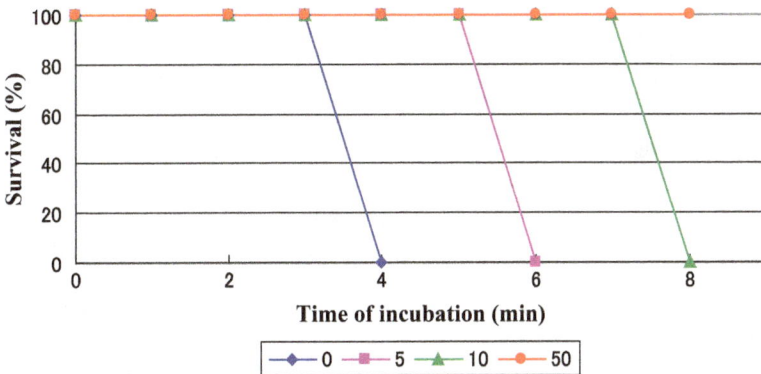

Figure 16.5: Detoxication effect of BSA against the Ag ions produced by Ag nanoparticle dispersion (21 nm). The Ag ion used was generated by dialysis treatment for 13 days. The numerical value (0, 5, 10, 50) in the graph represents the concentration of BSA (µg/mL). The vertical axis shows the percentage of surviving cells during the incubation time. The horizontal axis shows time of incubation (min).

16.7.3 Cytotoxicity assessment of multi-walled carbon nanotube

For high content multi-walled carbon nanotube (MWCNT) (50 mg/mL) dispersion during the non-proliferative phase, *Paramecium* began to incorporate MWCNT into food vacuoles immediately after addition (Figure 16.6A), and food vacuoles continued to form at the rate of one food vacuole per minute (Figure 16.6B) [17]. MWCNTs were concentrated in food vacuoles, and no NWCNTs invading the cytoplasm from food vacuoles were observed. All *Paramecium* used in the test survived during the 96-h-incubation period, and no physiological changes indicating toxicity were observed.

For low-content MWCNT (5–40 µg/mL) dispersions, in young *Paramecium*, no cytotoxicity was observed at any content of MWCNT. On the other hand, in old-aged *Paramecium*, the survival rate decreased in a dose-dependent manner (Figure 16.7).

The effects of MWCNTs on *Paramecium* during the proliferative phase were investigated in lettuce juice culture medium containing dietary bacteria.

In the experimental group in which one *Paramecium* cell was placed in the culture medium containing 50 mg/mL of MWCNT, the *Paramecium* once performed cell division in the first 24 h. However, the two daughter cells gradually lost their ability to swim, and after 48 h, 95% of the cells had died [17].

The reversibility of toxic effects of MWCNTs was evaluated by transferring cells cultured in MWCNT-containing medium for 48 h to MWCNT-free medium. Statistically, it was found that the proliferative capacity of all cells treated with 30 to 500 µg/mL of MWCNT dispersion was restored to the control level after 48 h of culture [17].

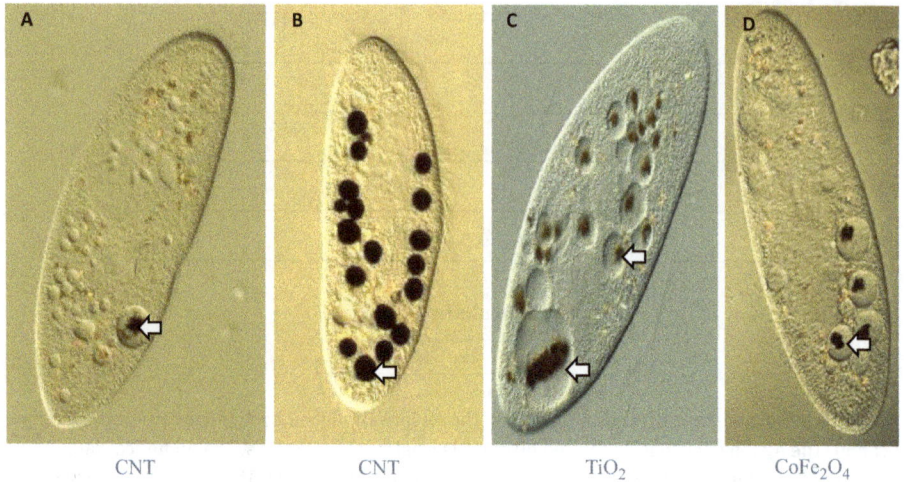

Figure 16.6: *Paramecium* that ate nanoparticles. Arrows indicate food vacuoles incorporating nanoparticles. (A) After 1 min incubation in CNT suspension. (B) After 30 min incubation in CNT suspension. (C) After 30 min incubation in TiO_2 suspension. (D) After 30 min incubation in $CoFe_2O_4$ suspension. The photographs of (B), (C), and (D) suggest that *Paramecium* preferentially incorporates CNT, TiO_2, $CoFe_2O_4$, and in that order.

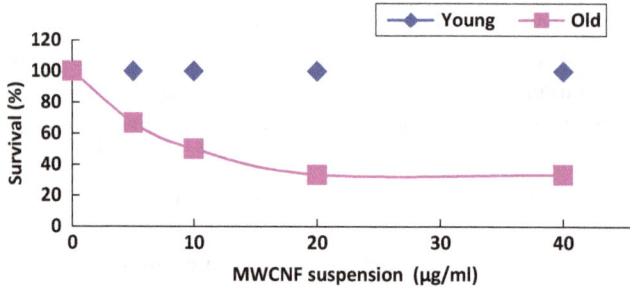

Figure 16.7: Survival rate of non-proliferative cells under the presence of MWCNT. The vertical axis shows the percentage of surviving cells during 24 h of culture. The horizontal axis shows the amount of MWCNT in the dispersion. Young represents cells that have divided about 100 times after conjugation and old represent cells that have divided about 500 times. There was no statistically significant difference between the amounts of MWCNT in the young. On the other hand, a dose dependency was observed in the old (Tukey, n = 10).

In summary, MWCNT was found to be cytotoxic to proliferating cells in both young and aged cells. However, in the non-proliferative state, it was not toxic to young cells, but it was toxic to old cells. No toxicity appeared in young or aged cells when MWCNT were injected intracellularly (Table 16.2). The molecular mechanism of MWCNT cytotoxicity is not yet understood. Producing the appropriate regulatory

Table 16.2: Toxic effect of MWCNT on young and old *Paramecium* cells.

Application	Cell cycle stage	Young	Old
		Conc. of 50% inhibition	
From outside	growth phase	50 µg/mL	50 µg/mL
From outside	stationary phase	No inhibition	50 µg/mL
Microinjection (5 pg/cell)	growth phase	No inhibition	No inhibition

guidelines of MWCNTs requires biochemical testing and identification of chemicals that interact with MWCNTs.

16.7.4 *Paramecium* cytotoxicity evaluation flowchart

We summarized the toxicity of nanoparticle dispersions on *Paramecium* survival, cell division, and progeny productivity (Table 16.3). The flow chart for assessing the cytotoxicity of nanomaterials in the *Paramecium* system is summarized in Figure 16.8.

Table 16.3: Summary of nanoparticle cytotoxicity.

Cell function	Material	Effect
Survival	Ag	Toxic
	Au	Non toxic
	MWCNT	Non toxic
Cell division	MWCNT	Toxic
Progeny production	CoFe2O4	Non toxic
	TiO2	Toxic

16.8 Current status and problems of nanotechnology in Japan

As of 2018, 322 Japanese products have been registered in the Nanotechnology Product Database (StatNano.com), but data for a comprehensive survey of all nanomaterial products have not yet been created. The current state of commercialization of nanocarbon materials has been disclosed by the Nanotechnology Business Promotion Council (NBCI). This section outlines the Japanese government's key initiatives in nanotechnology since 2007.

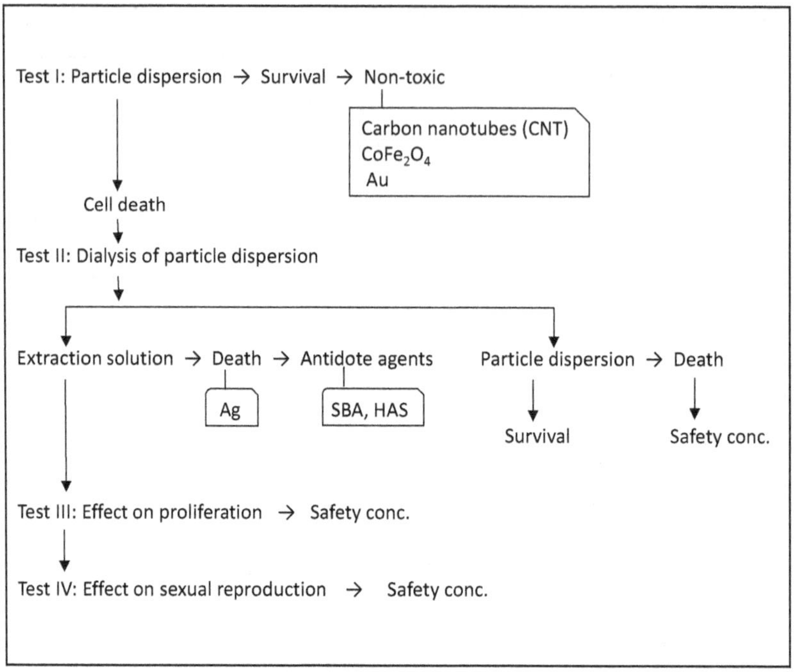

Figure 16.8: Flowchart for evaluation of nanoparticle cytotoxicity in *Paramecium* system.

At this time, the latest view of the Japanese government was published by the Nanotechnology and Materials Science and Technology Committee of the Nanotechnology Materials Science and Technology Strategy of the Ministry of Education, Culture, Sports, Science and Technology [18]. Below is an overview of the committee's basic policies.

The Japanese nanotechnology plays an important role in supporting the industrial base and is at high level internationally. As for future prospects, with the rapid progress of artificial intelligence/Internet of Things/big data technologies, the speed of material development that incorporates data-driven research and development (R&D) methods has begun. The government will put together R&D strategies in the field of nanotechnology and materials, disseminate them widely to society, and drive the realization of a future society. Therefore, the government will make efforts to change from conventional products to products with new functions and promote the creation of attractive materials that society supports. In this way, the government aims to realize a social revolution through materials(material revolution).

The Japanese government's historic initiatives are outlined below. In the Third Science and Technology Basic Plan (2006), two big projects, "Nanotechnology R&D promotion and basic development on social acceptance" in the group of collaborative measures of the Cabinet Office and the "Study on risk assessment and management method of new technology system: Risk assessment of nanomaterials" project were promoted in the National Institute of Advanced Industrial Science and Technology and the Safety Science Research Division [19]. The Second International Dialogue on Responsible Research and Development of Nanotechnology on 2006 proposed that 5–7% of the R&D budget should be allocated to ethical, legal, and social issues/environment, health and safety [20]. However, the Japanese government has not taken steps to achieve this goal.

In 2016, the Fifth Science and Technology Basic Plan advocated "Science and technology to co-create with society." The Ministry of Economy, Trade and Industry held the "Study group on safety measures for nanomaterials manufacturers" [21]. In the ministry's report, the ministry instructed manufacturers to take safety measures under their own control. The report also sets out the basic policy that the manufacturers and the Japanese government should collect and actively disseminate information on the scientific knowledge and applications regarding the safety of nanomaterials. Based on this basic policy, the Ministry of Economy, Trade and Industry announced the "Nanomaterials Information Collection and Dissemination Program" on the website in 2010. The ministry has compiled the status of voluntary risk information and safety measures by 31 nanomaterial manufacturers.

In March 2009, the Labor Standards Bureau of the Ministry of Health, Labor and Welfare issued a notice entitled "Precautions to Prevent Exposure to Nanomaterials." In March 2010, the Ministry of Health, Labor and Welfare's Pharmaceutical and Food Administration held a study meeting to discuss the issues for promoting safety measures for nanomaterials used in products for general consumers and the direction of safety measures, and issued a report.

In March 2010, the Ministry of the Environment announced "Guidelines for Preventing Environmental Impact of Industrial Nanomaterials" [22]. In this report, the ministry summarized the issues and points that governments and manufacturers should address when managing emissions of nanomaterials into air and water, waste management, and more.

In "Research and development of nanoparticle characterization method" conducted from 2006 to 2010, the New Energy and Industrial Technology Development Organization published a study on the comprehensive evaluation and management of the risk of industrially produced nanometer-scale particles of carbon nanotubes, fullerenes, and titanium dioxide as a chemical substance. The main items are the sample preparation method and measurement procedure for nano-substance toxicity test, nano-substance emission/exposure assessment report, nano-substance toxicity test method, nano-substance filter performance evaluation procedure, the composition of airborne particles of nanomaterials and social acceptance of nanotechnology.

In 2016, the Japanese government issued "Characteristics of working suspensions of nanomaterials for in vitro tests to evaluate toxicity specific to nanomaterials" in accordance with international standardization of hazard evaluation methods for nanomaterials. Since this report can be applied to the hazard assessment of carbon nanotubes, it was also used as a procedure for the safety management test method for nanocarbon materials. The Ministry of Health, Labor and Welfare added a specific carbon nanotube (MWNT-7) to the target of the carcinogenicity guideline, and provided the guideline on labor exposure [23].

From February 2017, the National Institute of Advanced Industrial Science and Technology's Safety Sciences Research Division has published a "Risk assessment report" on its website. In addition, the research results on the safety of nanocarbon materials were published as "NanoSafety Web Site." The ministry also publishes a summary of trends in laws, regulations, and guidelines on manufactured nanomaterials from various organizations around the world.

The nanoproduct manufacturing industry needs to establish new organizations to guarantee the safety of individual nanoproducts in the future. However, as of 2018, the "Chemical substance examination and manufacturing regulation law" (Chemical Substances Control Law) does not require nanomaterial manufacturers to report toxicity tests of nanomaterials. On the other hand, the Japanese government has issued an information provision sheet as a program to collect and transmit information on nanomaterials, requesting information from manufacturers of carbon nanotubes, carbon black, titanium dioxide, and the like. Since June 2018, the NBCI has posted a FAQ on "Regulations and Standardization of Nanomaterials" on its website [24].

What our research has revealed is that Japan's nanotechnology policy involves four Japanese government ministries and the Cabinet Office. These five organizations have separately set up councils and/or committees with private businesses to collect information and build databases, and have made measures. Nanoproduct manufacturers believe that the Japanese government needs to set national standards for product safety and quality control, as well as individual safety standards for consumers and producers of nanoproducts. Economic strategies aimed at maintaining a stable economy and responding to international competition require a philosophy that allows close cooperation between the four ministries and their leaders.

16.9 Future outlook

In Japan, there is no movement related to the regulation of nanomaterials by law. When releasing products in the global market, it is required to publish and disseminate safety evaluation data in accordance with the regulations of each country. Since the targets of scientific evaluation of nanoproducts are diverse, such as the composition, size, shape, and surface state of compounds, enormous amounts of time, money, and equipment are required. The future situation suggests that information on the safety of various nanomaterials and related products will be prepared by actual testing by individual companies and approved by relevant ministries. In countries without such schemes/systems, it will be difficult for public and private sectors to obtain safety information on nanomaterials. In order for the international community to enjoy the benefits of nanomaterials widely, there is a need for internationally established safety standards. Therefore, it is necessary to work under the framework of national initiative and international cooperation.

In vivo studies with animals are required for the safety assessment. Once the results of the *in vivo* test are no longer disclosed as trade secret data, countries that do not have a corresponding scheme/system (such as Japan) may be left behind in collecting safety information. As a countermeasure, it is possible to publicly make a database of safety information so that it can be used by a third party. By encouraging the use of this system, it will be possible to respond to new research and evaluation of new materials, verify the safety of imported and exported nanomaterials, and make regulatory approval decisions.

References

[1]   Crick F. On protein synthesis. Symp Soc Exp Biol 1958, XII, 139–63.
[2]   Crick F. Central dogma of molecular biology. Nature 1970, 227, 5258, 561–63.
[3]   Prusiner SB. Novel proteinaceous infectious particles cause scrapie. Science 1982, 216, 4542, 136–44.
[4]   Pan KM. et. al. Conversion of alpha-helices into beta-sheets features in the formation of scrapie prion protein. Proc Natl Acad Sci, USA, 1993, 90, 83, 10962–66.
[5]   Dixit VA. A simple model to solve a complex drug toxicity problem. Toxicol Res 2019, 8, issue 2, 157–71.
[6]   Wichterman R. The Biology of *Paramecium* 2nd ed. Nutrition, Growth, and Respiration. New York, Plenum Press, 1986, 197–210.
[7]   Haga N, Sato J. Induction of Mating-type change in a Non-selfer mutant of *Paramecium* by microinjecting wild-type genomic DNA. Euro J Protis 1996, 32, Sup, 132–36.
[8]   Takenaka Y, Yanagi A, Masuda H, Mitsui Y, Mizuno H, Haga N. Direct observation of histone H2B-YFP fusion protein and transport of their mRNA between conjugating Paramecia. Gene 2007, 395, 108–15.
[9]   Haga N, Hiwatashi K. A protein called immaturin controlling sexual immaturity in *Paramecium*. Nature 1981, 289, 177–79.
[10]   Hiwatashi K. Determination and inheritance of mating type in *Paramecium caudatum*. Genetics 1968, 58, 373–86.
[11]   Takagi Y, Yoshida M. Clonal death associated with the number of fissions in *Paramecium caudatum*. J Cell Sci 1980, 41, 177–91.
[12]   Haga N, Karino S. Microinjection of immaturin rejuvenates sexual activity of old *Paramecium*. J Cell Sci 1986, 86, 263–71.
[13]   Eckert R. Bioelectric control of ciliary action. Science 1972, 176, 437–81.
[14]   Eckert R, Naitho Y. Bioelectric control of locomotion in the ciliates. J Protozool 1972, 19, 237–41.
[15]   Drake PL, Hazelwsood KJ. Exposure-related health effects on silver and silver compounds: A review. Ann Occup Hyg 2005, 49, 575–85.
[16]   Abe T, Haneda K, Haga N. Silver nanoparticle cytotoxicity and antidote proteins against silver toxicity in *Paramecium*. Nano Biomed 2014, 6, 1, 35–40.

[17] Haga N, Haneda K, (2007) *Paramecium* as a bioassay system for elucidation of cytotoxicity and biocompatibility of nanoparticles: effect of carbon nanofibers on proliferation and survival.
[18] Nanotechnology Materials Science and Technology Research and Development Strategy of Ministry of Education, Culture, Sports, Science and Technology,"Nanotechnology and materials science and technology R&D strategy" Nanotechnology Materials Science and Technology Committee (August 2019) (Japanese) Press Releases.
[19] The 3rd Science and Technology Basic Plan (2006–2010)(Japanese).
[20] The 2nd International dialogue on responsible research and development of Nanotechnology (2006)(Japanese).
[21] The 5th Science and Technology Basic Plan (2016)(Japanese).
[22] Ministry of the Environment (2010) "Guidelines for Preventing Environmental Impact of Industrial Nanomaterials" (Japanese) Press Releases.
[23] National Institute of Advanced Industrial Science and Technology, Research Institute for Safety Science (2017) "Safety Test Comprehensive Procedure Manual for Nano Carbon Materials" (Japanese) Press Releases.
[24] Nanotechnology Business Promotion Council (NBCI), E-mail: info08@nbci.jp.

Rune Nydal

17 Revitalizing nanoethics: Nanotechnology at the center of nanoethics

Abstract: The chapter revisits the discussions on the need for nanoethics as a particular field of ethics. Many authors argued there was nothing morally specific to the field of nanotechnology. Nanoethics should consequently be developed alongside – and build on work done in – other branches of applied ethics (like bioethics). Nanoethics, however, came to be presented as a "new ethics": one that focused on process rather than product; an ethics that made a difference for the "science in the making." In this context, nanotechnology became an important site for the articulation of responsible research and innovation (RRI) initiatives whose identity in part was marked by a negative definition: the integration of ethics in nanotechnology is not to be done like it had been done in genomics. Nanotechnology should not continue the applied ethics pathway of scrutinizing ethical, legal, and social issues (ELSI). This paper suggests that the turn to RRI – away from ELSI – is part of the reason why nanoethics lost its initial momentum. As nanoethics turned into a field focusing on the "ethics of new and emerging technology," research situated in and dedicated to nanotechnology were gradually downplayed. As a result, nanotechnology – as a host for RRI innovations – has lost valuable substantial analysis from fields like theology, law, and philosophy. It is time to revitalize nanoethics. As nanotechnology is normalized, it is important to revitalize nanoethics by reintroducing careful scrutiny of nanotechnological issues into the sites where nanotechnology is fabricated. The point is not to devalue the importance of the "new ethics" of emerging technology – but to draw attention to what is lost.

Keywords: revitalizing nanoethics, RRI, ELSI, revisiting the call for nanoethics

17.1 Introduction

Do nanotechnology initiatives require nanoethics? Is there a need for nanoethics like there has been a need for bioethics? This chapter revisits this question discussed in reports worldwide during the first decade of this century as countries considered the need to invest in the field [1–7]. The aim of revisiting this question is to understand why nanoethics have come to be quite invisible in times where no technology is normalized and nanoethics is needed more than ever.

Rune Nydal, Programme for Applied Ethics, Department of philosophy and religious studies, Norwegian University of Science and Technology, Norway

https://doi.org/10.1515/9783110669282-017

It is striking how the early debate in nanoethics was marked by the question of the legitimacy of its own existence – questions that, for instance, were not asked in the same way in the context of genomics. The Human Genome Project of the 1990's raised ethical, legal, and social issues (ELSI) one could easily communicate to scientists, policymakers, and the public. The genome initiative simply actualized, enforced, and extended the scope of an already existing field, namely bioethics [8]. While there were disagreements concerning how to do bioethics, the question on nanoethics was a more fundamental one – it was a question of whether nanoethics was needed at all. Many authors argued that there simply were no nano-specific ethical issues that could justify such a term. Certainly, nanotechnology initiatives would raise ethical issues, but they were already taken care of so to speak. The issues were not of a different kind with respect to the ones already systemized and discussed in bioethics and other existing branches of applied ethics [9–17].

Nanoethics nevertheless arose as field in its own right. It was first of all a field that came to draw attention to a different set of questions, in part by explicitly distancing itself from bioethics and other branches of applied ethics. The need for nanoethics was presented as a need for a "new" or "better ethics": one that focused on process rather than product; an ethics that made a difference for the "science in the making" [18–22]. Nano initiatives, if successful, would transform society in unpredictable ways. The urgent "new" task for ethics was to take part in the work of steering these developments. It was the nature of this approach along with the theoretical perspectives that supported the analysis that came to define mainstream nanoethics rather than recognizable nanospecific ethical issues.

This chapter aims to show that this framing of nanoethics is part of the reason why nanoethics have lost its initial momentum. Nanoethics gradually turned into a field focusing on the ethics of emerging technology in general and eventually subsumed under the general heading of responsible research and innovation (RRI). The initial excitement and interdisciplinary mobilization that once sparked the birth of nanoethics gradually burned out as the nano-research activity itself faded away from the center of attention. From the perspective of the ethics of emerging technology, any ethically relevant distinction between nanotechnology and other technologies like biotechnology is leveled out. What was lost along the way was nanoethics situated in and dedicated to what is recognized as nanotechnological practice. As a result, nanoethics evolved as a field that gradually lost contributions from traditional humanist and social science studies of ethical, religious, and cultural issues.

It is time to revitalize nanoethics; to reintroduce the focus on particular ELSI raised by nanotechnological research projects and initiatives, even if the borderlines for nanotechnology remain fuzzy. The aim is not to downplay the distinctive contribution of "the ethics of new and emerging technology," but rather to draw attention to what have been lost in the field of nanotechnology. As nanotechnology is normalized, it is important to put nanotechnology at the center of attention and reintroduce the scholarly work of philosophers, theologians, and cultural analysts into the mix.

17.2 Nanoethics: a new and better ethics

Nanoethics emerged as a scholarly field during the first decade of this century. A number of books on nanoethics were published [23–27], a dedicated journal was established ("Nanoethics – Ethics for Technologies that Converge at the Nanoscale" in 2007) with close connections to the S-Net society ("Society for the Study of Nanoscience and Emerging Technologies" that first met in 2009) [28]. There had also been funding opportunities for nanoethics. Research projects and larger research centers had been funded from the EU framework program, the US National Science Foundation as well as by local national research programs in countries like my own, Norway. But nanoethics remained as fuzzy as the field of nanotechnology itself during the decade affecting the debate on nanoethics in two ways.

One strand focused on the question of what justified the term "nanoethics" as a distinctive field of applied ethics.Was there a need for specific courses in nanoethics, PhD training programs for scholars to specialize in nanoethics, scientific associations with designated conferences and journals discussing nanoethical issues? The answer in applied ethics circles seemed to be no – at least if establishing nanoethics as a field of its own presupposed the existence of novel set of nonspecific ethical issues. One should, for instance, expect issues like privacy and questions of human enhancement to be of critical importance for nanotechnology following new compact sensors and the creation of powerful interfaces between molecular biology and electronics. But such topics were already widely discussed in applied ethics in general and bioethics in particular. As for instance, Søren Holm argued, a rich knowledge base and tool case had already been established since the 1970's for these and other expected issues. Instead of imagining nanoethics as a distinctive new field, one would be better off using bioethics as a platform for developing a nanoethics. There would still be a need for specialization in nanoethics, "because good ethical analysis of nanotechnology requires in-depth knowledge of nanoscience, nanotechnology and the social field in which nanotechnology is becoming embedded" [9, p. 37].

The second strand drew attention to the way in which nanotechnology called for radical changes in existing research and innovation systems. Nanotechnology, despite its underdeveloped state, presented itself as a field that would radically change people's daily lives. Famously, the US National Nanotechnology Initiative declared extraordinary expectations: "[t]he impact of nanotechnology on the health, wealth, and lives of people could be at least as significant as the combined influences of microelectronics, medical imaging, computer-aided engineering, and man-made polymers developed in this century" [29, p. 205]. By 2004, governments worldwide were investing heavily in nanotechnology raising concerns about the unexamined social impact of these investments. A commentary in *Science* magazine stated, "Is the field moving so fast that it's destined to repeat the mistakes of earlier technological revolutions?" Given the combination of high expectations for societal transformation, unclear risks and the immaturity of the field, nanotechnology represented an historic opportunity of "getting it right from the start" [30].

The stated need for nanoethics came to be linked to this historic opportunity, rather than the identification of nano-specific ethical issues. The challenges of early ethical intervention were reinforced due to the way science was turning into novel social institutions. Two reports become central, the 2001 NSF report "Societal Implications of Nanoscience and Nanotechnology" and the 2004 EU report "Converging Technologies – Shaping the Future of European Societies" [1, 2]. These reports contextualized nanotechnology in a transatlantic debate on "the new technology wave" of particularly powerful technologies referred interchangeably to as "emerging," "enabling," or "converging" technologies [31] – terms that now have become part of our daily language. Importantly, these technologies were characterized by novel "modes of knowledge production" where research were to be steered toward the "context of application" calling attention to the importance of examining novel patterns of "coproduction" of science and society [32–36].

A key feature of the transatlantic literature on converging technologies, as Deborah Johnson [19] pointed out, concerned the way technology was reconceptualized as sociotechnical systems (rather than material objects). Analyzing technology as sociotechnical systems helped theorize the way in which technology and society are mutually constituted through a dynamic of "coproduction." Understanding the dynamics of coproduction provides a gateway for understanding how technology and values are intertwined: "Acknowledging the contingency of technological development and the many actors and groups that shape it is critical to understanding the potential of nanoethics" [19, pp. 23–24].

Johnson paper drew attention to an important assumption: the potential of nanoethics critically depend on acknowledging key theoretical insights under development within the field of science and technology studies (STS). Analyzing the phenomena of the new technology wave through the theoretical STS discourse – often loosely referred to under the heading of "coproduction" – came with an intriguing promise: a nanoethics capable of make a difference for "science in the making." Nanoethicists could "be among the many actors who shape the meaning and materiality of an emerging technology" [19, p. 21]. Nanotechnology became a key site for the discussion on how one could understand and adequately respond to ongoing changes taking place in the research and innovation systems. Given this context, nanoethics could be associated with the aspiration to make the nanotechnology initiative, as Armid Grunwald described, "a 'model' of dealing responsibly with new and emerging sciences and technologies" [37, p. 192].

Rhetorically, the "novel" nature of nanoethics could be presented as a shift of focus from "issues" to "interactions," reflected in proposals of new terms replacing the term "nanoethics." The proposed SEIN acronym (social and ethical interactions with nanotechnology and science) is particularly illustrative as it explicitly draws attention to the work of identifying and assessing interactions rather than issues. "Scientists, engineers, and all of the many stakeholders in our joint socio-nanotechnological future, need to engage in multi-directional discussions about societal values, needs, scientific/engineering prospects and probabilities. We envision such multi-directional discussions along the lines of a huge version of an old-fashioned town meeting, where we are all collectively constructing the 'nano society' of the future" [21, p. 105]. Many scholars now explored the opportunity nanotechnology provided to experiment on public and upstream engagement, finding means to facilitate and modulate interactions between technological and societal actors and interests promoting labels like Anticipatory governance, Real-time technology assessment (RTTA), or Constructive technology assessment (CTA) [18, 38–40].

Nanotechnology would still require the examining of ethical issues. But, as will be argued in the following, research activity focusing on philosophical, religious, and cultural issues became delegitimized following the call for a shift from issues to interactions. From an applied ethics point of view, the research task is to analyze morality in shifting sociotechnical systems [19, 20, 22, 31, 41, 42]. The task of nano-ethics should be seen as a broad one, as Nigel Cameron early pointed out: The task is not only to help modulate interactions between technological and societal actors but, as also to "articulate our values once again not only at the individual level [. . .] but also at the level of the ideals about our European societies" [31, p. 28]. As the field of nanoethics developed, normative issues tended to be kept unanalyzed.

17.3 Why and how nanoethics lost momentum

The framing of nanoethics as a new and better ethics is part of an unfortunate narrative of there being a need for a shift *away* from ELSI toward RRI. Nanotechnology itself – and thereby also nanoethics, as this chapter suggests – faded away from the center of attention as an effect of this shift.

By 2015, the "Society for the Study of Nanoscience and Emerging Technology" had become the "Society for the Study of New and Emerging Technology," and the subtitle of the journal *Nanoethics* had changed to "Studies of New and Emerging Technologies" replacing the previous "Ethics for Technologies that Converge at the Nanoscale" [28]. The shift of names reflect how nanotechnology gradually turned into a model case for dealing responsibly with new and emerging technology in general. Nanotechnology – as a specific research arena for ethics– did not have to get backgrounded following this contextualization of the field. But it did in part due to the way in which nanotechnology became a key arena for the articulation and development of RRI.

The story of RRI connects to the way in which RRI was presented as a shift from ELSI studies. When nanotechnological initiatives were discussed, ELSI had been established as a transatlantic funding category. The question of whether nanoethics was needed (like was the case with bioethics) soon took the form of a question of whether there should be a NELSI [23], that is an ELSI *of* nano like there had been an ELSI *of* genomics? One should expect, and welcome, discussions on what one could learn from these past efforts of including the work of the humanities and social scientist as "part of" a scientific initiative as it was done in genomics. But, as different scholars have pointed out, ELSI were misrepresented and came to represent what one should not do, and consequently why an alternative was needed [43–45].

A caricature was created, allowing sweeping characteristics informed by the dominant theories and self-understanding of science and technology studies. ELSI came to represent the past, simply taken to belong within the modernist "linear model" – the downstream-oriented paradigm. ELSI could function as a signifier for the type of approach one needed to overcome in order to give way to up- and mid-stream governance [see, for example, 21, 46–48]. "Over the years," Rebecca Walker and Clair Morrissey write, "criticisms of the ELSI Program [of the human genome project] have remained remarkably stable and include concern over a lack of independence from [. . .] the very scientists whose work is the subject of critical inquiry [. . .], ineffectiveness in generating policy in the area of human genomics, and a dearth of actual public engagement" [43, p. 482]. Their review of the ELSI literature contradicts this simple story as it demonstrates a complex literature, where a set of methodological approaches work along-side ethical analysis that often were undistinguished from social, legal, or political investigations.

The caricature of ELSI narratives matter for the argument this chapter makes. The story of nanoethics – one should remember – also involve disputes of disciplinary hegemony. In the process where RRI became the dominant term (rather than suggested terms like SEIN, NELSI, RTTA, or CTA), ELSI appeared as one of the code words. If you trashed the term you communicated you where concerned with the processes where our collective future are shaped. As a consequence – in contrast to ELSI researchers – one where geared towards collaboration and integration. But ELSI scholarship builds on a tradition that does put emphasis on collaboration – but for different reasons: to do better substantial ethical analysis. Ethical inquiry needs to be "context sensitive"; the research revolves around, and requires knowledge of, the specific research practice under investigation. RRI, in contrast, focuses on creating better research and innovation processes through involvement and engagement. A consequence of this difference – as will be discussed in the following – is nanotechnology lost key competence of applied ethics simply because nanotechnology itself faded away from the center of attention in the RRI discourse.

The ELSI program of the Human Genome Project builds on a two-decade long tradition of experimentations within the field of applied ethics. The field had already gained valuable experiences in researching emerging ethical issues that arose in the research and application context (like in vitro fertilization, diagnosis leading to selective abortion, stem cell research, euthanasia options, and so on) [42, 49]. A key entry point of applied ethics is the ethical concern or issue raised by the subject field under investigation. It is the issues that unite applied ethicists working in the field, whether they are trained in philosophy, theology, social or cultural – studies – or in the relevant technical field. Understanding and suggesting solutions called for ways to engage and interact with empirical research fields, medical and biological expertise, stakeholders, and contributors to the public debate. Applied ethics then arose as an interdisciplinary field where the ethicist's theoretical and methodological apparatus were expanded and reformulated [50, 51].

A statement from John Weckert, the first editor of the journal *Nanoethics*, illustrates how an applied ethics approach had marked his judgments during the first six years of the journal. "Much of the material published would not count as ethics for a purist, not a purist philosopher anyway. My approach has been that anything that contributes meaningfully to ethical discussions is legitimate subject matter. It didn't bother me too much if in some papers mention of ethics was scant providing that a careful reader could see how they contributed to that discussion" [52]. The inclusion criteria were broad but quite demanding: that the work "contributes meaningfully to ethical discussions" of the nanotechnological research field.

The point is a quite simple one. Knowledge of nanotechnology critically matters for applied ethics as it aims to identify what is at stake in the practice under investigation. Doing nanoethics involves having a conversation among scholars who are interested in questions like what are the instruments, history, financial structures, key literature, key journals, potentials and visions of nanolabs, and the dynamics of the convergence of fields? An ELSI *of* genomics would need to take place at different places and networks than an ELSI *of* nanotechnology. There is always an ELSI of x, where x puts different demands on the researcher.

The x gradually faded away as nanotechnology drifted into a paradigmatic site for RRI allowing nanotechnology to be substituted with "new and emerging technology." It is striking how any reference to x is typically missing in the highly cited RRI definitions like: "Responsible innovation means taking care of the future through collective stewardship of science and innovation in the present" [53], or "A transparent, interactive process by which societal actors and innovators become mutually responsive to each other with a view to the (ethical) acceptability, sustainability and societal desirability of the innovation process and its marketable products (in order to allow a proper embedding of scientific and technological advances in our society)" [54]. Such definitions illustrate the shift of focus from "issues" to "interactions." The key words of RRI are words like "collective stewardship" or "transparent, interactive process." The key words of ELSI approaches are typically constituted by a list of issues that arise in the x in question.

RRI drew attention away from the subject field as the question of methods appeared to start having a life of their own independent of any x. Policy documents in Norway drifted from an issue-based legitimation of the why's of ELSA to a theoretical legitimation for an RRI strategy. The policy documents clarifying the why's of

RRI draw attention to science topics like the discussions on the new social contract, linear model, division of moral labor, and shifts of modes of knowledge production [42].The x does not matter for RRI, whether it is an RRI of biotechnology or nanotechnology. RRI was presented as underdeveloped, typically calling for "experimentations" and developments of "RRI frameworks" in ways that did not invite participation from traditional normative fields.

A framework published by the Research Council of Norway is illustrative. The document builds on a UK research council's framework (of anticipation, reflexivity, inclusion, and responsiveness) in need of further development since "RRI is a figuration; it is open, not 'owned' by anyone and therefore invites and inspires experimentation, development activities and learning across established boundaries, sectors and disciplines." The document not only assumes there are serious deficiencies in the way the ethics of science is ordinary done but understands them as inadequate and in need of replacement. "RRI is motivated more by discontinuity than continuity in relation to tools/instruments that are becoming inadequate in the knowledge society" [55].

What is important for the perspective of this chapter concerns the question of what is lost in the turn from ELSI to RRI. Traditional normative oriented fields of study like philosophy, theology, and cultural studies are suppressed along with the traditions and tool cases developed in these fields. It is not a matter if these fields are not thriving elsewhere and need nanotechnology, it is rather a matter of what got lost in nanotechnology initiatives.

17.4 Nano at the center of nanoethics

Understood as an important birthplace for RRI, nanoethics have been quite successful. RRI initiatives have contributed to the normalization of socio-humanist research activity in the new wave of technologies. Researchers, whether they are engaged in ICT, nanotechnology or biotechnology initiatives, have become accustomed to the RRI ambition of building capacities for a responsible research and innovation. and RRI initiatives have enrolled and empowered scholarly communities that were less dominant in ELSI initiatives.

This chapter cares for what is lost, without disallowing what is gained. Nanotechnology, as the chapter has tried to show, has drifted away from being a specific research arena for nanoethics following connected shifts from nanotechnology to emerging technologies, issues to interactions, and ELSI to RRI. Part of the story of nanotechnology drifted out of sight is the loss of disciplinary capacities and competences of research communities researching substantial ethical, religious, and cultural issues of nanotechnology. Without work of identification and deliberation of issues, the ethics and policy of nanotechnology will be left "in the dark" as Nick Bostrøm pointed out in the early phase of nanoethics. Nanoethics should not only be concerned with "the structures of governance – with *who* should make the decisions, but rather on the terms of the discourse: what kind of considerations should be taken into account, and in what ways" [56, p. 144].

As nanotechnology is normalized and matured, it is time to revisit the ambition of "getting it right from the start." The quality of nanoethics should be measured in terms of its ability to do so by feeding into the nano-research, innovation, and application contexts in constructive ways. As Chris Toumay notes [57, 58], the field is underdeveloped in terms of its ability to provide recommendations for researchers, developers, or policymakers. What issues are likely to arise? What are the different positions, the stated concerns, and strengths and weakness of the arguments that circulates? Even if the borders of nano remain fuzzy, there are recognizable nano-research activities like seminars, conferences, training programs, funding programs as well as research and innovation centers. At a very practical level, ethical issues are handled in these arenas – more or less as a result of explicit and careful deliberation. The "NELSI landscape" portrayed from the start has not changed and much remains underdeveloped compared to genomics. Understanding the details of the ethical landscape – and how it shifts over time – has become even more relevant as nanotechnological activity is normalized, like the issues of hype, public relations, risks, choices of medical and military applications, nano-divide, and human enhancement [59].

It is time to revitalize nanoethics, leaving behind the unproductive choice between issues and interactions and reintroducing the characteristic research activity of applied ethics into the mix. This chapter has understood the "new ethics" of emerging technology as an important contribution to and extension of applied ethics rather than as an alternative to it. It seems necessary to state a simple point: philosophical, religious, and cultural analyses are needed to deliver a substantial normative analysis of what is at stake. Substantial analysis of issues supports the quality of the deliberation process that takes place in numerous interactions shaping our collective futures. New and emerging technologies call for ongoing substantial ethical and political analysess simply because of the way in which technologies reshape normative discourses. Such research requires nanoethics that revolve around the sites of nanotechnological research and innovation.

References

[1]   Roco MC, Bainbridge W, eds. Societal implications of nanoscience and nanotechnology. Report from the workshop held at the National Science Foundation, 28–29. September, 2000. USA, National Science Foundation, 2001.
[2]   Nordmann A, rapporteur. Converging technologies – shaping the future of European societies. Geo-Eco-Urbo-Orbo-Macro-Micro-Nano. High level expert group: Foresighting the new technology wave. European Commission, 2004.

[3]   Royal Society & Royal Academy of Engineering. Nanoscience and Nanotechnologies: Opportunities and Uncertainties. London, UK, Royal Society & Royal Academy of Engineering, 2004.

[4]   Royal Netherlands Academy of Arts and Sciences. How big can small actually be? Some remarks on research at the nanometre scale and the potential consequences of nanotechnology. Prepared for the Dutch Minister of Education, Culture and Science. Amsterdam, Netherlands, Royal Netherlands Academy of Arts and Sciences, 2004.

[5]   The Research Council of Norway. Nanoteknologierognyematerialer: Helse, miljø, etikkogsamfunn. Oslo, Norway, Norgesforskningsråd, 2005.

[6]   UNESCO. The Ethics and Politics of Nanotechnology. Paris, France, United Nations Educational, Scientific and Cultural Organization (UNESCO), 2006.

[7]   National Academics Forum. Environmental, social, legal and ethical aspects of the development of nanotechnologies in Australia. A report for The National Nanotechnology Strategy Taskforce. Department of Industry Tourism and Resources. Melbourne, Australia, 2006.

[8]   Walker R, Morrissey C. Bioethics methods in the ethical, legal and social implications of the human genome project literature. Bioethics, USA 2014, 28, 9, 481–90.

[9]   Holm S. Does Nanotechnology Require a New "Nanoethics"? Gordijn B, Cutter AM eds, In Pursuit of Nanoethics: Transatlantic Reflections on Nanotechnology. Dordrecht, Netherlands, Springer, 2014, 31–39.

[10] Ebbesen M, Andersen S, Basenbacher F. Ethics in nanotechnology: Starting from scratch? Bull Sci Technol Soc, USA 2006, 26, 6, 451–62.

[11] Allhoff F. On the Autonomy and Justification of Nanoethics. Allhoff F, Lin P eds, Nanotechnology & Society: Current and Emerging Ethical Issues. Dordrecht, Netherlands, Springer, 2009, 185–210.

[12] Grunwald A. Nanotechnology – a new field of ethical inquiry? Sci Eng Ethics, USA 2005, 11, 187–201.

[13] Allhoff F, Lin P. What's so special about nanotechnology and nanoethics? Int J Appl Philos, USA 2006, 20, 2, 179–90.

[14] Ferrari A. Developments in the debate on nanoethics: Traditional approaches and the need for new kinds of analysis. Nanoethics, Netherlands 2010, 4, 1, 27–52.

[15] Brownsword R. Nanoethics: Old wine, new bottles? J Consum Policy, USA 2009, 32, 355–79.

[16] Litton P. "Nanoethics"? What's new? Hastings Cent Rep, USA 2007, 32, 1, 22–25.

[17] Keiper A. Nanoethics as a discipline? The new atlantis. J Technol Soc, USA 2007, Number, 16, 55–67.

[18] Guston D, Sarewitz D. Real-time technology assessment. Technol Soc, Netherlands 2002, 24, 93–109.

[19] Johnson DG. Ethics and technology 'in the making': An essay on the challenge of nanoethics. Nanoethics, Netherlands 2007, 1, 21–30.

[20] Moor J. Why we need better ethics for emerging technologies. Ethics Info Technol, Netherlands 2005, 7, 111–19.

[21] Baird D, Vogt T. SEIN: Social and ethical interactions with nano. Nanotechnol Law Bus Germany 2006, 1, 391–96.

[22] Swierstra T, Rip A. Nano-ethics as NEST-ethics: Patterns of moral argumentation about new and emerging science and technology. Nanoethics, Netherlands 2007, 1, 3–20.

[23] Cameron M, Mitchell E, eds. Nanoscale: Issues and Perspectives for the Nano Century. USA, John Wiley & Sons, 2007.

[24] Allhoff F, Lin P, Moor J, Weckert J, eds. Nanoethics. The Ethical and Social Implications of Nanotechnology. New Jersey, USA, John Wiley & Sons, 2007.

[25] Fisher E, Selin C, Wetmore JM, eds. The Yearbook of Nanotechnology in Society. Volume 1: Presenting Futures. USA, Springer, 2008.

[26] Allhoff F, Lin P, eds. Nanotechnology & Society. Current and Emerging Ethical Issues. USA, Springer, 2009.

[27] Kjølberg K, Wickson F, eds. Nano meets Macro. Social Perspectives on Nanoscale Sciences and Technologies. UK, Pan standford publishing, 2010.

[28] Coenen C. S.NET and Nanoethics: A pair of siblings. Nanoethics, Netherlands 2015, 9, 197–98.

[29] National Nanotechnology Initiative. Leading to the next industrial revolution. Microscale Thermophys Eng, UK 2000, 4, 3, 205–12.

[30] Service RF. Nanotechnology grows up. Is the field moving so fast that it's destined to repeat the mistakes of earlier technological revolutions? Science, USA 2004, 304, 1732–34.

[31] De S, Cameron NM. Ethics, Policy, and the Nanotechnology Initiative: The Transatlantic Debate on "Converging Technologies." De S, Cameron NM, Mitchell E eds, Nanoscale: Issues and Perspectives for the Nano Century. USA, John Wiley & Sons, 2007, Chapter 3, 27–41.

[32] Gibbons M, Limoges C, Nowotny H, Schwartzman S, Scott P, Trow M. The New Production of Knowledge. UK, Sage, 1994.

[33] Ziman J. Real Science. What It Is and What It Means. UK, Cambridge University Press, 2000.

[34] Nowotny H, Scott P, Gibbons M. Rethinking Science. UK, Polity Press, 2001.

[35] Jasanoff S, ed. States of Knowledge: The Coproduction of Science and the Social Order. UK, Routledge, 2004.

[36] Carrier M, Normann A, eds. Science in the Context of Application. Boston Studies in the Philosophy of Science. Germany, Springer, 2010.

[37] Grunwald A. Responsible Research and Innovation: An Emerging Issue in Research Policy Rooted in the Debate on Nanotechnology. Arnaldi S, Ferrari A, Magaudda P, Marin F eds, Responsibility in Nanotechnology Development. Dordrecht, Netherlands, Springer, 2014, 191–205.

[38] Schot J, Rip A. The past and future of constructive technology assessment. Technol Forecast Soc Change, Netherlands 1996, 54, 251–68.

[39] Toumay C. Science and democracy. Nat Nanotechnol, USA 2006, 1, 6–7.

[40] Fisher E, O'Rourke M, Evans R, Kennedy EB, Gorman ME, Seager TP. Mapping the integrative field: Taking stock of socio-technical collaborations. J Responsible Innov, USA 2015, 2, 1, 39–61.

[41] Nydal R, Strand R. God nanoetikk – God nanoetikkutvikling. Nord J Appl Ethics, Norway 2008, 2, 2, 33–51.

[42] Nydal R, Myhr AI, Myskja BK. From ethics of restriction to ethics of construction. ELSA research in Norway. Nord J Sci Technol Stud, Norway 2014, 3, 1, 34–45.

[43] Walker R, Morrissey C. Bioethics methods in the ethical, legal and social implications of the human genome project literature. Bioethics, USA 2014, 28, 9, 481–90.

[44] Zwart H, Landeweerd L, Van Rooij A. Adapt or perish? Assessing the recent shift in the European research funding arena from 'ELSA' to 'RRI'. Life Sci Soc Policy, UK 2014, 10, Article 11. https://doi.org/10.1186/s40504-014-0011-x.

[45] Myskja BK, Nydal R, Myhr AI. We have never been ELSI researchers–there is no need for a post-ELSI shift. Life Sci Soc Policy, UK 2014, 10, Article 9. https://doi.org/10.1186/s40504-014-0009-4.

[46] Winner L. Testimony to the committee on science of the U.S.S House of Representatives on the societal implications of nanotechnology. Wednesday, April 9, 2003. (Accessed April 2021 at https://homepages.rpi.edu/~winner/testimony.htm)

[47] Yesley MS. What's ELSI got to do with it? Bioethics and the human genome project. New Genet Soc, UK 2008, 27, 1, 1–6.

[48] Balmer AS, Calvert J, Marris C, Molyneux-Hodgson S, Frow E, Kearnes M, Bulpin K, Schyfter P, Mackenzie A, Martin P. Taking roles in interdisciplinary collaborations: Reflections on working in post-ELSI spaces in the UK synthetic biology community. Sci Technol Stud, UK 2015, 28, 3, 3–25.

[49] Meslin EM, Thomson EJ, Boyer JT. Bioethics inside the beltway. The Ethical, legal, and social implications research program at the national human genome research institute. Kennedy Inst Ethics J, USA 1997, 7, 3, 291–98.

[50] Toulmin S. How medicine saved the life of ethics. Perspect Biol Med, USA 1982, 25, 736–50.

[51] Musschenga AW. Empirical Ethics, context-sensitivity, and contextualism. J Med Philos: Forum Bioethics Philos Med, UK 2005, 30, 5, 467–90.

[52] Weckert J. Editorial. Nanoethics, Netherlands 2012, 6, 3, 153.

[53] Stilgoe J, Owens R, Macnaghten P. Developing a framework for responsible innovation. Res Policy, Netherlands 2013, 42, 1568–80.

[54] Von Schomberg R. A Vision of Responsible Innovation. Owen R, Heintz M, Bessant J eds, Responsible Innovation. UK, John Wiley, 2013, 51–74.

[55] The Research Council of Norway. A framework for Responsible Innovation-under BIOTEK2021, IKTPLUSS, NANO2021 and SAMANSVAR Version 1.0, 2015. (Accessed April 2021 at www.for skningsradet.no/contentassets/1975cf4657c24ffea33d274adfff0319/rri-rammeverk.pdf)

[56] Bostrom N. Technological Revolutions: Ethics and Policy in the Dark. De S, Cameron NM, Mitchell E eds, Nanoscale: Issues and Perspectives for the Nano Century. USA, John Wiley & Sons, 2007, Chapter 10, 129–152.

[57] Toumay C. Early voices for ethics in nanotechnology. Nat Nanotechnol, USA 2019, 14, 304–05.

[58] Toumay C. Later voices on ethics in nanotechnology. Nat Nanotechnol, USA 2019, 14, 636–37.

[59] Mekel M, De S, Cameron NM. The NELSI Landscape. De S, Cameron NM, Mitchell E eds, Nanoscale: Issues and Perspectives for the Nano Century. USA, John Wiley & Sons, 2007, Chapter 20, 359–376.

Marcel Van de Voorde and Gunjan Jeswani
Conclusion

> The use of scientific knowledge needs the guiding light of ethical wisdom. Such is the wisdom that inspired the Hippocratic Oath, the 1948 Universal Declaration of Human Rights, the Geneva Convention, and other laudable international codes of conduct. Hence religious and ethical wisdom, by answering questions of meaning and value, play a central role in professional formation. And consequently, those universities where the quest for truth goes hand in hand with the search for what is good and noble, offer an indispensable service to society.[1]

> *Pope Benedict XVI,*
> *Blessing of the cornerstone of*
> *Madaba University of the Latin Patriarchate,*
> *in Jordan, May 9, 2009*

Nanotechnology is clearly a breakthrough technology because it unleashes the power to manipulate matter on the nanoscale, tailoring it to individual and societal needs. Indeed, current nanotechnology research is aimed at the development, analysis, and application of a wide variety of incredible nanostructures such as buckminsterfullerene, nanotubes, nanoparticles, molecular motors, a range of biomolecules, quantum dots and quantum wires, nanocapsules, nanopores, and self-assembling nano-robots. Such applications promise to improve the standard of living, health care, and nutrition of people worldwide. Nanotechnology also promises to minimize or perhaps eliminate pollution, mitigate existing ecological imbalance, and eradicate various diseases by offering protection against harmful bacteria and viruses, including the dreadful plague of Covid-19. In the future, nanotechnology is expected to bring effective innovations to enhance health care and in critical resources: water and energy supply, agriculture, and industrial engineering.

At present, both excitement and apprehension surround nanoscale technologies. Uncertainty about nanotechnology's possible adverse effects on human health and the environment cannot be overlooked at present. Many experts fear that the power of knowledge provided by nanotechnologies, most often combined with artificial intelligence and localization techniques, might disproportionately benefit private companies rather than the general public of individuals in society at the cost of important privacy concerns. Optimists believe that to solve the problem of war and terrorism and foster peace, nanotechnology is urgently needed. Pessimists fear that nanotechnology will be used for unethical purposes, such as creating weapons of

---

[1] On the Vatican official Website : www.vatican.va/content/benedict-xvi/en/speeches/2009/may/documents/hf_ben-xvi_spe_20090509_pietra-madaba.html

---

Marcel Van de Voorde, University of Technology Delft, The Netherlands
Gunjan Jeswani, Faculty of Pharmaceutical Sciences, SSTC, SSGI, Bhilai, India

https://doi.org/10.1515/9783110669282-018

mass destruction that are used to start wars in times of peace. Uncertainty surrounds many threats and risks posed by nanotechnology. There is a great debate regarding the methodological challenge to risk management. Without input from informed stakeholders, it is impossible to determine the appropriate benchmarks that will detect, predict, control risks, or choose which trade-offs are made for the good of civil society and human civilization.

Balanced and inclusive development of nanotechnology requires international interdisciplinarity and international cooperation to coordinate training, research, and technology transfer between different regions and countries and also fill the gap between developed and developing countries. Given the extraordinary economic and social potential of nanotechnology, attempting to stifle scientific and technological developments in nanotechnology would be immoral. But, developing this enormous potential without integrated ethical conduct would be equally immoral. Therefore, development of this disruptive technology requires forethought. Facing numerous promises and fear, it is important to decide the direction of development, its extent and methodology, in tune with ethical reflection. Nanoethics cannot lag behind nanotechnology.

Therefore, this book on "nanoethics" provides a vital forum for informed discussion on ethical, legal, psychological, economic, cultural, and social concerns related to nanotechnology. Some major points to ponder include: Shouldn't the orientation of research and development projects be geared toward the public good since the massive amounts of funds are contributed by the public? Is it not the aim of nanotechnology to improve the well-being of the society? In different chapters, solutions are given to the queries, and at the same time, important questions are asked, for example, what are the numerous stakeholders searching for? Immediate financial profit? Personal development and a contribution to science through nanoscale discovery and new engineered applications? Do they have the common good as their primary goal? And if so, what are their working assumptions about the common good? Are these assumptions focused on the benefits for all or a few? Do they minimize dangers for the present generation and consider future generations? And, notably, have they considered these questions?

In this book, attempts have been made to find solutions to some of the most pressing queries while raising questions that require further examination:

1. Who will be benefited from advances in nanotechnology, and at what cost?
2. Who will take the responsibility for the potential adverse effects of these devices and systems that are invisible to the human eye?
3. How does international competition affect the development of nanotechnology? How will world powers introduce and regulate nanotechnology applications?
4. Is the public aware of the potential and associated risks of nanotechnology? How can the developments of nanotechnology be made more transparent to the public?

The present situation calls for shared and widespread recognition of nanotechnological innovations that may come with possible risks. Therefore, it is utmost important to identify and assess the risks regarding their acceptability in light of gains.

The following key conclusions are drawn from the topics discussed in the various chapters in the book. Nanoethics offers a rigorous study of philosophical, scientific, ethical, social issues, and political problems. It allocates responsibility and enables funding conditions that impact the ability of decision-makers to pause and include ethics in their plans for nanotechnology research, in light of the social problems. It also gives ethics its rightful place in nanotechnology.

Nanotechnology obviously enhances other technologies, that is, computer and material science and so on. Therefore in rigorous discussion of social and ethical issues, scientists and researchers from those disciplines as well as from nanotechnology itself must be involved. This critical issue requires an interdisciplinary perspective. Hence, nano-scientists, engineers, ethicists, and, more generally, philosophers, social scientists, psychologists, economists, science communicators, ecologists, agriculturalists, and health workers must be brought together, depending on the particular problem being explored. Therefore, public and private funding organizations must take serious account of the ethical reflections put forward in various aspects of this book. Academic and corporate training must contribute to this as an integral part of a humanistic education that cultivates all human dimensions, arouses curiosity, and awakens a sense of individual and communal responsibility.

Solving current ethical problems will hopefully incentivize ongoing interdisciplinary discussions that will include stakeholders reflecting a multidirectional nanoscene ethical debate must continue until questions regarding the potential, capacity, and controllability of nanotechnology coexist.

The development of nanotechnology is in a phase where it is still possible to choose the road to more inclusive applications and balance the power of different social groups. Nanotechnology is a vehicle for applying or discarding the traditional paradigm. It is expected that this comprehensive resource can function as the first step toward inclusive debate that will equitably benefit all stakeholders. The reader is provided with solutions and recommendations suggested by eminent scientists and philosophers working on nanotechnology and its implications. Let us give the last word to "one of the most courageous minds of our time, Andrei Sakharov," as Mrs von der Leyen said at the beginning of her European Union speech:

He always spoke of his unshakeable faith in the hidden power of the human spirit.

May this strength, found within the hidden power of material at the nanoscale, make nanotechnology the lever for a future commensurate with the highest of human aspirations, and may ongoing ethical reflection be our moral compass as we traverse uncharted nanoethics using this book as our roadmap.

Marcel Van de Voorde
Gunjan Jeswani

Index

https://doi.org/10.1515/9783110669282-019

www.ingramcontent.com/pod-product-compliance
Lightning Source LLC
Chambersburg PA
CBHW080649220326
41598CB00033B/5150